Coupled Site and Soil-Structure Interaction Effects with Application to Seismic Risk Mitigation

NATO Science for Peace and Security Series

This Series presents the results of scientific meetings supported under the NATO Programme: Science for Peace and Security (SPS).

The NATO SPS Programme supports meetings in the following Key Priority areas: (1) Defence Against Terrorism; (2) Countering other Threats to Security and (3) NATO, Partner and Mediterranean Dialogue Country Priorities. The types of meeting supported are generally "Advanced Study Institutes" and "Advanced Research Workshops". The NATO SPS Series collects together the results of these meetings. The meetings are co-organized by scientists from NATO countries and scientists from NATO's "Partner" or "Mediterranean Dialogue" countries. The observations and recommendations made at the meetings, as well as the contents of the volumes in the Series, reflect those of participants and contributors only; they should not necessarily be regarded as reflecting NATO views or policy.

Advanced Study Institutes (ASI) are high-level tutorial courses intended to convey the latest developments in a subject to an advanced-level audience

Advanced Research Workshops (ARW) are expert meetings where an intense but informal exchange of views at the frontiers of a subject aims at identifying directions for future action

Following a transformation of the programme in 2006 the Series has been re-named and re-organised. Recent volumes on topics not related to security, which result from meetings supported under the programme earlier, may be found in the NATO Science Series.

The Series is published by IOS Press, Amsterdam, and Springer, Dordrecht, in conjunction with the NATO Public Diplomacy Division.

Sub-Series

A.	Chemistry and Biology	Springer
B.	Physics and Biophysics	Springer
C.	Environmental Security	Springer
D.	Information and Communication Security	IOS Press
E.	Human and Societal Dynamics	IOS Press

http://www.nato.int/science
http://www.springer.com
http://www.iospress.nl

Series C: Environmental Security

Coupled Site and Soil-Structure Interaction Effects with Application to Seismic Risk Mitigation

edited by

Tom Schanz
Laboratory of Foundation Engineering,
Soil and Rock Mechanics
Ruhr-Universität, Bochum, Germany

and

Roumen Iankov
Institute of Mechanics,
Bulgarian Academy of Sciences
Sofia, Bulgaria

Springer
Published in cooperation with NATO Public Diplomacy Division

Proceedings of the NATO Advanced Research Workshop on
Coupled Site and Soil-Structure Interaction Effects with Application
to Seismic Risk Mitigation, Borovets, Bulgaria
30 August–3 September 2008

Library of Congress Control Number: 2009926512

ISBN 978-90-481-2709-2 (PB)
ISBN 978-90-481-2696-5 (HB)
ISBN 978-90-481-2697-2 (e-book)

Published by Springer,
P.O. Box 17, 3300 AA Dordrecht, The Netherlands.

www.springer.com

Printed on acid-free paper

PREFACE

The purpose of NATO ARW 983188 *Coupled Site and Soil-Structure Interaction Effects with Application to Seismic Risk Mitigation*, held in Borovets, Bulgaria, 30 August—3 September 2008, was to present state-of-the-art, on-site, soil-structure interaction effects (SSSI), as manifested in the broader area of south and south-eastern Europe, which is the most seismically prone region of the European continent. Another objective was to attempt to define the seismic risk posed to the built environment in this area and to present modern methods for seismic risk mitigation.

The ARW was very successful and generated an interdisciplinary-type information exchange between the three main groups of participants: geophysicists, geotechnical engineers, and structural engineers. The presentations during the workshop can be grouped into four subject areas: (1) site conditions and their role in seismic hazard analyses, (2) soil-structure interaction, (3) the role of site effects and of soil-structure interaction in the design of structures, and (4) general and related subjects. The following fields were addressed during the presentations and the discussions: strong ground motion (near-field effects, seismic-wave propagation, free-field motion); geotechnical engineering (slopes, foundations, lifelines, dams, and retaining walls); and structural engineering (buildings, bridges, field measurements, and protective systems).

The work presented in this volume includes contributions from engineers and scientists, mainly from south-eastern Europe and the neighbouring regions of the Near East. The arrangement of contributions in different chapters is not rigorous, and many papers present similar material, which includes broad coverage and different disciplines, since earthquake engineering is by its nature an interdisciplinary subject.

Recommendations for Future Research

The conclusions reached by the workshop participants can be summarized as follows.

1. It is important to create an extensive strong-motion database for major urban areas in the seismically prone regions of Europe, and to document

local soil and geological site conditions. These data are essential for all aspects of earthquake engineering research and applications and should be made available to the research community through Web sites.

2. Development of hybrid methods for computer simulations of free-field strong ground motion are of paramount importance if reliable artificial time histories are to be produced "on demand" for the aforementioned regions.
3. It is important to develop and implement protective systems for special classes of structures in the earthquake-prone regions of Europe.
4. It is hoped that in the future the cost of protective systems and the placement of technology will become economically feasible to the point that they can be implemented in a routine fashion in the large groups of conventional structural systems.
5. The ultimate goal is a high level of protection of the built environment to earthquakes and the availability of low-cost insurance.

The roundtable discussions during the final day of the workshop addressed a large number of topics. The following represents a summary of the principal and most important observations and recommendations.

SITE CONDITIONS

New characterizations of site conditions in the near field should be developed that include all relevant components of the forces acting on a structure. With large amplitudes of strong motion, surface soil experiences large, nonlinear response, and ultimately soil failure and liquefaction can lead to large transient and permanent motions. Examples of ground failure that can follow liquefaction are lateral spreading, ground oscillations, flow failure, and loss of bearing strength. *Lateral spreads* involve displacements of surface blocks of sediment facilitated by liquefaction in a subsurface layer. This type of failure may occur on slopes up to 3° and is particularly destructive to pipelines, bridge piers, and other long and shallow structures situated in flood plain areas adjacent to rivers. *Ground oscillations* occur when the slopes are too small to result in lateral spreads following liquefaction at depth. The overlying surface blocks break, one from another, and then oscillate on liquefied substrate. *Flow failures* are a more catastrophic form of material transport and usually occur on slopes greater than 3°. The flow consists of liquefied soil and blocks of intact material riding on and with liquefied substrate, on land or under the sea. *Loss of bearing strength* can occur when the soil liquefies under a structure. The building can settle, tip, or float upward if the structure is buoyant. The accompanying motions can lead to large transient and permanent displacements and rotations, which so far have been neither evaluated through simulation nor recorded by strong-motion instruments.

INFRASTRUCTURE LOCATED IN THE NEAR-FIELD

Consequently, any structure, and in particular all extended structures (e.g., long buildings, bridges, tunnels, dams), in the area where such large nonlinearities in the soil occur, will, in addition to the horizontal components of inertial forces caused by strong earthquake shaking, experience large differential motions and large differential rotations of their foundation(s). Bridge peers or foundations of long buildings supported by soil, which the earthquake has separated into blocks by strong shaking, will be forced to deform, accompanied by large differential motions (translations and rotations) of soil blocks, and they will experience both the inertial and pseudo-static aspects of those motions. At present, we can only speculate about how much larger these motions will be relative to the tilts and angular accelerations and velocities we can estimate from the linear-wave theory. Few observations, however, suggest that those can be orders of magnitude larger than the predictions based on the linear theory. For successful design, it will be necessary to prescribe the resulting forcing functions, which will include, in a balanced way, the simultaneous action of all components of possible motion. The description of how to scale those balanced forcing functions can start from principles similar to what we use today for the design of structures crossing an active fault. Because the complexity of such motions and the multiplicity of possible outcomes will increase with amplitudes of incident strong-motion waves, specification of the driving forces for design may best be formulated in terms of their distribution functions. This will require systematic and long-range research programs focusing on two key tasks: (1) development of advanced numerical simulation models, and (2) the recording of all six components of strong motion, in the near field, as well as their analysis and interpretation. Such description of the near-field motion will have to be used in the selection of design forces within distances that are equal to about one source dimension (e.g., up to 20–50 km in California) away from the fault. In the far field, we should be able to continue to use the traditional local site parameters to describe the effects of the local site conditions for most design applications.

SEISMIC HAZARD ASSESSMENT

Beyond the near field (say, for distances greater than several tens of km) the classical empirical scaling of strong-motion amplitudes and of their duration seem to work well in all regions where sufficient strong-motion data are available and where the corresponding empirical equations have been developed. In the Balkans, there is a good strong-motion database for the territory of former Yugoslavia, and a fair number of strong motion records are now becoming available for Greece and Bulgaria. For all seismic hazard calculations and for the development of the hazard maps for use in the design codes, the locally developed empirical scaling equations must be used. Some previous studies in this region have used a mixture of attenuation equations from different parts of the world. These results are not valid and lead to erroneous hazard maps. Therefore, a systematic effort should be made to further develop and refine the existing region specific attenuation and scaling equations, and to use the existing attenuation and scaling relations, which have already been developed specifically for the Balkan region and for the Mediterranean basin.

SOIL-STRUCTURE INTERACTION

Nonlinear phenomena that accompany soil-structure interaction (nonlinear response in the soil and separation of the foundation from the soil) are powerful sinks of earthquake shaking energy. The ability of these phenomena to prevent the wave energy from entering structures should be studied and quantified so that their passive absorption capacity can be included in the design—which takes advantage of the phenomena that are nearly always present during large ground shaking seismic events.

The soil-structure interaction of extended structures (dams, bridges, tunnels) is also very sensitive to the differential motions and rotations of individual supports (as illustrated above during lateral spreading, ground oscillations, flow failure, and loss of bearing strength). Future research should aim to quantify all such differential excitations and to develop design tools to account for their action. The ramifications of complex ground phenomena associated with strong ground shaking of marginal soil deposits on the dynamic response of existing buildings is an important research topic that may have repercussions in future revisions of the EC8 seismic design code.

DESIGN OF STRUCTURES

The traditional approach to the design of earthquake-resistant structures is based on the vibrational solutions of the problem. With almost no exceptions, this approach is converted to equivalent horizontal action on the structure in the current simplified code-design procedures. This approach works reasonably well in the far field, but it breaks down in the near field, where strong motion is characterized by powerful pulses associated with large peak velocities and large point rotations. New research is needed to formulate design procedures, which will be based on wave propagation principles. In this approach, the structures will be designed for the largest strains (drift) caused by *design* strong-motion pulse entering the structure as a wave and propagating through the structure in linear or nonlinear fashion, depending upon the amplitude of the pulse. Furthermore, new research is necessary for the development of the design methods in the near field, which takes into consideration large differential displacements and rotations, which, as discussed above, accompany nonlinear motions of individual soil blocks.

PROTECTION OF STRUCTURES

It is no longer sufficient in modern post-industrial societies to apply routine maintenance to the built environment. In seismically prone regions of Europe, it is essential to monitor and check large categories of structures (high-rise buildings, industrial units, lifelines, infrastructure, etc.) for their response to micro-tremors. Also, it is necessary to selectively reinforce structures in order to improve their resistance to possible earthquake shaking, using economically feasible techniques. Finally, regarding new construction, it is important to implement primarily passive energy-absorption devices (e.g., bracing systems, dampers, base isolators) that will add an extra safety margin to the structural skeleton at an affordable cost. This will have the added advantage of stimulating European industry to produce and to implement these devices that can be exported to other earthquake-prone regions of the world at a later date, once their efficiency has been established.

SEISMIC RISK MANAGEMENT ISSUES

We close by perceiving that the managing of seismic risk, which is the ultimate goal of all aspects of earthquake engineering, is a multi-disciplinary endeavour, requiring the cooperation of many researchers from many different fields, of which the principal ones are seismology, geotechnical engineering, and design of structures. The success in all of these areas depends upon access to modern data banks with spatial and temporal attributes of ground seismicity, including soil and geological site properties, and requires the development of advanced analysis tools for design and testing of structures. The researchers in earthquake engineering and seismology, in turn, must produce information for local and national civil authorities, as well as for the private sector, for decision-making purposes on how to best handle the seismic risk.

Acknowledgements

The organizing committee and the directors of the workshop would like to acknowledge the NATO Scientific Committee for the generous grant that made it possible to organize the workshop and Mrs. Liz Cowan from SPS Section at NATO Public Diplomacy Division for her invaluable help during the preparation of the workshop and the final and financial reports. We thank the staff at the Institute of Mechanics of the Bulgarian Academy of Sciences, and personally the director, Dr. Emil Manoach, for providing excellent assistance in the organization and realization of the workshop.

The editors of these proceedings also wish to recognize the fruitful interaction with Springer publishers while preparing this volume, especially the assistance of Ms Annelies Kersberger from NATO Publishing Unit. Her help with submission of the manuscripts is highly appreciated. Finally, the excellent typesetting was provided by Mr. Venelin Chernogorov (wily@fmi.uni-sofia.bg) of Sofia University, Faculty of Mathematics and Informatics.

We are most grateful to all of the participants for taking their time to come and participate. Finally, we express our gratitude to those who submitted their papers for this publication.

Maria Datcheva, Roumen Iankov,
Institute of Mechanics, Bulgarian Academy of Sciences, Sofia, Bulgaria

George Manolis,
Aristotle University, Thessaloniki, Greece

Mihailo Trifunac,
University of Southern California, USA

Tom Schanz,
Ruhr-Universität Bochum, Germany

March 2009

LIST OF WORKSHOP PARTICIPANTS

Sabit F. Zeyniyev
Science and Research Institute of Erosion and Irrigation,
M. Hadi St. 103/59, AZE-1129 Baku, Azerbaijan
sabitzeyniyev@mail.ru

Ivan Brlek (key speaker)
Federal Hydro-meteorological Institute, Center for Seismology,
Bardakcije 12, 71000 Sarajevo, Bosnia & Herzegovina
ivanbrlek@yahoo.com

Zdravko Bonev
University of Architecture, Civil Engineering and Geodesy,
Hr. Smirnenski Blvd. 1, 1046 Sofia, Bulgaria
zbp_uacg@abv.bg

Petia Dineva (key speaker)
Institute of Mechanics, Bulgarian Academy of Sciences,
Acad. G. Bonchev St., Bl. 4, 1113 Sofia, Bulgaria
petia@imbm.bas.bg

Roumen Iankov
Institute of Mechanics, Bulgarian Academy of Sciences,
Acad. G. Bonchev St., Bl. 4, 1113 Sofia, Bulgaria
iankovr@yahoo.com

Radan Ivanov Ivanov
VSU "Lyuben Karavelov", 175 Suhodolska St., 1373 Sofia, Bulgaria
r_ivanov@vsu.bg

Mihaela Kouteva-Guentcheva
CLSMEE, Bulgarian Academy of Sciences,
Acad. G. Bonchev St., Bl. 3, 1113 Sofia, Bulgaria
mkouteva@geophys.bas.bg

Tsviatko Rangelov
Institute of Mathematics and Informatics, Bulgarian Academy of Sciences,
Acad. G. Bonchev St., Bl. 8, 1113 Sofia, Bulgaria
rangelov@math.bas.bg

Marijan Herak (key speaker)
Department of Geophysics, Faculty of Sciences and Mathematics,
University of Zagreb, Horvatovac 95, 10000 Zagreb, Croatia
herak@irb.hr

Behrooz Gatmiri (key speaker)
Ecole Nationale des Ponts et Chaussées, Paris,
6 et. 8, Av. Blaise Pascal, 77455 Champs-Sur-Marne, France
gatmiri@cermes.enpc.fr

Bettina Albers (key speaker)
Technische Universität Berlin,
Fachgebiet Grundbau und Bodenmechanik, Sekr. TIB 1-B7,
Gustav-Meyer-Allee 25, 13355 Berlin, Germany
albers@grundbau.tu-berlin.de

Gottfried Schmidt
Bauhaus-Universität Weimar, Coudraystr. 11C, 99423 Weimar, Germany
Gottfried.schmidt@bauing.uni-weimar.de

Frank Wuttke
Bauhaus-Universität Weimar, Coudraystr. 11C, 99423 Weimar, Germany
frank.wuttke@uni-weimar.de

Andreas Kappos
Aristotle University of Thessaloniki, Department of Civil Engineering,
54124 Thessaloniki, Greece
ajkap@civil.auth.gr

George Manolis (key speaker)
Department of Civil Engineering, Aristotle University,
Thessaloniki, GR-54006, Greece
gdm@civil.auth.gr

Prodromos N. Psarropoulos
Department of Infrastructure Engineering, Hellenic Air-force Academy,
Themistocleous St. 43, 16674 Athens, Greece
prod@central.ntua.gr

Anastasios G. Sextos
Aristotle University Thessaloniki, Department of Civil Engineering,
Aristotle University Campus, 54124 Thessaloniki, Greece
asextos@civil.auth.gr

Yiannis Tsompanakis
Division of Mechanics, Department of Applied Sciences,
Technical University of Crete,
University Campus, GR-73100, Chania, Crete, Greece
jt@science.tuc.gr

Varvara Zania
Division of Mechanics, Department of Applied Sciences,
Technical University of Crete,
University Campus, GR-73100, Chania, Crete, Greece
zaniab@tee.gr

Claudio di Prisco (key speaker)
Department of Structural Engineering,
Politecnico di Milano, Piazza Leonardo da Vinci 32, 20133 Milano, Italy
claudio.diprisco@polimi.it

Giuseppe Oliveto
Department of Civil and Environmental Engineering, University of Catania,
Viale Andrea Doria, 6, 95125 Catania, Italy
goliveto@dica.unict.it

Vlado Gicev
Department of Computer Science, Goce Delcev University,
Toso Arsov St. 14, 2000 Stip, F.Y.R. of Macedonia
vgicev@gmail.com

Andrei Bala
National Institute for Earth Physics,
12 Calugareni St., 077125 Bucharest-Magurele, Romania
bala@infp.ro

Mihaela Lazarescu
National R&D Institute for Environmental Protection, ICIM Bucharest,
Spl. Independentei 294, sector 6, 060031 Bucharest, Romania
mihaela_lazarescu@hotmail.com

S. Umit Dikmen
Istanbul Kültür University, Department of Civil Engineering,
Atakoy Kampusu, Bakirkoy, 34156 Istanbul, Turkey
u.dikmen@iku.edu.tr

Dmytro Rudakov
National Mining University,
Department of Hydrogeology and Engineering Geology,
K. Marx av., 19, 49005 Dnipropetrovs'k, Ukraine
dmi3rud@mail.ru

Roel Snieder (key speaker)
Center for Wave Phenomena and Department of Geophysics,
Colorado School of Mines, Illinois Street 1500,
CO 80401-1887 Golden, USA
rsnieder@mines.edu

Mihailo Trifunac (key speaker)
University of Southern California,
Department of Civil Engineering, KAP 216D, mc 2531,
90089-2531 Los Angeles, California, USA
trifunac@usc.edu

Sadillakhon Umarkhonov
Namangan Engineering-Pedagogical Institute,
Namangan region, K. Karvan St. 116, 717200 Chust city, Uzbekistan
saidullo@yahoo.com

Co-directors

Tom Schanz
Laboratory of Foundation Engineering, Soil and Rock Mechanics,
Faculty of Civil Engineering, Ruhr-Universität Bochum,
Universitätsstraße 150, D-44780 Bochum, Germany
tom.schanz@rub.de

Radomir Folić
University of Novi Sad, Faculty of Technical Sciences,
Department of Civil Engineering,
Trg D. Obradovica 6, 21000 Novi Sad, Serbia
folic@uns.ns.ac.yu

CONTENTS

Preface v

List of Workshop Participants xi

I SITE CONDITIONS AND THEIR ROLE IN SEISMIC HAZARD ANALYSES

M. D. Trifunac. The Nature of Site Response During Earthquakes 3

M. Kouteva-Guentcheva, I. Paskaleva, and G. F. Panza. Earthquake Source and Local Geology Effects on the Seismic Site Response 33

T. Rangelov and P. Dineva. Wave Propagation in the Anisotropic Inhomogeneous Half-Plane 43

T. Schanz, F. Wuttke, and P. Dineva. Hybrid Simulation of Seismic Wave Propagation in Laterally Inhomogeneous Media 53

B. Albers. On the Influence of Saturation and Frequency on Monochromatic Plane Waves in Unsaturated Soils 65

P. N. Psarropoulos. Local Site Effects and Seismic Response of Bridges 77

B. Gatmiri. Local Site Effect Evaluation in Seismic Risk Mitigation 89

A. Bala, S. F. Balan, J. Ritter, and D. Hannich. Seismic Site Effect Modelling Based on *In Situ* Borehole Measurements in Bucharest, Romania 101

II SOIL-STRUCTURE INTERACTION

Y. Tsompanakis. Issues Related to the Dynamic Interaction of Retaining Walls and Retained Soil Layer 115

V. Zania, Y. Tsompanakis, and P. N. Psarropoulos. The Effect of Soil-Structure Interaction and Site Effects on Dynamic Response and Stability of Earth Structures 127

C. di Prisco, A. Galli, and M. Vecchiotti. Cyclic and Dynamic Mechanical Behaviour of Shallow Foundations on Granular Deposits 139

V. Gicev. Soil-Structure Interaction in Nonlinear Soil 151

M. I. Todorovska. Separation of the Effects of Soil-Structure Interaction in Frequency Estimation of Buildings from Earthquake Records 169

B. Folić and R. Folić. Analysis of Seismic Interactions Soil-Foundation–Bridge Structures for Different Foundations 179

III THE ROLE OF SITE EFFECTS AND OF SOIL STRUCTURE INTERACTION IN DESIGN OF STRUCTURES

A. J. Kappos and A. G. Sextos. Seismic Assessment of Bridges Accounting for Nonlinear Material and Soil Response, and Varying Boundary Conditions 195

G. D. Manolis and A. M. Athanatopoulou. Structural Response to Complex Synthetic Ground Motions 209

A. G. Sextos and O. Taskari. Single and Multi-Platform Simulation of Linear and Non-Linear Bridge-Soil Systems 225

Z. Bonev, G. Necevska-Cvetanovska, E. Vaseva, R. Apostolska, and D. Blagov. Design Seismic Response Evaluation of Wall Systems Including Foundation Flexibility 241

T. Schanz, Z. Bonev, F. Wuttke, R. Iankov, and V. Georgiev. Design Seismic Performance of R/C Frame Structures Taking into Account Foundation Flexibility 255

S. U. Dikmen, A. M. Turk, and G. Kiymaz. Effect of Depth of Ground Water on the Seismic Response of Frame Type Buildings on Sand Deposits. Part I: Soil Response 273

IV GENERAL AND RELATED SUBJECTS

R. Snieder. Extracting the Time-Domain Building Response from Random Vibrations 283

M. Herak and D. Herak. Recent Measurements of Ambient Vibrations in Free-Field and in Buildings in Croatia 293

I. O. Sadovenko, D. V. Rudakov, and V. I. Timoschuk. Analysis of Dynamic Impact on a Ground Slope During Destruction of an Emergency House 305

Author Index 313

Part I

Site Conditions and Their Role in Seismic Hazard Analyses

THE NATURE OF SITE RESPONSE DURING EARTHQUAKES

Mihailo D. Trifunac (trifunac@usc.edu)*
Department of Civil Engineering, University of Southern California, Los Angeles, California 9089-2531, U.S.A.

Abstract. The traditional approach for empirical scaling of the amplitudes of strong earthquake ground motion revolves around the linear representation of the amplification of seismic waves when they propagate through soft surface sediments and soil. However, in the near field, when the amplitudes of shaking become large, the soil experiences nonlinear strains, and tensile cracks, fissures, and pounding zones form, resulting in highly nonlinear response characteristics. This means that the characteristic site response, and the patterns of amplifications measured via small earthquake records, or by analysis of microtremors, will disappear, departing from the linear amplification characteristics completely. This leads to chaos and creates a problem for seismic zoning because the nonlinear response is strongly dependent upon the amplitudes and on the time history of shaking, so that it becomes virtually impossible to predict the distribution of amplification from the local site conditions. If we assume that the observed damage distribution is a useful indication of the distribution and of the nature of shaking amplitudes, we can conduct a full-scale experiment every time a moderate or large earthquake leads to some damage. Analyses of these patterns, combined with detailed maps of the properties of the soil and of surface geology, suggest that there are reappearing patterns of nonlinear site response from one earthquake to the next. We show one such example for two earthquakes in the Los Angeles metropolitan area. This example implies that the relative movement along the boundaries of the blocks of soil, and along the cracks formed by previous strong shaking, may recur during future earthquakes. The implication is significant for all engineering analyses of response and for engineering design in the near field because it means that in the vicinity of these cracks the complexity of strong shaking is further increased by large differential motions and by large transient and permanent strains and tilts.

Keywords: effects of site response during earthquakes, local soil site conditions, local geologic site conditions, nonlinear site response, site response in near field

1. Introduction

Studies of the effects that local site conditions have on the characteristics of strong earthquake ground motion are as old as earthquake engineering. Descriptions of early investigations can be found in the papers of Reid (1910) and Sezawa, Kanai, and their co-workers (Duke, 1958). These studies first

*http://www.usc.edu/dept/civil_eng/Earthquake_eng/

T. Schanz and R. Iankov (eds.), *Coupled Site and Soil-Structure Interaction Effects with Application to Seismic Risk Mitigation,* NATO Science for Peace and Security Series C: Environmental Security,

emerged from observations of damage, which showed considerable spatial variations and complexities. The levels of observed damage could be correlated with the available information on the site conditions, which were extracted from maps of surface soil and surface geology. Concurrent theoretical studies of linear-wave propagation, which showed amplification of amplitudes, as seismic waves emerged from "hard" into "soft" surface deposits, contributed to the formation of a view that the strong-motion shaking is amplified in the soft surface soils and sediments. This simplified view prevailed for many years, and it is evident in the formulation of early design codes (Freeman, 1932) and in the guidelines for the design of important structures (Coulter et al., 1973). A perusal of Kanai's descriptions of the patterns of damage to Japanese wooden houses, for example, reveals his appreciation for the details of many seemingly conflicting observations (Kanai, 1983), although in the end the simplifications needed for the development of design codes prevailed. The absence of recorded strong-motion accelerograms by dense arrays, and the lack of three-dimensional soil and geological characterizations of sites, eventually led to simplified site descriptions, many of which continue to be in use today.

Looking back at numerous studies of site effects, certain characteristics and trends emerge. First, many studies were carried out by prominent seismologists (e.g., Gutenberg, 1957), who usually work only with linear waves with long-period motions (say, longer than ~1 s), small wave amplitudes, and large epicentral distances (e.g., more than ~100 km). Second, engineering contributions to the studies of site effects, in the beginning, used only the amplitudes of peak acceleration (i.e., they did not consider the frequency content of ground motion) and tended to use only the site characterization in terms of the surface soil conditions (with dimensions rarely exceeding ~200 m) (e.g., Seed et al., 1976; Ambraseys et al., 1996; Lee, 2007). This trend continues today. It should not be so, but it is rationalized by the fact that it is expensive and difficult to include deeper site characterization and to use a wider zone surrounding the site (e.g., on the scale of several hundred meters to several kilometers). Third, with few exceptions, most studies of site effects are based on forward modelling and regression analyses, and they rarely test the significance of the computed regression coefficients and do not test for cross-correlations among the parameters of the model. The soil site-condition variables (which should be important for short-period motions) and the geologic site-condition variables (which are important for intermediate and long-period motion) are correlated by the nature of their formation, and they are usually not considered simultaneously in most regression models. The result is that most scaling methods, which are based on the site conditions and consider only soil-site classification, average out the effects of the geological site conditions and are characterized by large uncertainty in the prediction of spectral amplitudes. Fourth, because most strong-motion data are available for fault-to-station distances in the range from about 25 to 100 km, essentially all published regression models reflect the trends in the data for this distance range. Since the significant damage to structures occurs mainly within several tens of kilometres from the fault (in the near field), the nature of the site effects and the extent to which they influence the ground motion will be different from what is determined from the regression analyses of the distant recordings, in that they will describe essentially linear and almost-linear site response. Fifth, it is assumed that the site effects are repeatable from one earthquake to the next and that they do not depend significantly on the azimuth, angle of incidence, and amplitudes of seismic waves. However, studies of multiple earthquake recordings at the same strong-motion stations show that this assumption holds at best only about 50% of time, and only at some recording stations (Trifunac et al., 1999; Trifunac and Ivanović, 2003a, b). Sixth, it is very rarely asked whether the parametrization of the site conditions should have been done differently, on the basis of some rational physical considerations (Todorovska and Trifunac, 1998), so that it could be correlated with, and shown to be significant in terms of, the end result (e.g., distribution of damage).

More recently, licensing pressures resulting from the need for consensus building among ground-motion experts, at first in the design of nuclear power plants and then in the revision of the design codes, have resulted in the emergence of group efforts for the development of scaling equations of strong ground motion. On the positive side, this has resulted in increased exchange of ideas and more discussions among the researchers who work on the effects of local site conditions. However, this has also reduced the role of original, individual approaches and has led to the adoption of scaling models, which favour the "average" view rather than the search for the "best" physical models. This consensus building may help to speed up the licensing process, but nature will follow its course, and what individual researchers may not be able to change in the consensus, future earthquakes certainly will. In the following, I will not be guided by any "consensus" views but rather will try to outline how local site effects have been addressed in the past and how we might improve their representation, based on what is known to date.

In summary, the shortcomings of the studies dealing with the effects of site conditions on the amplitudes of strong ground motion are that (1) what are adopted as "site conditions" are often not based on the physical nature of the problem—i.e., on a careful study of the nature of wave propagation through geologic and soil layers—but rather on the heuristic description of the information that a geologist and an engineer can gather from published maps and through field observation; (2) the form of the regression equations that are used to describe the trends is often not based on the nature of the problem but rather on mathematical forms that lead to manageable regression analyses; and (3) the formulation is essentially linear (Trifunac, 1990). Consequently, the results and lesions from such studies are valid only at a certain distance from the earthquake faults, where nonlinearities in the site response are absent or small. In the near field, where large motions cause damage and destruction of structures, and where the soil experiences large, nonlinear deformations, these results cease to predict the outcome, and new methods must be developed to provide characterization of strong-motion amplitudes for engineering applications. In this paper, I will discuss some of these alternatives and give examples of the phenomena that need to be modelled, using examples from selected earthquake studies.

2. The Linear Approach

The linear (transfer-function) representation of strong ground motion can be viewed in the frequency domain as

$$O(f) = E(f)P(f)S(f), \tag{1}$$

where f is frequency, $O(f)$ and $E(f)$ are, respectively, the Fourier spectra of the motion at a site and at the earthquake source, and $P(f)$ and $S(f)$ are the transfer functions of the propagation path and of the local site effects. This representation is meaningful for epicentral distances that are large relative to the source dimensions, when the earthquake source can be approximated by a point source. In the near field, the small distance between the site and the large area of the rupturing fault results in geometrical nonlinearities, which are caused by the spatial distribution of wave arrivals from different segments of the fault surface. Thus, in the near field, Eq. (1) ceases to be valid because $E(f)$, $P(f)$, and $S(f)$ become complex, geometrically nonlinear functions of the space coordinates. While $O(f)$ could be represented by an equation related to Eq. (1), it would have to be in the form of an integral over the fault surface, with $P(f)$ and $S(f)$ being functions that depend upon the geologic environment and on the site location. Further, $E(f)$ would have to include contributions from near-field terms in the representation of the source radiation ($1/r^2$ and $1/r^4$ terms, where r is the distance between the site and a point on the fault surface; Haskell, 1969; Trifunac, 1974). With $\varphi_j r_i R_{i,j} R/a \rightarrow \infty$, where R is the epicentral distance and a is some representative size of asperities on the fault surface, Eq. (1) asymptotically becomes linear (geometrically, since there is no need to integrate over the fault surface) and can represent the site and the propagation effects well.

For two sites having different site conditions and a separation distance that is small relative to a large epicentral distance, it is reasonable to assume that their motions will differ mainly due to the differences in $S(f)$, while their $P(f)$ can be assumed to be nearly the same. This reasoning has evolved into a framework for most theoretical and empirical studies of the effects of site conditions on the amplitudes of strong ground motion (Kanai, 1983; Trifunac, 1990). In the following, this approach will be illustrated through several representative studies.

In equation (1) $S(f)$ models the site effects in general and can represent the geological site effects, the soil site effects, both of those together, or the surface topography, and it may include other site characteristics that may be relevant. In this paper, I discuss the role of $S(f)$ only as representing the geological site effects, soil site effects, or both of those together, and I do not consider examples of any other aspect of site dependence. While using this approach, it is important to define precisely and a priori what is included in $S(f)$ to avoid ambiguity in interpreting the end results. It is remarkable how many papers, even some written by very experienced researchers, use imprecise site descriptions (e.g., by mixing the geological and soil site conditions), only to arrive at wrong conclusions (Aki, 1988).

2.1. GEOLOGICAL SITE CONDITIONS

Considering the size of geological inhomogeneities, the distances travelled by strong-motion waves, and the wavelengths associated with the frequencies of interest in earthquake engineering (0.05 to 50 Hz), it is clear that the local geologic conditions play a prominent role in determining the local site amplifications (Trifunac, 1976a, 1978, 1979; Trifunac and Anderson, 1977, 1978a,b).

In this paper, I use "geological site conditions" to represent the binary interpretation of the site conditions as can be determined from geological maps ($s = 0$ for sites on sediments, and $s = 2$ for sites on the basement rock). Trifunac and Brady (1976) show examples of how the geological site descriptions can be converted to $s = 0$ or 2, and to $s = 1$ for "in-between" sites, which are near the contact of sediments with basement rock, or which are in a complex setting that does not allow unequivocal and simple site description. Sites on sediments ($s = 0$) can further be described by their thickness (h) above the basement rock (Trifunac and Lee, 1978, 1979). The nature of the geological site conditions, as described by s and/or h, involves a scale that is measured in kilometers (Trifunac, 1990).

Before the advent of digital computer, analyses of the amplification of ground motion, were performed by manually measuring the recorded peaks of instrument response. Periods of motion were evaluated from the frequency of zero crossing or by approximating individual peaks by half-sine pulses. For example, Reid (1910) found amplification of 1 to 2 for sandstone, 2 to 4 for sand, and 4 to 12 for man-made fill and marsh. For seismometer response to local earthquakes, with periods of the peaks in the range from 0.5 to 1 s, Gutenberg (1957) analysed recordings from 25 temporary stations on sediments and one reference station on basement rock. He found amplification of about 2 to 3 on deep sediments. Similar trends were later observed by Borcherdt (1970), Borcherdt and Gibbs (1976), and Campbell and Duke (1974). After digital data processing became possible and the recorded strong-motion accelerations could be corrected and integrated to give velocities and displacements, Trifunac and Brady (1976) and Trifunac (1976a,b) extended this work to all peaks of strong ground motion and found excellent agreement with the results of Gutenberg (1957) for periods longer than about 0.5 s and for peak velocities and peak displacements. However, they found a reversal of this trend for peak accelerations (i.e., for strong-motion amplitudes at high frequencies) and showed that the peak accelerations recorded on basement rock are comparable to or larger than the peaks recorded on sediments and on alluvium. The work of Trifunac and Brady (1976) brought out the significance of the frequency-dependent nature in the amplification by local site effects.

The age of sediments (and of rock) under the recording station also influences the amplitudes and the duration of strong motion. It can be shown to correlate well with the geologic site classification ($s = 0$, 1, and 2) and can be used as an additional variable in the regression models. Studies of how the age of site deposits interacts with other site parameters and contributes to the overall duration of strong motion are described in Novikova and Trifunac (1995). Studies of how the age of local site deposits contributes to spectral amplitudes in regression equations that also include the geologic and soil site parameters have not been performed thus far.

2.2. SOIL SITE CONDITIONS

Characterization of the soil site conditions involves a depth scale, which originally extended to about 200 m (deep, cohesionless soils, as in Seed et al., 1976, but which in more recent studies has been reduced to only 30 m below the surface (Chiou et al., 2008). Because of this small thickness, soils can be expected to contribute mainly to the high-frequency, linear changes in the incident seismic waves, but because of their low stiffness and nonlinear behaviour they can play a significant role at all frequencies of the observed motions. The soil site conditions introduced by Seed et al. (1976) involve four groups: “rock” ($s_L = 0$, for sites with a shear-wave velocity of less than

800 m/s and a thickness of less than 10 m), stiff soil sites ($s_L = 1$, with a shear-wave velocity of less than 800 m/s and a soil thickness of less than 75 to 100 m), deep soil sites ($s_L = 2$, with a shear-wave velocity of less than 800 m/s and a thickness of between 100 and 200 m), and soft-to-medium clay and sand ($s_L = 3$) (where the notation $s_L = 0, 1, 2, 3$ is as introduced and used by Trifunac, 1987, and Lee, 1987).

Categorical variables, which describe the shallow soil site conditions in terms of the average shear-wave velocity $\overline{v}$ in the top 30 m of soil, were at first defined as: A for $\overline{v} > 750$ m/s, B for $360 < \overline{v} < 750$ m/s, C for $180 < \overline{v} < 360$ m/s, and D for $\overline{v} < 180$ m/s. With minor variations, these categorical variables continue to be refined as more data become available (Chiou et al., 2008).

Trifunac (1987) showed that the local soil and geologic site conditions must be considered simultaneously in the empirical scaling of strong-motion spectral amplitudes, and he presented a family of such scaling equations. Lee (1987) extended this work to the scaling of pseudo-relative velocity spectra. In searching for the most stable equations, and in order to find the type of regression analysis that is most suitable for such scaling, eight different models were considered, two pairs for direct scaling in terms of the local geologic conditions modelled by the depth of sediments, and two pairs for scaling in terms of the simple geologic site conditions ($s = 0$, 1, and 2). Each pair consisted of one set of equations for scaling in terms of earthquake magnitude and one set for scaling in terms of the site intensity. Corresponding to these four models, in which the simultaneous effects of both local soil and local geologic conditions were considered, a set of four other models with two-stage regression was also analysed, first with respect to all scaling parameters, including the local geologic conditions, and then with respect to the residuals in terms of the local soil conditions only. These regression analyses are too complex to review here, but for the purpose of this paper it is sufficient to note that all local soil and geologic site effects can be described by the coefficient functions of the period of motion T. These functions, representing amplification, typically are small or negative for short periods and positive for intermediate and long periods.

It is noted here that both the derived scaling functions for site amplification in terms of the geological site parameters (s and h) and the soil site parameters (s_L), as well as the corresponding parameters in the site database, are correlated. This is to be expected because of the nature of the creation, transport and the deposition of soil materials. For the data set used by Trifunac (1987), there were many (33%) deep-soil sites ($s_L = 2$) over sediments ($s = 0$, or $h > 0$) and 10% "rock"-soil sites ($s_L = 0$) over basement rock ($s = 2$, or $h = 0$). There were, however, also many (27%) stiff-soil sites

($s_L = 1$) over sediments ($s = 0$, or $h > 0$) and 8% "rock"-soil sites ($s_L = 0$) over intermediate geologic sites ($s = 1$) (Trifunac, 1990). Consequently, the use of regression models, which describe the site conditions in terms of only soil or geological site parameters, averages out the dependence upon the site parameter, which is not used in the analysis. This leads to erroneous prediction of the amplification by local site conditions, and, using the distribution of the site conditions in the study by Trifunac (1987) as an illustration, these erroneous predictions occur about 40% of the time. In view of this, it is remarkable how many studies still continue to develop scaling equations using only the soil site classification variables (e.g., Abrahamson and Silva, 1997; Ambraseys et al., 2005a,b; Boore et al., 1997), as if all strong-motion data has been recorded under identical geologic site conditions!

Here, I discuss only the results based on the s, h, s_L, and $\bar{v}$ (or A, B, C, and D) site parameters. Examples of other site-specific parameters that have been considered in the analysis of the local site effects on the amplitudes of strong motion are described in Rogers et al. (1985). They studied the role of nine geotechnical parameters (mean percentage of silt and clay, thickness of Quaternary, age, thickness of Holocene, depth to water table, textural type, depth to crystalline basement, depth to cementation, and mean shear-wave velocity) and found that in addition to void ratio and shear-wave velocity the thickness of unconsolidated sediment and the depth to basement rock are significant parameters controlling overall site effects. In another, related study, Goto et al. (1982) described the relationship between the site effects and the blow counts (N-value).

2.3. $R_{I,J}$, R_I AND φ_J

The above-discussed geological and soil site conditions represent a characterization of the recording site as a "point on the ground surface" and ignore the horizontal extent and geometry of those conditions. In a series of papers on the duration of strong ground motion, Novikova and Trifunac (1993a,b, 1994a,b, 1995) introduced the additional site- and earthquake-specific variables $R_{i,j}$, r_i and φ_j. $R_{i,j}$ represents the effective horizontal distance (in km) from the site (j) to a basement outcrop (i), which is at distance r_i from the earthquake epicentre and is capable of reflecting strong-motion waves from the source back toward the site (j), thus contributing to prolongation of strong ground motion (Fig. 1). φ_j is the angle containing those outcrops, as seen from the recording station (j), and is evaluated separately for each earthquake. Both $R_{i,j}$ and φ_j were found to contribute significantly to duration of strong motion and were therefore adopted as new site-specific variables in the empirical scaling of the duration of strong shaking. Through prolongation

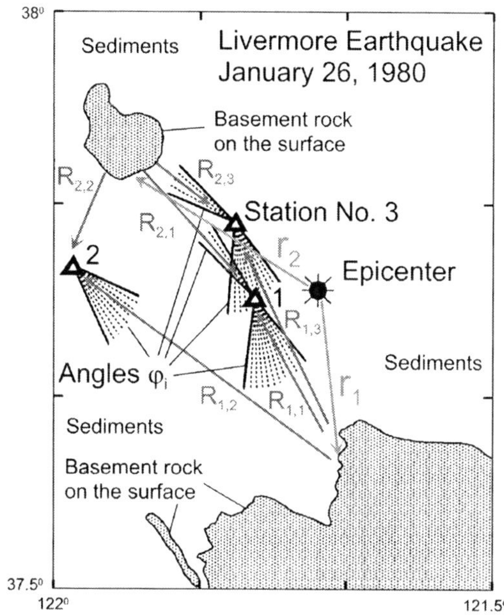

Figure 1. Parameters for horizontal reflections: the angles φ_j subtended at the recording stations by the surface of the rocks from which reflections occur, and the distances r_i and R_{ij} (from Novikova and Trifunac, 1993b).

of shaking, these site parameters will also affect the spectral amplitudes of strong motion, but the empirical studies for their inclusion in the scaling models of spectral amplitudes have yet to be carried out.

2.4. PERCENTAGE OF DISTANCE TRAVELLED THROUGH BASEMENT ROCK – P

Between the source and the recording station, the strong-motion waves encounter different configurations and a number of sedimentary basins (Fig. 2). At each interface, complex reflections and refractions occur, and many new waves are generated. To characterize such effects on the amplitudes and on the duration of strong shaking, one can begin by considering the percentage of the wave path, from epicentre to the recording site, covered by the basement rock, for each path type separately. Then, $p = 100$ represents a path entirely through rock, and $p = 0$ is for the path only through sediments. It has been shown that p is a significant variable and that the scaling equations can be developed for a family of different paths (Lee and Trifunac, 1995; Lee et al., 1995; Novikova and Trifunac, 1995).

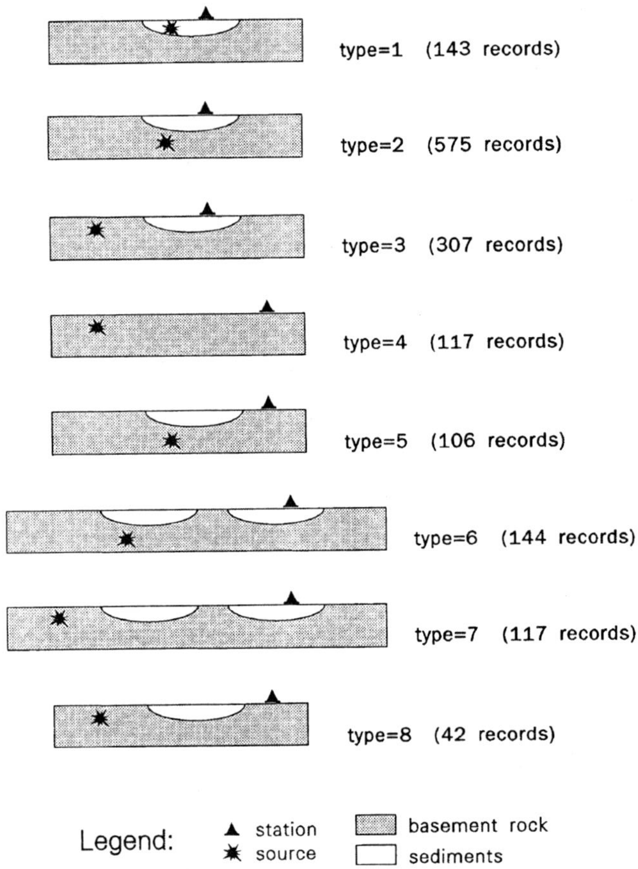

Figure 2. A schematic representation of the propagation path types. For each type the number of acceleration records, which could be used in the regression analyses is shown in the brackets (from Novikova and Trifunac, 1995).

2.5. EXPERIMENTAL METHODS

One approach for estimation of the effects the local soil and geological site conditions have on the amplitudes of strong motion assumes that those effects can also be seen during other forms of excitation. This has led to studies of microtremors, microseisms and of small earthquakes preceding and following (i.e., aftershocks) the damaging earthquakes.

2.5.1. *Microtremors*

During the 1930s and 1940s, Kanai (1983) promoted the measurement of microtremors as a vehicle for experimental estimation of local site effects.

Specifically, he used microtremors to estimate the "predominant site period" and proposed procedures for estimation of the local site effects. Numerous papers have been published about this approach, but successful procedures capable of predicting the amplification during strong earthquake shaking are yet to be formulated. Comparisons of earthquake and microtremor measurements, of the distribution of strong-motion amplitudes, and of the site-predominant periods in California did not produce useful results (Udwadia and Trifunac, 1973). A comparison of the spatial distribution of strong-motion amplitudes and the distribution of damage following the 1994 Northridge, California earthquake with the distribution of amplitudes of long-period microtremors was also not successful (Trifunac and Todorovska, 2000a). In contrast, the use of microtremors in the measurement of structural properties has been very successful (e.g., Ivanović et al., 2000), which suggests that more advanced analysis procedures may yet be developed to make microtremors useful in the estimation of the amplification properties of local site conditions.

2.5.2. *Small earthquakes and aftershocks*

Because destructive earthquakes occur infrequently, many attempts have been made to use the recordings from smaller earthquakes and from aftershocks to predict the amplification of waves by local site conditions. Most of this work is based on the wave amplitudes, which are one-to-several orders of magnitude smaller than the amplitudes of strong shaking. An example of what can be learned from several comprehensive aftershock studies of one earthquake can be found in the paper by Trifunac and Todorovska (2000b). They review three studies of amplification based on the recordings of aftershocks of the Northridge earthquake (Gao et al., 1996; Hartzell et al., 1996; Field and Hough, 1997) and one study of amplification based on four local earthquakes (1971 San Fernando, 1987 Whittier-Narrows, 1991 Sierra Madre, and 1994 Northridge, all in California) by Harmsen (1997). Trifunac and Todorovska conclude that (1) the aftershock studies could not consider longer-period motions (0.2–2 Hz), which contribute to the damaging energy, and (2) that within the current (linear) methods of analysis of aftershock data the results are not useful for prediction of site amplification and of the nonlinear and damaging nature of strong motion within 25 to 30 km from the Northridge fault. For sites further than about 30 km from the fault, where the peak ground velocity was smaller than 15 to 20 cm/s, predictions of the amplification by the local site conditions based on small earthquake and aftershock studies led to fair agreement with the amplifications observed during the main event.

3. Nonlinear Site Response

There are many different signs that the large strong-motion amplitudes in the near field lead to nonlinear response of soil and sedimentary deposits near the surface. The evidence can be seen in the records of strong motion, which show saturation of peak amplitudes, shifting, broadening, and amplitude reduction of the spectral peaks. It can also be seen in the near field, for example, as permanent deformation of surface soil, movement of soil blocks, landslides, and liquefaction. In the following, we discuss a few examples that are mainly associated with the evidence based on recorded motions.

3.1. SATURATION OF PEAK AMPLITUDES

We illustrate the saturation of peak amplitudes by the recorded motions during the 1994 Northridge, California earthquake. Figure 3 shows the nonparametric attenuation functions for peak accelerations at "soft" (C) and "hard" (A and B) soil sites for horizontal (solid lines) and vertical (dashed lines) peak amplitudes, derived by smooth interpolation through the recorded values and plotted versus shortest distance to the map view of the rupture surface. It shows that the horizontal peak accelerations on "soft" sites became saturated in the range between 0.4 and 0.6 g for distances less than about 25 km. It also

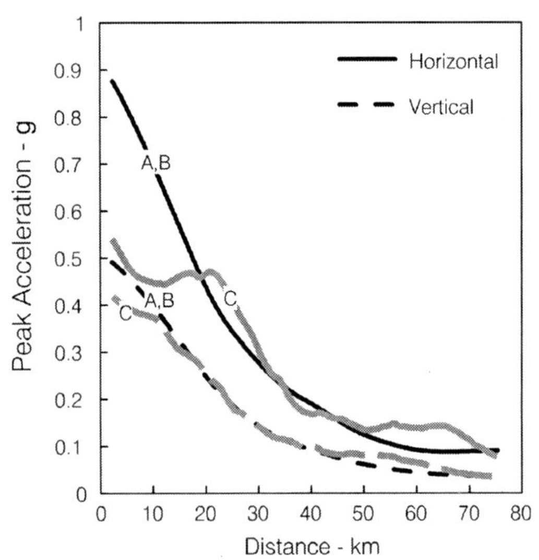

Figure 3. onparametric attenuation functions for peak acceleration at "soft" (C) sites, and "hard" sites (A and B), for Northridge earthquake, and horizontal (solid lines) and vertical (dashed lines) components of motion.

shows that the horizontal peaks at "hard" sites, as well as the vertical peaks at "soft" and "hard" sites, did not reach saturation during this earthquake. For the sediments and soils in the San Fernando Valley (Trifunac and Todorovska, 1998a), this shows that a noticeable reduction of recorded horizontal peak accelerations occurs when the strain in the soil exceeds 10^{-3}, at sites with $\bar{v} < 360$ m/s (C sites). In the San Fernando Valley, during the Northridge earthquake, the area where the recorded strain exceeded 10^{-3} was limited to distances less than 15–20 km from the fault (Trifunac and Todorovska, 1996). Within the same distance range from the fault, there were numerous and unambiguous signs of large nonlinear soil response (EERI, 1995).

3.1.1. *Recurrence and shifting of predominant peaks*

To predict site-specific ground motion during future earthquakes at the site of a structure, analyses of the soil and geology surrounding the site are carried out. The site is usually modelled by parallel layers, with physical properties measured by different *in situ* methods. These models are next used to estimate the site-specific transfer functions (for linear response) or to evaluate nonlinear site response (for large strong-motion amplitudes) via numerical simulation. It is assumed that the site properties do not change with time or with the direction of wave arrival. Also, it is usually assumed that the overall amplification can be modelled by vertically incident shear waves in a stratum with parallel layers, even though it is known that a significant part of the recorded strong-motion energy is propagated to the site by surface waves (Trifunac, 1971a).

Two- and three-dimensional (2-D, 3-D) inhomogeneities at the site lead to shifting, disappearance, and reoccurrence of the spectral peaks in the site-specific linear transfer functions. This is caused by interference, focussing, scattering, and diffraction of waves in the irregular medium surrounding the site (Trifunac, 1971b). Even when the problem may be described by linear material properties, the irregular site geometry contributes to complex changes in the transfer functions, which depend in a nonlinear manner upon the incident angle and the azimuth of wave arrivals. These changes depend also upon the epicentral distance and the 3-D geological inhomogeneities along the propagation path.

Sands tend to settle and densify when subjected to strong shaking (Lee and Albaisa, 1974; Tokimatsu and Seed, 1987). If the sand is saturated and there is little or no drainage, the earthquake shaking can lead to excess pore pressure, and settlement follows as the excess pore pressure dissipates. The settlement can occur instantaneously or within about a day following the shaking. Settlement from earthquake shaking also occurs in dry sands. One of the consequences is compaction, which is accompanied by an increase in the effective shear modulus. The implication for analyses of strong ground motion is that, after settlement, the site-specific peaks in the spectra of recorded motions can shift toward shorter periods. Dynamic compaction of soil following strong shaking will thus result in a "stiffer" site for shaking by waves from the aftershocks. Aftershocks have a small source area, and consequently the pencils of wave arrivals at the site are narrow. All of this may lead to more-coherent high-frequency motions and result in larger high-frequency amplitudes of spectra of recorded motions. In contrast, the main seismic events have extended source areas, are the result of the fracture of many asperities, which are randomly distributed in time and space, and produce waves that propagate along different paths toward the site. This will lead to less-coherent high-frequency signals and apparent "reduction" of the high-frequency spectral amplitudes, which may be misinterpreted as resulting from nonlinear soil response (Hartzell, 1998).

The rare occurrence of intermediate and strong earthquakes rules out the possibility of evaluating site-specific transfer functions for design directly from representative strong-motion earthquake recordings. As already noted, this lack of real data has led to the idea that recording and analysing weak motions (from microtremors and microseisms) will help estimate the site-specific transfer functions experimentally (Kanai, 1983). However, field tests in El Centro, California did not show any similarity of spectra of recorded earthquakes and of measured microtremors at a site because the recorded waves (1) are of a different type, and (2) have different propagation paths (Udwadia and Trifunac, 1973). One-dimensional, equivalent, linear numerical-simulation studies have also concluded that "the use of small earthquake records as the basis for evaluating site response during strong earthquakes may be misleading" (Idriss and Seed, 1968).

In spite of this negative evidence, site-specific response is often investigated by comparing recorded strong motions during an earthquake with weak motions during subsequent aftershocks. The basic premise is that Fourier amplitude spectra of recorded motions can be represented by a product, as in Eq. (1). If a site repeatedly amplifies PSV amplitudes at certain frequencies, structures at the site with similar natural frequencies will also experience larger response. Therefore, to find out how often these peaks recur during excitation by different earthquakes it would be useful to search for the repeated occurrence of local peaks in the Fourier amplitude spectra that also appear as strong and well-defined peaks in the PSV spectra (Trifunac et al., 1999; Trifunac and Ivanović, 2003a,b).

Figure 4a shows by solid dots ("obvious" peaks) and open circles ("not-so-obvious" peaks) the periods of spectral peaks that can be identified during 41 events (listed according to the amplitude of their peak velocity, shown on

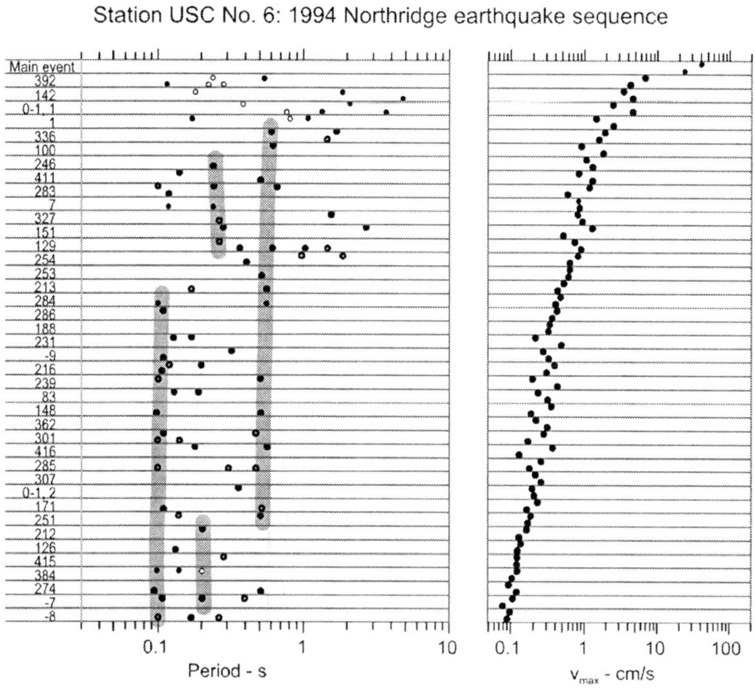

Figure 4a. Periods of identified peaks of Fourier spectra (left) and of peak ground velocity (right) for 41 records of the 1994 Northridge earthquake and its aftershocks, recorded at station USC 6 (from Trifunac et al., 1999).

the right) recorded at station USC 6 in the San Fernando Valley (Anderson et al., 1981) during the 1994 Northridge earthquake (main event) and 40 of its aftershocks that triggered the accelerograph at this station. Wide gray lines mark the periods near 0.10, 0.20, and 0.55 s, which reappear in many records. Two important characteristics of this plot should be noted. First, the site-characteristic peaks are not present in all recordings. Considering all of the sites studied in this manner thus far (Trifunac et al., 1999; Trifunac and Ivanović, 2003a,b), the site peaks occur again at most about 50% of time, but usually less often. Second, the spectral peaks shift to longer periods or completely disappear for motions with peak ground velocity larger than about 10 cm/s. Figure 4b shows a similar plot, but with the contributing events arranged in the chronological order following the main (Northridge) event. It shows a clear shift of the period of the peak, from about 1 s (during the main event) toward 0.3 s 10 min later, during events 9 and 20. The right side of this plot shows the approximate values of strain in the ground, in the range between 10^{-5} and 10^{-3}. To illustrate the long-term (seven years) variations

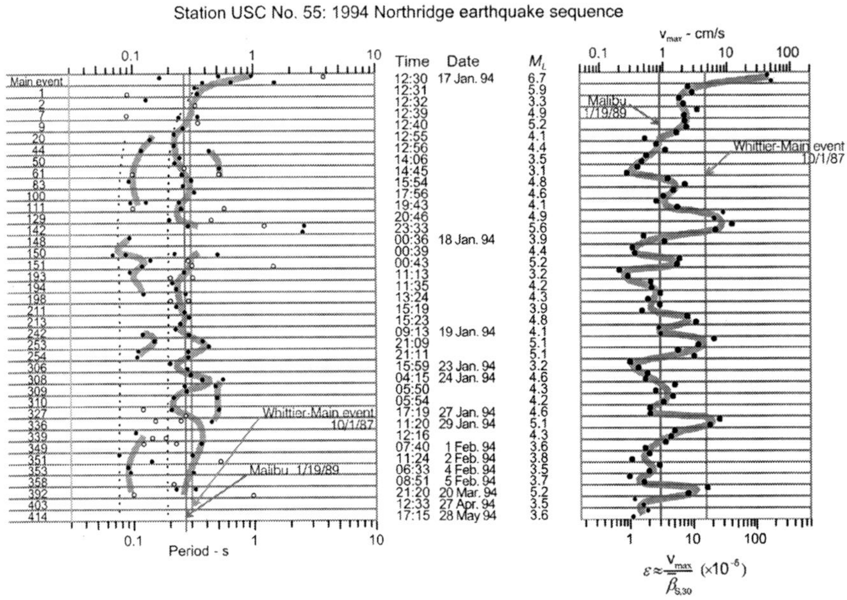

Figure 4b. Periods of identified peaks of Fourier amplitude spectra (left) and of peak horizontal ground velocity v_{max} (right) for 41 records of the 1994 Northridge earthquake and its aftershocks, recorded at station USC 6, arranged in chronological order. The bottom-right scale shows an estimate of peak strain $v_{max}/\beta_{s,30}$, where $\beta_{s,30}$ is the average shear-wave velocity in the top 30 m of soil. The periods and the peak velocities for two preceding earthquakes are shown by vertical lines (1987 Whittier Narrows and 1989 Malibu) (from Trifunac et al., 1999).

in the site strain amplitudes, the strains during the Malibu (1/19/89) and Whittier-Narrows (10/1/87) earthquakes are also shown. It can be seen that at this station the layer stiffness returns to its original value during several early aftershocks.

Figures 4a and b, together with other such figures we have studied (Trifunac et al., 1999; Trifunac and Ivanović, 2003a,b), show that it is possible to measure the site-characteristic peaks by analysis of multiple recordings at a station when the motions are small (i.e., peak velocity is less than 5–10 cm/s). However, the resulting peaks do not appear with every excitation, and for the cases we studied they appear at most about 50% of time. With peak ground velocity exceeding 10 cm/s, site-characteristic peaks begin to disappear, and as the peak ground velocity approaches and exceeds 100 cm/s (Fig. 5) essentially all peaks disappear (Trifunac and Ivanović, 2003a). This is consistent with the conclusions of Gao et al. (1996), Hartzell et al. (1996), and Trifunac and Todorovska (2000b).

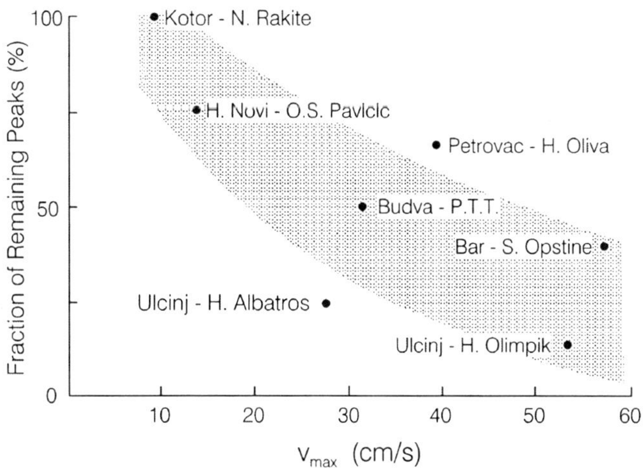

Figure 5. Fraction of identified site-specific peaks that remain in the data set, as peak ground velocity increases from 10 to 60 cm/s. At 60 cm/s and above, 60% to 90% of all site-specific peaks disappear due to nonlinear response of the soil (from Trifunac and Ivanović, 2003a).

3.2. MOVEMENT OF SOIL BLOCKS

Many observations in the epicentral regions (cracks in the pavement, buckled curbs, and concentrations of breaks in the pipes of the water distribution system) show that the near-surface soil does not move as a continuum but rather as a collection of blocks of material moving one relative to the other. This suggests that a radically different and new approach to modelling the effects of the local soil on strong ground motion and damage—and consequently for microzonation of metropolitan areas—is needed to predict the effects of damaging earthquakes.

Trifunac and Todorovska (1998b) studied simultaneously the spatial distribution of damaged (red-tagged) buildings (RTBs) and of pipe breaks following the 1994 Northridge earthquake, and they discovered that the areas with RTBs do not overlap with the areas with a large concentration of pipe breaks, except where the ground shaking was very severe (i.e., peak ground velocity exceeding about 150 cm/s). Their interpretation is that typical buildings (i.e., wood-frame buildings, which represented 84% of all buildings that received red tags) suffered less damage where the soil response was not linear. They defined so-called "gray zones" with somewhat fuzzy boundaries, but such that, wherever possible, they included the RTBs and excluded the pipe breaks. A model that could predict the location of these gray zones has not yet been formulated, but the possibility that such zones exist, in which buildings are more prone to damage because of specific features of the site geology and soil, is very significant for seismic hazard mapping and deserves detailed

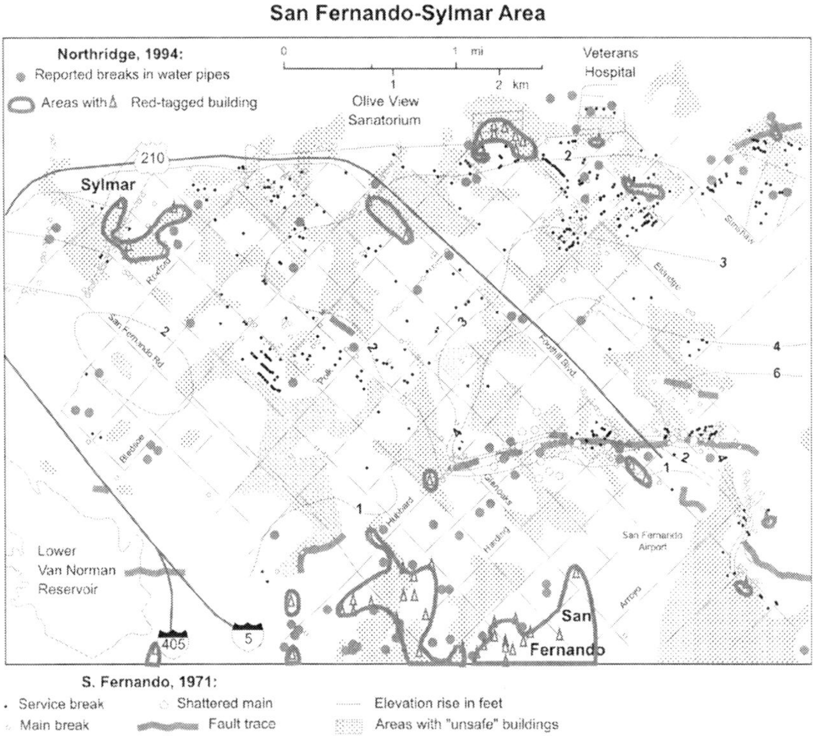

Figure 6. Sylmar–San Fernando area: overlay of the "gray zones" (for all "unsafe" buildings) and locations of pipe breaks for the 1971 San Fernando earthquake with the "gray zones" for the 1994 Northridge earthquake (from Trifunac and Todorovska, 2004).

further investigation. The authors then studied the distribution of RTBs during the 1971 San Fernando earthquake and discovered that it is possible to construct the gray zones so that they include the damaged buildings from both earthquakes while excluding the sites of the pipe breaks, also during both earthquakes (Trifunac and Todorovska, 2004). An example illustrating this is shown in Fig. 6. It can be seen that the buildings severely damaged by the Northridge earthquake occurred essentially within the gray zones defined for the San Fernando earthquake, which had occurred 23 years earlier. This figure also shows the gray zones for both earthquakes. It can be seen that in the San Fernando–Sylmar area the shaking from the Northridge earthquake "extended" the boundaries of the gray zones drawn for the San Fernando earthquake, but in a manner consistent with the principle that the gray zones do not include sites with breaks in the water pipes.

The aim of Trifunac and Todorovska's (2004) paper was to find (1) whether the gray zones (first discovered for the 1994 Northridge earthquake) existed also for the San Fernando earthquake, (2) whether and to what degree the gray zones for both earthquakes overlapped, and (3) what determines the location of the gray zones—e.g., the patterns and distribution of strong-motion amplitudes, the distribution of weaker buildings, or some other site characteristics. For the same population of buildings, two earthquakes with similar size and mechanism, and occurring within the same area, would be expected to produce similar effects. However, the Los Angeles metropolitan area grew between 1971 and 1994, and these two earthquakes neither had the same focus nor the same source mechanism. Thus, comparing the damage from these two earthquakes was not a simple task. Nevertheless, Trifunac and Todorovska's (2004) paper shows that the overall trends for both earthquakes appear to be stable, significant, and consistent. The conclusion reached is that the formation of the gray zones is mainly governed by the local soil and geologic conditions at the site, which do not change significantly during the life of a typical building (50–100 years). The implications of these observations are important, both for the future development of seismic zoning methods and for the characterization of site-specific models, with the goal being the prediction of strong motion in the near field when a local site experiences large, nonlinear deformations.

The above examples of the separation of the gray zones (with damaged buildings) from the areas with the breaks in the water pipe system, for San Fernando 1971 and Northridge 1994 earthquakes is not unique. This type of separation can and should be analysed and interpreted following any earthquake for which sufficiently detailed data exists. We only have to search for such data and interpret it (Trifunac, 2003).

3.3. NUMERICAL MODELS

Numerical methods (finite-element and finite-difference) have been used for studies of the irregular geometry of sediments and soil layers and to explore the characteristics of nonlinear response. The majority of the published papers address only one-dimensional wave propagation in simple models (e.g., Gičev and Trifunac, 2008). These studies show the complexity and the multitude of possible outcomes, which are difficult to describe with a few parameters, and thus it is difficult to incorporate them into the engineering regression analyses of recorded strong motion. As with most problems that involve large, nonlinear deformations, the number of possible outcomes becomes large and complex (Trifunac, 2009). It appears at present that we will continue to learn a great deal about the nature of nonlinear site response by investigation of the results obtained by numerical models, but the potential simple breakthroughs for robust engineering predictions can come only from many more recordings of nonlinear motions in the near field.

4. How should the Local Site Conditions be modeled?

The above-reviewed methods for description of the effects the local site conditions have on the amplitudes and spectral content of strong earthquake ground motion can all be categorized into the same group—ad hoc forward representations. In all approaches, an assumption is made with regard to how the local site effects can be modelled (this includes parametric representation for use in regression analyses and representations for numerical response simulations), and the model parameters are selected by trial and error or by a regression analysis. However, after a model has been developed, the relevance of the model is almost never addressed. This lack of the critical tests is common in the selection of the model parameters and assumptions and in the verification of the entire modelling approach. The use of Kolmogorov-Smirnoff and χ^2 tests (Lee, 2002, 2007; Trifunac and Anderson, 1977), for example, is alarmingly rare even in the most recent papers on this subject, and the question of whether the assumptions and the models are relevant with respect to the observed damage from earthquake shaking is almost never present. Many modelling approaches to selecting model parameters to represent the effects of the local site conditions appear logical and pragmatic within the limited selection framework. However, if we are to develop reliable and robust engineering tools, we must also ask the question: Is the result significant and relevant? I will illustrate this by two examples.

4.1. IS THE AVERAGE SHEAR-WAVE VELOCITY IN THE TOP 30 M A RELEVANT SCALING PARAMETER?

Lee et al. (1995), in their regression analyses of peak accelerations of strong ground motion, studied the significance of the shear-wave velocity parameter in two different ways. First, they used the average shear-wave velocity in the top 30 m of soil, and then they considered the categorical variables A, B, C, and D. Simultaneously, they considered the soil-type parameter s_L (= 0, 1, and 2). They used the student t-statistic and found that the soil-type classification (s_L) is significant, while the velocity-type classification (either average shear-wave velocity or the categorical variables A, B, and C) is not significant. They concluded that further use of the average shear-wave velocity in the top 30 m of soil, or of the corresponding categorical variables A, B, C, and D, is not indicated, while the soil-type classification variable s_L is significant and should be included in all regression models of linear strong motion. They commented that the apparent physical explanation of why s_L is significant and why the average shear-wave velocity is not is that s_L included information on the soil depth well beyond the top 30 m.

Novikova and Trifunac (1995) investigated different regression models for prediction of the duration of strong ground motion, with dependence upon (1) the local soil and geologic site parameters, (2) the geometry of the site conditions, and (3) the age of the materials under the recording station. They found that the age of local deposits is a significant site variable and that it should be included in the empirical prediction equations. However, they also found that the contribution of the average shear-wave velocity variable, $\overline{v}$, in the top 30 m is not significant for frequencies below 2.5 Hz and is significant only for the higher frequencies. The variable $\overline{v}$ describes the properties of only a very thin soil layer, the influence of which on the linear amplitudes of waves longer than 30 m should be small.

Castellaro et al. (2008) revisited old data on the relationship between Fourier spectrum amplitudes of recorded acceleration and $\overline{v}$, and discussed the requirements for meaningful regression and the significance tests of the results. They concluded that "in spite of its almost universal adoption as a key parameter in seismic site classification, $\overline{v}$ appears a weak proxy to seismic amplification".

4.2. WHAT CHARACTERIZATION OF SITE CONDITIONS IS RELEVANT?

An important, often-overlooked principle is that a prediction should be evaluated by a comparison of the actual outcome against a prediction published before the event. Post-facto detailed studies do augment our knowledge, but the only true test is a comparison of the outcome with a prediction made previously (Trifunac, 1989; Trifunac et al., 1994). Thus, a model proposed for prediction of the effects that the local site conditions have on the amplitudes of shaking, or better yet on some measure of structural response, should be evaluated by comparison with some future actual outcome. To illustrate this, we correlate a normalized measure of damage with nonlinear site response and consider different descriptions of the local site properties (measured or postulated), as shown in Fig. 7. In this figure, we plot the number of red-tagged (solid points represent seriously damaged) and yellow-tagged (open circles represent moderately damaged) buildings per 1,000 housing units, normalized relative to the area average versus the number of pipe breaks per 1,000 housing units per area average. In simple terms, we are plotting a measure of damage versus a measure of the strain amplitude in the local soil, as seen through a filter of surface geology, average shear-wave velocity in the soil, and two different liquefaction criteria.

In parts (a), (b), (c), and (d) of Fig. 7, we consider four different site characteristics: (1) surface geology, (2) average shear-wave velocity in the top 30 m of soil, (3) liquefaction susceptibility using L.A. maps, and (4) liquefaction susceptibility using U.S. Geological Survey (USGS) maps. For surface

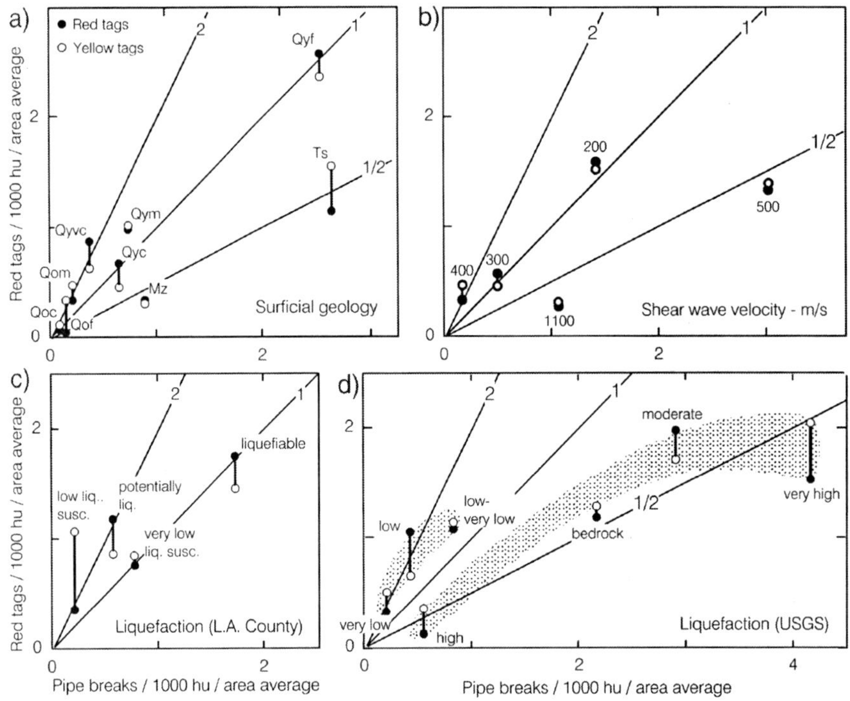

Figure 7. Occurrence of red-tagged buildings (solid points) and yellow-tagged buildings (open circles) versus pipe breaks (both normalized to unit average for the total area of the map) relative to (a) surface geology, (b) surface shear-wave velocity, (c) liquefaction susceptibility based on L.A. County maps, and (d) liquefaction susceptibility based on USGS maps (from Todorovska and Trifunac, 1998).

geology (Fig. 7a), we consider the following: Qyf (fine-grained Holocene alluvium), Qym (medium-grained Holocene alluvium), Qyc (coarse-grained Holocene alluvium), Qyvc (very-coarse-grained Holocene alluvium), Qof (fine-grained Pleistocene alluvium), Qom (medium-grained Pleistocene alluvium), Qoc (coarse-grained Pleistocene alluvium), Ts (Tertiary and pre-Tertiary sedimentary rock), and Mz (Mesozoic and pre-Mesozoic rocks) (Tinsley and Fumal, 1985; Trifunac and Todorovska, 1998a). For average shear-wave velocity in the top 30 m of soil (in Fig. 7b), we consider 200, 300, 400, 500, and 1,100 m/s. In Fig. 7c, we use the liquefaction susceptibility categories as defined in the maps for Los Angeles County: Liquefiable, Potentially Liquefiable, Low Liquefaction Susceptibility, and Very-Low Liquefaction Susceptibility (Leighton and Associates, Inc., 1990)). In Fig. 7d, we use the liquefaction susceptibility categories in the USGS maps: Very High, High, Moderate, Low, Low-Very Low, Very Low, and Bedrock (Tinsley et al., 1985).

Figures 7a and b show that only for Ts and Mz rock sites, and for shear-wave velocities in the soil equal to 500 and 1,100 m/s, there were more pipe breaks than damaged buildings (compared with the respective total area averages). This is due to hillside ground conditions at most of the sites contributing to the data set and the occurrence of landslides. Figure 7d shows that for the sites with "moderate", "high", and "very high" liquefaction susceptibility there were proportionally fewer damaged buildings than pipe breaks (compared with the respective total area averages), by approximately a factor of two. This is in excellent agreement with the mechanism for the formation of the "gray zones" (as discussed above) and the passive isolation of single-family, wood-frame dwellings from the incident-seismic-wave energy (Trifunac and Todorovska, 1998b). It can be seen that neither in terms of surface geology nor in terms of the average shear-wave velocity in the top 30 m does the site characterization correlate with the damage to wood-frame residential buildings in the near field. The site characterization in terms of the liquefaction susceptibility (USGS) as described by Tinsley et al. (1985) is the only site characterization in this group of four that is indicated as a useful and significant site-characterization parameter for damaging levels of strong motion. Perusal of the liquefaction susceptibility criteria in Tinsley et al. (1985) shows that the ultimate categorical variables, like "very high" or "moderate", are derived on the basis of multiple site characteristics and therefore can also describe the relevant site properties for the purposes of amplitude scaling. It can be seen that in the near field, for damaging levels of strong motion, local geological and soil ($\overline{v}$) site conditions cease to be good predictors of the damage to wood-frame structures, while the composite site characterization in terms of the liquefaction susceptibility, as defined in the maps of Tinsley et al. (1985), works reasonably well.

5. Discussion and Conclusions

I have illustrated the contemporary approaches for inclusion of the effects that local site conditions have on the amplitudes of strong ground motion and how those approaches are essentially based on concepts that have evolved from classical linear-wave-propagation theory. While this approach works in the far field, I showed examples of how it ceases to apply in the near field, where the buildings get damaged and where the soil experiences large nonlinear and permanent deformations. I hinted, using an example of a refined site characterization that correlates well with the observed damage (Fig. 7d), that better and more physically meaningful site characterizations can continue to be developed, but this would still leaves us within the traditional "linear" approach for the scaling of strong-motion amplitudes. To go beyond this linear approach and to predict the nature of strong motion in the near-field region that describes the forces on the engineering structures, we must change the entire approach and formulate a new one. This new approach must include all relevant components in the description of the forces acting on a structure. The first step in this direction will require that we abandon the traditional scaling, which is based on only one scalar quantity (e.g., peak acceleration, amplitude of a response spectrum, peak strain, or peak differential displacement) to describe the strong-motion effects on the response of structures. To accomplish this goal, we will have to work with multi-parametric representation and include all relevant components of all forces that act in the near field and that contribute significantly to the response. In the following, we illustrate how this could be done.

With large amplitudes of strong motion, surface soil experiences large, nonlinear response, and ultimately soil failure and liquefaction can lead to large transient and permanent motions. We illustrate this by examples of ground failure that can follow liquefaction: lateral spreading, ground oscillations, flow failure, and loss of bearing strength. *Lateral spreads* involve displacements of surface blocks of sediment facilitated by liquefaction in a subsurface layer. This type of failure may occur on slopes up to 3° and is particularly destructive to pipelines, bridge piers, and other long and shallow structures situated in flood plain areas adjacent to rivers. *Ground oscillations* occur when the slopes are too small to result in lateral spreads following liquefaction at depth. The overlying surface blocks break, one from another, and then oscillate on liquefied substrate. *Flow failures* are a more catastrophic form of material transport and usually occur on slopes greater than 3°. The flow consists of liquefied soil and blocks of intact material riding on and with liquefied substrate, on land or under the sea. *Loss of bearing strength* can occur when the soil liquefies under a structure. The building can settle, tip, or float upward if the structure is buoyant. The accompanying motions can lead to large transient and permanent displacements and rotations, which so far have been neither evaluated through simulation nor recorded by strong-motion instruments.

Consequently, any structure, and in particular all extended structures (e.g., long buildings, bridges, tunnels, dams), in the area where such large nonlinearities in the soil occur, will, in addition to the horizontal components of inertial forces caused by strong earthquake shaking, experience large differential motions and large differential rotations of their foundation(s). Bridge peers or foundations of long buildings supported by soil, which the earthquake has separated into blocks by strong shaking, will be forced to deform, accompanied by large differential motions (translations and rotations) of soil blocks, and they will experience both the inertial and pseudo-static aspects of those motions. At present, we can only speculate about how much larger these motions will be relative to the tilts and angular accelerations and velocities we can estimate from the linear-wave theory. Few observations, however, suggest that those can be orders of magnitude larger than the predictions based on the linear theory (Trifunac, 2008a). For successful design, it will be necessary to prescribe the resulting forcing functions, which will include, in a balanced way, the simultaneous action of all components of possible motion. The description of how to scale those balanced forcing functions can start from principles similar to what we use today for the design of structures crossing an active fault (Todorovska et al., 2007; Trifunac, 2008b). Because the complexity of such motions and the multiplicity of possible outcomes will increase with amplitudes of incident strong-motion waves, specification of the driving forces for design may best be formulated in terms of their distribution functions. This will require systematic and long-range research programs focusing on two key tasks: (1) development of advanced numerical simulation models, and (2) the recording of all six components of strong motion, in the near field, and their analysis and interpretation. Such description of the near-field motion will have to be used in the selection of design forces within distances that are equal to about one source dimension (e.g., up to 20 to 50 km in California) away from the fault. In the far field, we should be able to continue to use the traditional local site parameters to describe the effects of the local site conditions for most design applications.

References

Abrahamson, N. A. and Silva, W. J. (1997) Empirical response spectral attenuation relations for shallow crustal earthquakes, *Seism. Res. Lett.* **68**(1), 94–127.

Aki, K. (1988) Local site effects on strong ground motion. In *Proc. of Earthquake Engineering and Soil Dynamics II*, GT Div./ASCE, Park City, Utah.

Ambraseys, N. N., Douglas, J., Sarma, S. K., and Smit, P. M. (2005a) Equations for the estimation of strong ground motions from shallow crustal earthquakes using data from Europe and the middle east: Horizontal peak ground acceleration and spectral acceleration, *Bull. Earthquake Eng.* **3**, 1–53.

Ambraseys, N. N., Douglas, J., Sarma, S. K., and Smit, P. M. (2005b) Equations for the estimation of strong ground motions from shallow crustal earthquakes using data from Europe and the middle east: Vertical peak ground acceleration and spectral acceleration, *Bull. Earthquake Eng.* **3**, 55–73.

Ambraseys, N. N., Simpson, K. A., and Bommer, J. J. (1996) Prediction of horizontal response spectra in Europe, *Earthquake Eng. Structural Dyn.* **25**, 371–400.

Anderson, J. G., Trifunac, M. D., Teng, T. L., Amini, A., and Moslem, K. (1981) Los Angeles vicinity strong motion accelerograph network, Report CE 81-04, Department of Civil Engineering, Univ. of Southern Calif., Los Angeles, CA.

Boore, D. M., Joyner, W. B., and Fumal, T. (1997) Equations for estimating horizontal response spectra and peak acceleration from western north american earthquakes: A summary of recent work, *Seism. Res. Lett.* **68**(1), 128–153.

Borcherdt, R. D. (1970) Effects of local geology on ground motion near San Francisco Bay, *Bull. Seism. Soc. Am.* **60**, 29–61.

Borcherdt, R. D. and Gibbs, J. F. (1976) Effects of local geological conditions in the San Francisco Bay region on ground motions and intensities of the 1906 earthquake, *Bull. Seism. Soc. Am.* **66**, 467–500.

Campbell, K. and Duke, C. M. (1974) Bedrock intensity attenuation and site factors from San Fernando earthquake records, *Bull. Seism. Soc. Am.* **64**, 173–185.

Castellaro, S., Mulargia, F., and Rossi, P. L. (2008) Vs30: Proxy for seismic amplification? *Seism. Res. Lett.* **79**(4), 540–543.

Chiou, B., Darragh, R., Gregor, N., and Silva, W. (2008) NGA project strong-motion database, *Earthquake Spectra* **24**(1), 23–44.

Coulter, H. W., Waldron, H. H., and Devine, J. F. (1973) Seismic and geologic siting considerations for nuclear facilities. In *Proc. of the 5th World Conf. on Earthquake Engineering*, Rome, Italy.

Duke, C. M. (1958) Bibliography of effects of soil conditions on earthquake damage, Berkeley, CA, Earthquake Engineering Research Institute.

Earthquake Engineering Research Institute (1995) Northridge earthquake of January 17, 1994, Reconnaissance Report, Vol. 1. In *Earthquake Spectra* **11**, Suppl. C.

Field, E. H. and Hough, S. H. (1997) The variability of PSV response spectra across a dense array deployed during the Northridge aftershock sequence, *Earthquake Spectra* **13**(2), 243–257.

Freeman, J. R. (1932) *Earthquake Damage and Earthquake Insurance*, New York, McGraw-Hill.

Gao, S., Liu, H., Davis, P. M., and Knopoff, L. (1996) Localized amplification of seismic waves and correlation with damage due to the Northridge earthquake: Evidence for focusing in Santa Monica, *Bull. Seism. Soc. Am.* **86**(1B), S209–S230.

Gičev, V. and Trifunac, M. D. (2008) Transient and permanent rotations in a shear layer excited by strong earthquake pulses, *Bull. Seism. Soc. Amer.* (submitted). doi:10.1785/0120080066.

Goto, H., Kameda, H., and Sugito, A. (1982) Use of N-value profiles for estimation of site dependent earthquake motions, *Collected Papers* **317**, pp. 69–78, Japanese Society of Civil Engineering (in Japanese).

Gutenberg, B. (1957) Effects of ground on earthquake motion, *Bull. Seism. Soc. Am.* **47**, 221–250.

Harmsen, S. C. (1997) Determination of site amplification in the Los Angeles urban area from inversion of strong motion records, *Bull. Seism. Soc. Am.* **87**, 866–887.

Hartzell, S. (1998) Variability of nonlinear sediment response during Northridge, California earthquake, *Bull. Seism. Soc. Am.* **88**(6), 1426–1437.

Hartzell, S., Leeds, A., Frankel, A., and Michael, J. (1996) Site response for urban Los Angeles using aftershocks of the Northridge earthquake, *Bull. Seism. Soc. Am.* **86**(1B), S168–S192.

Haskell, N. A. (1969) Elastic displacements in the near-field of a propagating fault, *Bull. Seism. Soc. Am.* **59**, 865–908.

Idriss, I. M. and Seed, H. B. (1968) An analysis of ground motions during the 1957 San Francisco earthquake, *Bull. Seism. Soc. Amer.* **58**, 2013–2032.

Ivanović, S. S., Trifunac, M. D., and Todorovska, M. I. (2000) Ambient vibration tests of structures—A review, *Bull. Indian Soc. Earthquake Tech.* **37**(4), 165–197.

Kanai, K. (1983) *Engineering Seismology*, Tokyo, Univ. of Tokyo Press.

Lee, K. L. and Albaisa, A. (1974) Earthquake induced settlements in saturated sands, *J. Geotechnical Eng.* **100**(GT4), 387–406.

Lee, V. W. (1987) Influence of local soil and geologic site conditions on pseudo relative velocity response spectrum amplitudes of recorded strong motion accelerations, Report No. CE 87-05, Department of Civil Engineering, Univ. of Southern California, Los Angeles, CA.

Lee, V. W. (2002) Empirical scaling of earthquake ground motion. Part I: Attenuation and scaling response spectra, *ISET J.* **39**(4), 219–254.

Lee, V. W. (2007) Empirical scaling and regression methods for earthquake strong-motion spectra—A review, *ISET J.* **44**(1), 39–69.

Lee, V. W. and Trifunac, M. D. (1995) Frequency dependent attenuation function and fourier amplitude spectra of strong earthquake ground motion in California, Report No. CE 95-03, Dept. of Civil Eng., Univ. of Southern Cal., Los Angeles, CA.

Lee, V. W., Trifunac, M. D., Todorovska, M. I., and Novikova, E. I. (1995) Empirical equations describing attenuation of the peaks of strong ground motion, in terms of magnitude, distance, path effects and site conditions, Report No. CE 95-02, Dept. of Civil Eng., Univ. of Southern California, Los Angeles, CA.

Leighton and Associates, Inc. (1990) Technical appendix to the safety of the Los Angeles County General Plan, Vol. 1, Prepared for Los Angeles County Board of Supervisors, Regional Planning Comm., Dept. of Regional Planning.

Novikova, E. I. and Trifunac, M. D. (1993a) Modified Mercalli intensity and the geometry of the sedimentary basin as the scaling parameters of the frequency dependent duration of strong ground motion, *Soil Dyn. Earthquake Eng.* **12**(4), 209–225.

Novikova, E. I. and Trifunac, M. D. (1993b) Duration of strong earthquake ground motion: Physical basis and empirical equations, Report No. CE 93-02, Dept. of Civil Eng., Univ. Southern California, Los Angeles, CA.

Novikova, E. I. and Trifunac, M. D. (1994a) Duration of strong ground motion in terms of earthquake magnitude epicentral distance, site conditions and site geometry, *Earthquake Eng. Structural Dyn.* **23**(6), 1023–1043.

Novikova, E. I. and Trifunac, M. D. (1994b) Influence of geometry of sedimentary basins on the frequency dependent duration of strong ground motion, *Earthquake Eng. and Eng. Vibration* **14**(2), 7–44.

Novikova, E. I. and Trifunac, M. D. (1995) Frequency dependent duration of strong earthquake ground motion: Updated empirical equations, Report No. CE 95-01, Dept. of Civil Eng., Univ. Southern California, Los Angeles, CA.

Reid, H. F. (1910) The California earthquake of April 18, 1906. In *The Mechanics of the Earthquake*, **2**, Report of the State Earthquake Investigation Commission, Carnegie Institute of Washington, Publ. 87, Washington, DC.

Rogers, A. M., Tinsley, J. C., and Borcherdt, R. D. (1985) Predicting relative ground response. In J. L. Ziony (ed.), *Evaluating Earthquake Hazards in the Los Angeles Region*, U.S.G.S. Professional Paper 1360, pp. 221–248.

Seed, H. B., Ugas, C., and Lysmer, J. (1976) Site-dependent spectra for earthquake-resistant design, *Bull. Seism. Soc. Am.* **66**, 221–243.

Tinsley, J. C. and Fumal, T. E. (1985) Mapping quaternary sedimentary deposits for areal variation in shaking respnse. In *Evaluating Earthquake Hazards in the Los Angeles Region—An Earth Science Perspective*, U.S. Geological Survey Pofessionl Paper 1360, Washington, DC.

Tinsley, J. C., Youd, T. L., Perkins, D. M., and Chen, A. T. F. (1985) Evaluating liquefaction potential, In *Evaluating Earthquake Hazards in the Los Angels Region—An Earth Science Perspective*, U.S.G.S. Professional Paper 1360, Washington, DC.

Todorovska, M. I. and Trifunac, M. D. (1998) Discussion of "The role of earthquake hazard maps in loss estimation: A study of the Northridge Earthquake," by R. B. Olshansky, *Earthquake Spectra* **14**(3), 557–563.

Todorovska, M. I., Trifunac, Todorovska M. D., and Lee, V. W. (2007) Shaking hazard compatible methodology for probabilistic assessment of permanent ground displacement across earthquake faults, *Soil Dyn. Earthquake Eng.* **27**(6), 586–597.

Tokimatsu, K. and Seed, H. B. (1987) Evaluation of settlements in sands due to earthquake shaking, *J. Geotechnical Eng., ASCE* **113**(8), 861–878.

Trifunac, M. D. (1971a) Response envelope spectrum and interpretation of strong earthquake ground motion, *Bull. Seism. Soc. Am.* **61**, 343–356.

Trifunac, M. D. (1971b) Surface motion of a semi-cylindrical alluvial valley for incident plane SH waves, *Bull. Seism. Soc. Am.* **61**(6), 1755–1770.

Trifunac, M. D. (1974) A three-dimensional dislocation model for the San Fernando, California, earthquake of 9 February 1971, *Bull. Seism. Soc. Am.* **64**, 149–172.

Trifunac, M. D. (1976a) Preliminary analysis of the peaks of strong earthquake ground motion dependence of peaks on earthquake magnitude, epicentral distance and recording site conditions, *Bull. Seism . Soc. Am.* **66**, 189–219.

Trifunac, M. D. (1976b) A note on the range of peak amplitudes of recorded accelerations, velocities and displacements with respect to the modified Mercalli intensity, *Earthquake Notes* **47**(1), 9–24.

Trifunac, M. D. (1978) Response spectra of earthquake ground motion, *J. Eng. Mech. Div. ASCE* **104**, 1081–1097.

Trifunac, M. D. (1979) Preliminary empirical model for scaling Fourier amplitude spectra of strong motion acceleration in terms of modified Mercalli intensity and geologic site conditions, *Earthquake Eng. Structural Dyn.* **7**, 63–74.

Trifunac, M. D. (1987) Influence of local soil and geologic site conditions on Fourier spectrum amplitudes of recorded strong motion accelerations, Report No. CE 87-04, Dept. of Civil Eng., Univ. of Southern California, Los Angeles, CA.

Trifunac, M. D. (1989) Threshold magnitudes which cause the ground motion exceeding the values expected during the next 50 years in a metropolitan area, *Geofizika* **6**, 1–12.

Trifunac, M. D. (1990) How to model amplification of strong earthquake ground motions by local soil and geologic site conditions, *Earthquake Eng. Structural Dyn.* **19**(6), 833–846.

Trifunac, M. D. (2003) Nonlinear soil response as a natural passive isolation mechanism. Paper II—The 1933, Long Beach, California earthquake, *Soil Dyn. Earthquake Eng.* **23**(7), 549–562.

Trifunac, M. D. (2008a) The role of strong motion rotations in the response of structures near earthquake faults, *Soil Dyn. Earthquake Eng.* **29**(20), 382–393, doi:10.1016/j.soildyn.2008.04.001.

Trifunac, M. D. (2008b) Design of structures crossing active faults. In *Monograph Celebrating 85th anniversary of the birth of Prof. Milan Djurić*, Gradjevinski Fakultet u Beogradu, Katedra za Tehničku Mehaniku i Teoriju Konstrukcija, Beograd.

Trifunac, M. D. (2009) Nonlinear problems in earthquake engineering. In *Springer's Encyclopedia of Complexity and System Science* (in press).

Trifunac, M. D. and Anderson, J. G. (1977) Preliminary empirical models for scaling absolute acceleration spectra, Report No. 77-03, Dept. of Civil Eng., Univ. of Southern California, Los Angeles, CA.

Trifunac, M. D. and Anderson, J. G. (1978a) Preliminary empirical models for scaling pseudo relative velocity spectra, Report No. 78-04, Dept. of Civil Eng., Univ. of Southern California, Los Angeles, CA.

Trifunac, M. D. and Anderson, J. G. (1978b) Preliminary models for scaling relative velocity spectra, Report No. 78-05, Dept. of Civil Eng., Univ. of Southern California, Los Angeles, CA.

Trifunac, M. D. and Brady, A. G. (1976) On the correlation of seismic intensity scales with the peaks of recorded strong ground motion, *Bull. Seism. Soc. Am.* **66**, 139–162.

Trifunac, M. D., Hao, T. Y., and Todorovska, M. I. (1999) On reoccurrence of site specific response, *Soil Dyn. Earthquake Eng.* **18**(8), 569–592.

Trifunac, M. D. and Ivanović, S. S. (2003a) Reoccurrence of site specific response in former Yugoslavia. Part I: Montenegro, *Soil Dyn. Earthquake Eng.* **23**(8), 637–661.

Trifunac, M. D. and Ivanović, S. S. (2003b) Reoccurrence of site specific response in former Yugoslavia. Part II: Friuli, Banja Luka, and Kopaonik, *Soil Dyn. Earthquake Eng.* **23**(8), 663–681.

Trifunac, M. D. and Lee, V. W. (1978) Dependence of the Fourier amplitude spectra of strong motion acceleration on the depth of sedimentary deposits, Report No. 78-14, Dept. of Civil Eng., Univ. of Southern California, Los Angeles, CA.

Trifunac, M. D. and Lee, V. W. (1979) Dependence of the pseudo relative velocity spectra of strong motion acceleration on the depth of sedimentary deposits, Report 79-02, Dept. of Civil Eng., Univ. of Southern Cal., Los Angeles, CA.

Trifunac, M. D. and Todorovska, M. I. (1996) Nonlinear soil response—1994 Northridge California, earthquake, *J. Geotech. Eng. ASCE* **122**(9), 725–735.

Trifunac, M. D. and Todorovska, M. I. (1998a) Damage distribution during the 1994 Northridge, California, earthquake in relation to generalized categories of surficial geology, *Soil Dyn. Earthquake Eng.* **17**(4), 239–253.

Trifunac, M. D. and Todorovska, M. I. (1998b) Nonlinear soil response as a natural passive isolation mechanism—the 1994 Northridge earthquake, *Soil Dyn. Earthquake Eng.* **17**(1), 41–51.

Trifunac, M. D. and Todorovska, M. I. (2000a) Long period microtremors, microseisms and earthquake damage: Northridge, CA, earthquake of 17 January, 1994, *Soil Dyn. Earthquake Eng.* **19**(4), 253–267.

Trifunac, M. D. and Todorovska, M. I. (2000b) Can aftershock studies predict site amplification? Northridge, CA, earthquake of 17 January, 1996, *Soil Dyn. Earthquake Eng.* **19**(4), 233–251.

Trifunac, M. D. and Todorovska, M. I. (2004) 1971 San Fernando and 1994 Northridge, California, earthquakes: Did the zones with severely damaged buildings reoccur? *Soil Dyn. Earthquake Eng.* **24**(3), 225–239.

Trifunac, M. D., Todorovska, M. I., and Ivanović, S. S. (1994) A note on distribution of uncorrected peak ground accelerations during the Northridge, California earthquake of 17 January 1994, *Soil Dyn. Earthquake Eng.* **13**(3), 187–196.

Udwadia, F. E. and Trifunac, M. D. (1973) Comparison of earthquake and microtremor ground motions in El Centro, California, *Bull. Seism. Soc. Am.* **63**(4), 1227–1253.

EARTHQUAKE SOURCE AND LOCAL GEOLOGY EFFECTS ON THE SEISMIC SITE RESPONSE

Mihaela Kouteva-Guentcheva (mkouteva@geophys.bas.bg)
Central Laboratory for Seismic Mechanics and Earthquake Engineering, Bulgarian Academy of Sciences, and ESP-ICTP, Italy

Ivanka Paskaleva (paskalev@geophys.bas.bg)
Central Laboratory for Seismic Mechanics and Earthquake Engineering, Bulgarian Academy of Sciences

Giuliano F. Panza (panza@dst.units.it)
Department of Earth Sciences, University of Trieste and ESP-ICTP, Italy

Abstract. Strong shallow and intermediate-depth scenario earthquakes for two major cites in Bulgaria are discussed. The contribution of the earthquake source and the local site geology to the seismic input is illustrated. Due to the lack of strong motion records a neo-deterministic seismic hazard assessment procedure is used to generate synthetic seismic signals. After some parametric analyses the computed signals are validated against the few available data. Prognostic estimates of the dynamic coefficient for the target sites are performed with respect to the defined scenario earthquakes and local site models, corresponding to the Eurocode 8 (EC8) ground types A, B and C. The obtained results show that: (1) the seismic source influence on the seismic input at a given site is comparable with that of the local site geology; (2) the dynamic coefficients, computed for accelerograms (observed and computed) due to strong intermediate-depth Vrancea earthquakes overestimate significantly the values recommended by the EC8 for periods $T > 1$ s.

Keywords: scenario earthquakes, site response, neo-deterministic seismic hazard assessment, dynamic coefficient

1. Introduction

The seismic hazard of Bulgaria is controlled by seismic sources located in country and also in the territory of the neighbouring countries (Romania, Greece, Turkey, Serbia and the F.Y.R. of Macedonia). Maximum intensity I = IX (MSK, Medvedev, 1977) is expected for the main Bulgarian city, Sofia. The seismic hazard in NE Bulgaria with the major town Russe, the biggest Bulgarian port on the Danube River, is controlled by the Vrancea seismic zone, located in Romania.

T. Schanz and R. Iankov (eds.), *Coupled Site and Soil-Structure Interaction Effects with Application to Seismic Risk Mitigation,* NATO Science for Peace and Security Series C: Environmental Security,

The intermediate-depth Vrancea earthquake sources are of practical and scientific interest due to their social and economic impact on the territory of the adjacent countries. A brief analysis of the available instrumental records of the strong intermediate-depth Vrancea earthquakes, with Mw ≥ 6.5 (Nenov et al., 1990; Ambraseys et al., 2002), has shown the significant effect of the earthquake source mechanism on the seismic input at sites, which clearly differ in local geological conditions and epicentral distances. Another Vrancea peculiarity is the much stronger frequency-dependent attenuation effect toward NW for higher frequencies (≥1 Hz) than the attenuation toward SE (Radulian et al., 2006). The unusually small attenuation at low frequency has important consequences on the seismic hazard assessment not only in Romania, but also in the neighboring countries (Bulgaria, Rep. of Moldova, Ukraine and even Russia). The main purposes of this study are to:

- provide strong scenario earthquakes (Mw ≥ 7.0), that can be used for prognostic estimates of the seismic input;
- to validate the synthetic seismic signals against the available data and to supply seismic input for the chosen scenarios.

2. The Neo-Deterministic Seismic Hazard Assessment Procedure and the Modeling of the Seismic Input at Sofia and Russe

To obtain the seismic input at Sofia and Russe a neo-deterministic procedure for earthquake ground motion modelling has been applied (Panza et al., 2001). To model the seismic input in Sofia the hybrid neo-deterministic approach (Fah et al., 1993; Panza et al., 2001) is used. It combines the modal summation technique used to describe the seismic wave propagation in the inelastic bedrock structure (Panza and Suhadolc, 1987; Panza et al., 2001) with finite difference method (Virieux, 1984; Virieux, 1986; Levander, 1988) used for the computation of wave propagation in the inelastic, laterally inhomogeneous sedimentary media (Stein and Wysession, 2003). To model the seismic input in Niš and Russe the analytical neo-deterministic approach based on the mode coupling technique, is applied (Romanelli et al., 1996; Romanelli et al., 1997; Panza et al., 2001).

The major advantages of this neo-deterministic procedure are: (1) the simultaneous treatment of the contribution of the seismic source and of the seismic wave propagation through inelastic media to the seismic motion at the target site/region and therefore (2) the application of this procedure does not require the use of any attenuation relation. Applying this procedure site response estimates are provided simultaneously in frequency and space domain. The irregular pattern of the site amplification in the frequency–space domain, obtained even when considering rather simplified geological settings, as for the Russe case study (e.g., Kouteva et al., 2004), confirms the complicated properties of the so-called "site effect". They are due to the complex evolution of the seismic wavefield (while it propagates through the laterally heterogeneous, geological media) that cannot be captured by standard convolutive methods (e.g., Reiter, 1990). The traditionally used attenuation relations, extracted from the available strong motion databases, represent the functional dependency of the random spectral acceleration on the random variables, magnitude, distance and measurement error, and thus the source of systematic error in the seismic hazard assessment that might be introduced by the attenuation relations (Klugel, 2007; Panza et al., 2008) is avoided when using the neo-deterministic approach. The problem how crustal properties affect the attenuation and the effect due to inelasticity is taken into account analytically by the neo-deterministic procedure, using variational techniques (Panza et al., 2001, and references therein).

3. Scenario Earthquakes

The scenario event represents different combinations of parameters, thus the scenario earthquakes can be different in what concerns source location, magnitude and parameters describing the geometry and the kinematics of the seismic source. Usually for an earthquake prone area, scenario earthquakes with different levels of severity are considered: moderate, severe and extreme earthquakes. Widely accepted in international practice in earthquake engineering analysis, including EUROCODE 8, is the return period of 475 years.

3.1. SHALLOW LOCAL EARTHQUAKES—THE CITY OF SOFIA

The return period of the maximum macroseismic intensity at Sofia, I_o = IX (MSK), is about 150 years (Christoskov et al., 1982), i.e., it could correspond to the strong earthquake scenario. The shallow scenario earthquakes considered in this study are listed in Table I. For more details see Paskaleva et al. (2008) and references therein.

3.2. STRONG INTERMEDIATE-DEPTH VRANCEA EARTHQUAKES

Considering the specific natural conditions, the various categories of elements and systems of risk and the Vrancea earthquake record, the suitable scenario earthquakes, for this area, should correspond to return periods ranging from some 50 to 200 years, when severe or extreme magnitudes are considered,

TABLE I. Scenario earthquakes—local shallow quakes, Sofia City*

Scenario	Geologic Profile	Mw	Strike Angle	Dip Angle	Rake Angle	Closest Distance to the Fault	Focal Depth
Sce1 all	M1, M2, M3	7.0	340°	77°	285°	10 km	10 km
Sce1 3a	M3	7.0	0°	44°	309°	10 km	10 km

City sketch* and details on the geology of the considered geological profiles was published by Paskaleva et al. (2004).

TABLE II. Scenario earthquakes—strong intermediate-depth Vrancea quakes*

Scenario	Latitude	Longitude	Mw	Strike Angle	Dip Angle	Rake Angle	Focal Depth
Sce 1	45.76°N	26.53°E	7.2	240°	72°	97°	132.7 km
Sce 2	45.80°N	26.70°E	7.8	225°	60°	80°	150.0 km

Sce1 seismic source corresponds to the 1986 Vrancea earthquake, August 30, VR86, (Dziewonsky et al., 1991) and Sce2 seismic source corresponds to the 1940 Vrancea quake, November 10, VR40 (Radulian et al., 2000 and references therein; Lungu et al., 2004).

respectively. Suitable Vrancea scenario events can be considered the quakes in the magnitude range from 7.2, severe earthquakes, to 7.8, extreme earthquakes (Georgescu and Sandi, 2000). The chosen scenario earthquakes are listed in Table II.

4. The Synthetic Strong-Motion Database

The seismic input in Sofia and Russe was computed applying the neo-deterministic procedure for seismic wave propagation modelling (Panza et al., 2001).

4.1. SHALLOW LOCAL EARTHQUAKES: CASE STUDY—SOFIA CITY

The City of Sofia is situated in the central southern part of the Sofia kettle, a continental basin in southern Bulgaria, filled with Miocene–Pliocene sediments. The bedrock is represented by heterogeneous (in composition) and different (in age) rocks, which outcrop within the depression. The Sofia kettle is filled with Neogene and Quaternary sediments and its thickness reaches

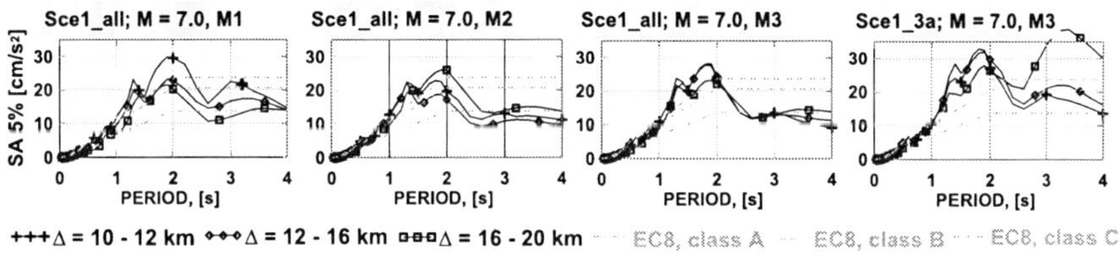

Figure 1. Dynamic coefficients of the computed signals versus the EC8 recommended curves (Paskaleva et al., 2008).

1200 m. From the structural point of view, the Sofia kettle represents a complex, asymmetric block structure graben, located in the West Srednogorie region, with an average altitude of about 550 m (Ivanov et al., 1998). Details on the tectonics and the local seismicity of the region, and on the construction the structural velocity computation models are provided by Paskaleva et al. (2004) and references therein.

Synthetic ground motions along three geological cross sections have been computed and validated by Paskaleva et al. (2004, 2008). The signals have been grouped in three ranges of epicentral distances: 10–12, 12–16 and 16–20 km. For each group mean ground motion spectral quantities are computed. The dynamic coefficients, computed from the synthetic seismic signals, and the EC8 recommended curves, are plotted in Fig. 1. This figure shows that the variation of the seismic source mechanism (Sce1all, M3 and Sce3a, M3) can cause significant change of the site response, comparable with the influence of the soil conditions (e.g., Sce1all for models M1, M2 and M3).

4.2. STRONG VRANCEA INTERMEDIATE-DEPTH EARTHQUAKES: CASE STUDIES OF RUSSE (BULGARIA) AND NIŠ (SERBIA)

4.2.1. *Validation of the seismic input*

The profile Vrancea-Russe passes through the Carpathians and the Moesian Platform, where Pliocene and significant Quaternary deposits are present. Moving away from the Vrancea zone a gradual transition from hard rocks to unconsolidated rocks and progressively softer soils is observed. For both cases the same bedrock model Roma06.str (Radulian et al., 2000) is used. For the validation of the computed signals at Russe and Niš , three local models, corresponding to EC8 ground type C ($V_{s,30}$ = 325 m/s) have been used: (a) deep model, top layer (of type C) 150 m thick, (b) intermediate model, top layer 60 m thick and (c) shallow model, top layer 30 m thick.

The frequency–time analysis of the available records (Nenov et al., 1990; Ambraseys et al., 2002) of strong intermediate-depth Vrancea earthquakes has shown that the frequencies up to 5 Hz have the major contribution to the seismic loading of practical importance. A typical example of the long period far reaching effect is the Vrancea March 4, 1977 earthquake. The available data on the Vrancea earthquakes of March 4, 1977 (VR77) and May 30, 1990 (VR901) are used to validate the computations at Niš (ep. distance ~ 500 km) and Russe (ep. distance ~ 220–230 km), respectively.

The values, including their uncertainties, of the parameters, describing the geometry and the motion at the earthquake source, are available from the GCMT Catalogue (`http://www.globalcmt.org/CMTfiles.html`), Radulian et al. (2000) and the Romplus catalogue (Oncescu et al., 1999). The seismic input in Russe and Niš has been computed and validated using the parameters given in Table III.

The comparison of the elastic acceleration response spectra, computed for 5% damping, of the synthetic and observed signals, considering VR77 (Niš) and VR901 (Russe), are shown in Figs. 2 and 3, respectively. Both figures clearly show the significant influence of the seismic source on the earthquake loading at the target sites.

The change of the focal depth of 84 and 94 km (CMT1 and CMT2, plotted in Fig. 2) to 100 km (CMT1m, CMT2m) for the Niš case shows a shift of the maximum amplitudes of the acceleration response spectra to larger periods.

TABLE III. Strong intermediate-depth Vrancea earthquakes. Data used for parametric studies and for validation of the seismic input

Source	Latitude	Longitude	Mw	Strike Angle	Dip Angle	Rake Angle	Focal Depth
Niš: VR77							
CMT1	45.23°N	26.17°E	7.5	236°	62°	92°	84 km
CMT2	45.23°N	26.17°E	7.5	50°	28°	86°	94 km
Other	45.80°N	26.80°E	7.5	225°	70°	110°	100 km
CMT1m	45.23°N	26.17°E	7.5	236°	62°	92°	100 km
CMT2m	45.23°N	26.17°E	7.5	50°	28°	86°	100 km
Russe: VR901							
CMT	45.92°N	26.81°E	6.9	236°	63°	101°	74 ± 6 km
NIEP	45.92°N	26.81°E	6.9	236°	63°	101°	90 km

VR77: Vrancea, March 4, 1977 and VR901: Vrancea, May 30, 1990; CMT1, CMT2, CMT (Global Centroid Moment Tensor Catalogue); CMT1m, CMT2m correspond to CMT1, CMT2 respectively, focal depth $H = 100$ km considered for both cases; NIEP corresponds to Radulian et al. (2000).

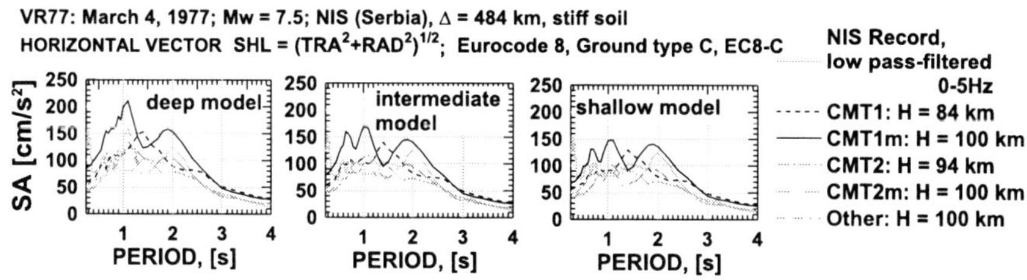

Figure 2. Niš site: Elastic acceleration response spectra, computed for 5% damping. Synthetics against observation (solid grey line).

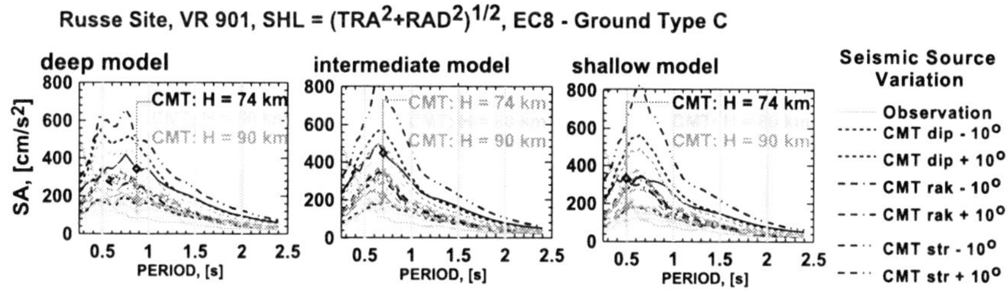

Figure 3. Russe site: Elastic acceleration response spectra, computed for 5% damping. Synthetics against observation (solid grey line).

A comparison of the obtained results for Niš, following Table III, shows that among the considered parameters, the focal depth seems to have a controlling impact on the seismic input.

The results of the theoretical modelling of the seismic input in Russe, including some parametric studies, are shown in Fig. 2. Among the earthquake source parameters, the focal depth has, here too, the most significant influence on the seismic loading at the target site.

4.2.2. *Scenario estimates*

The results of the computations made considering the chosen scenario earthquakes (Table II) and different local models are shown in Fig. 4.

Both scenarios, SCE1 (top) and SCE2 (bottom) in Fig. 4, show an obvious change, with varying epicentral distance, of the frequency content of the spectral site response, no matter which local model was considered. At epicentral distances ≥ 400 km, the dynamic coefficient at periods $T \geq 1.5$ s appear visibly higher than the EC8 recommendations. For the considered frequency content, 0–5 Hz, the shallow local models give dynamic coefficients that are closer to the EC8 recommendations.

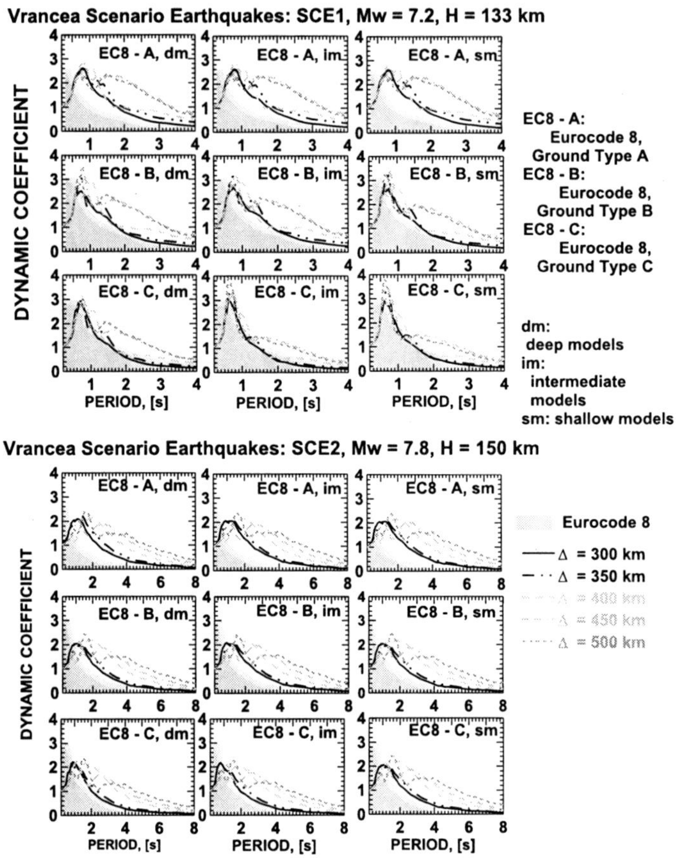

Figure 4. Vrancea scenario earthquakes (Table II): SCE1—strong event (top), SCE2—extreme event (bottom).

5. Concluding Remarks

The seismic input at a given site incorporates the coupled effects of the seismic source and of the inelastic media through which seismic wave propagates. Due to the lack of real strong motion records, synthetic seismic signals have been generated applying a neo-deterministic seismic hazard assessment procedure for given shallow and intermediate-depth scenario earthquakes. The computed signals are validated against the few available observations. The major outcome of this study can be summarized as follows:

- the computed synthetic seismic input for shallow earthquakes is consistent with the Eurocode 8 requirements;
- for the intermediate-depth earthquakes, local models with a thin top layer of type C supply synthetic seismic signals, that are quite close to the observed ones;

- the site response due to both, shallow and intermediate-depth, earthquakes is significantly influenced by the earthquake source mechanism;
- the dynamic coefficients, computed for accelerograms (observed and computed) due to strong intermediate-depth Vrancea earthquakes exceed significantly the values recommended by the EC8 for periods $T > 1$ s.

Acknowledgements

The financial support from the NATO SfP Project N980468, INTAS-Moldova 200505-104-7584; CEI Projects *Deterministic seismic hazard analysis and zoning of the territory of Romania, Bulgaria and Serbia* and *Geodynamical Model of Central Europe For Safe Development Of Ground Transportation Systems*, and the CEI university network are gratefully acknowledged.

References

Ambraseys, N., Smit, P., Sigbjornsson, R., Suhadolc, P., and Margaris, B. (2002) Internet-Site for European Strong-Motion Data, Technical Report, European Commission, Research-Directorate General, Environment and Climate Programme, Brussels.

Christoskov, L., Georgiev, T., Deneva, D., and Babachkova, B. (1982) On the seismicity and seismic hazard of Sofia valley. In *Proc. of 4th Int. Symp. on the Analysis of Seismicity and Seismic Risk*, Vol. IX, Bechyne castle, CSSR, pp. 448–454.

Dziewonsky, A. M., Ekstrom, G., Woodhouse, J. H., and Zwart, G. (1991) Centroid moment tensor solutions for April–June 1990, *Phys. Earth Planetary Interiors* **66**, 133–143.

EUROCODE 8 (1994) Basis of Design and Actions on Structures, CEN.

Fah, D., Iodice, C., Suhadolc, P., and Panza, G. F. (1993) A new method for the realistic estimation of seismic ground motion in megacities: The case of Rome, *Earthquake Spectra* **9**, 643–668.

Georgescu, E. S. and Sandi, H. (2000) Towards earthquake scenarios under the conditions of Romania. In *Proc. of 12th WCEE*, Auckland, New Zealand, No. 1699.

Ivanov, P., Frangov, G., and Yaneva, M. (1998) Engineering geological characteristics of quaternary sediments in the Sofia graben. In *Proc. of 3rd WG Meeting*, Sofia, pp. 33–37, Dec. 2–5.

Klugel, J. U. (2007) Error inflation in probabilistic seismic hazard analysis, *Eng. Geol.* **90**, 186–192.

Kouteva, M., Panza, G. F., Romanelli, F., and Paskaleva, I. (2004) Modelling of the ground motion at Russe site (NE Bulgaria) due to the Vrancea earthquakes, *J. Earthquake Eng.* **8**, 209–229.

Levander, A. R. (1988) Fourth-order finite-difference P-SV seismograms, *Geophysics* **53**, 1425–1436.

Lungu, D., Aldea, A., Demetriu, S., and Craifaleanu, I. (2004) Seismic strengthening of buildings and seismic instrumentation – two priorities for seismic risk reduction in Romania. In

Proc. of 1st Int. Conf. Science and Technology for Safe Development of Lifeline Systems, Natural Risks: Developments, Tools and Techniques in the CEI Area, Slovakia, Balkema.

Medvedev, S. V. (1977) Seismic intensity scale MSK-76, *Publ. Inst. Geophys. Pol. Acad. Sci.* **117**, 95–102.

Nenov, D., Paskaleva, I., Georgiev, G., and Trifunac, M. (1990) CATALOG of strong earthquake ground motion data in EQINFOS: Accelerograms recorded in Bulgaria between 1981 and 1987, Technical Report CE 90-02, Southern California University.

Oncescu, M. C., Marza, V. I., Rizescu, M., and Popa, M. (1999) The Romanian earthquake catalogue between 984–1997. In F. Wenzel, D. Lungu, and O. Novak (eds.), *Vrancea Earthquakes: Tectonics, Hazard and Risk Mitigation*, CISM Courses and Lectures, Kluwer Academic Publ., Dordrecht, Netherlands, pp. 43–47.

Panza, G. F., Kouteva, M., Vaccari, F., A., P., Cioflan, C. O., Romanelli, F., Paskaleva, I., Radulian, M., Gribovszk, i. K., Herak, M., Zaichenco, A., Marmureanu, G., Varga, P., and Zivcic, M. (2008) Recent achievements of the neo-deterministic seismic hazard assessment in the CEI region. In A. Santini and N. Moraci (eds.), *Proc. of 2008 Seismic Eng. Conf. Commemorating the 1908 Messina and Reggio Calabria Earthquake, Reggio Calabria, Italy, 8–11.07.2008*, Vol. 1020, Melville, New York, pp. 402–408, AIP.

Panza, G. F., Romanelli, F., and Vaccari, F. (2001) Seismic wave propagation in laterally heterogeneous anelastic media: theory and applications to the seismic zonation, *Adv. Geophys* **43**, 1–95.

Panza, G. F., and Suhadolc, P. (1987) Complete strong motion synthetics. In *Seismic Strong Motion Synthetics*, B. A. Bolt (ed.), Academic Press, Orlando, Computational Techniques 4, 153–204.

Paskaleva, I., Kouteva, M., Vaccari, F., and Panza, G. F. (2008) Application of the neo-deterministic seismic microzonation procedure in Bulgaria and validation of the seismic input against Eurocode 8. In A. Santini and N. Moraci (eds.), *Proc. of 2008 Seismic Eng. Conf. Commemorating the 1908 Messina and Reggio Calabria Earthquake, Reggio Calabria, Italy, 8–11.07.2008*, Vol. 1020, Melville, New York, pp. 394–401, AIP.

Paskaleva, I., Panza, G. F., Vaccari, F., and Ivanov, P. (2004) Deterministic modelling for microzonation of Sofia – an expected earthquake scenario, *Acta Geologica Geodetica Hungarica (AGGH)* **39**, 275–295.

Radulian, M., Panza, G. F., Popa, M., and Grecu, B. (2006) Seismic wave attenuation for Vrancea events revisited, *J. Earthquake Eng.* **10**, 411–427.

Radulian, M., Vaccari, F., Manderscu, N., Panza, G. F., and Moldoveanu, C. (2000) Seismic hazard of Romania: deterministic approach, *Pure Appl. Geophys.* **157**, 221–247.

Reiter, L. (1990) *Earthquake Hazard Analysis*, Columbia University Press, New York, North-Holland.

Romanelli, F., Bekkevold, J., and Panza, G. F. (1997) Analytical computation of coupling coefficients in non-poissonian media, *Geophys. J. Int.* **129**, 205–208.

Romanelli, F., Bing, Z., Vaccari, F., and Panza, G. F. (1996) Analytical computation of reflection and transmission coupling coefficients for Love waves, *Geophys. J. Int.* **125**, 132–138.

Stein, S. and Wysession, M. (2003) *An Introduction to Seismology, Earthquakes, and Earth Structure*, Blackwell Publishing, Oxford, UK.

Virieux, J. (1984) SH-wave propagation in heterogeneous media: Velocity-stress finite-difference method, *Geophysics* **49**, 1933–1957.

Virieux, J. (1986) P-SV wave propagation in heterogeneous media: Velocity-stress finite-difference method, *Geophysics* **51**, 889–901.

WAVE PROPAGATION IN THE ANISOTROPIC INHOMOGENEOUS HALF-PLANE

Tsviatko Rangelov (rangelov@math.bas.bg)
Institute of Mathematics and Informatics, Bulgarian Academy of Sciences, Acad. G. Bonchev St., Bl. 8, 1113 Sofia, Bulgaria

Petia Dineva (petia@imbm.bas.bg)
Institute of Mechanics, Bulgarian Academy of Sciences, Acad. G. Bonchev St., Bl. 4, 1113 Sofia, Bulgaria

Abstract. This study presents closed-form solutions for free-field motions in an anisotropic continuously inhomogeneous half-plane that includes contributions of incident waves as well as of waves reflected from the traction free horizontal surface. A state of plane strain holds and both pressure and vertically polarized shear waves are considered. Anisotropic material characteristics vary quadratically with respect to the depth coordinate. The method of solution is a hybrid approach based on the plane wave decomposition technique, augmented by appropriate functional transformation relations for the displacement vector. The existence of a unique analytical solution as a superposition of the incident P- or SV-wave and the corresponding reflected P- and SV-waves is proved under some restrictions on the incident wave direction. The simulation study reveals the influence of the material inhomogeneity and anisotropy on the displacement free field wave motion.

Keywords: anisotropic half-plane, quadratic inhomogeneity, free field motion

1. Introduction

An elastic medium is anisotropic when its physical properties at a given location depend on the direction in which they are considered. Anisotropy can be caused by the preferred orientation of anisotropic mineral grains or the preferred orientation of the shapes of isotropic minerals. It can also be caused by a stack of isotropic layers thicknesses smaller than the seismic wavelength, in which case the stack of layers can be treated as a single anisotropic medium (Backus, 1962). Another common case of anisotropy is the anisotropy due to the existence of fractures and cracks in the material (Thomsen, 1986).

Heterogeneity is the dependence of the physical properties of a medium on the position vector. The influence of both anisotropy and continuous inhomogeneity of soil deposits is often neglected for the sake of simplicity.

T. Schanz and R. Iankov (eds.), *Coupled Site and Soil-Structure Interaction Effects with Application to Seismic Risk Mitigation,* NATO Science for Peace and Security Series C: Environmental Security,

Many studies employing both body and surface waves in the earth concluded that there is significant departure from the homogeneous isotropic model, with anisotropy being dominant in deep rock deposits and inhomogeneity featuring prominently in near-surface soil deposits (Wetzel, 1987). Heterogeneity on the small scale (smaller than the seismic wavelength) appears as anisotropy on the large scale in Thomsen (1986). Wave propagation in inhomogeneous isotropic half-space is treated in the works of Manolis et al. (2007) and Dineva et al. (2006, 2008). Wave propagation in anisotropic media has been the subject of many publications (Daley and Hron, 1977; Thomsen, 1986; Graebner, 1992; Carcione, 2001; Tsvankin, 2005; Yakhno and Akmaz, 2007).

The knowledge of the displacement and traction free wave—fields in anisotropic and inhomogeneous media is valuable, since it serves as input to many classes of boundary-value problems describing wave scattering phenomena in vertically inhomogeneous geological media, which in turn is a prelude for studying site effects. The main aim of this work is to present closed-form solutions for free-field wave motions in an anisotropic continuously inhomogeneous half-plane that include contributions of incident waves as well as of waves reflected from the traction free horizontal surface. The problem is stated under plane strain condition. Both pressure and vertically polarized shear waves are considered. Anisotropic material characteristics vary quadratically with respect to the depth coordinate. The method of solution is a hybrid approach based on the plane wave decomposition technique, augmented by appropriate functional transformation relations for the displacement vector (Manolis et al., 2007; Manolis and Shaw, 1996). The existence of a unique analytical solution as a superposition of the incident P- or SV-wave and reflected correspondingly P- and SV-waves is proved under some restrictions on the incident wave direction. The obtained solution for the free-field motion depends on: (a) the parameters of the incident wave-its type, frequency, wave propagation direction and incident angle; (b) the reference anisotropic properties and mass density; (c) the inhomogeneity characteristics as direction and magnitude of the material gradient. Finally, a numerical simulation study is conducted to investigate the effect of the incident wave angle and frequency, type of material inhomogeneity and anisotropy on the free-field wave motions.

2. Problem Statement

In a rectangular Cartesian coordinate system Ox_1x_2 in R^2 consider an inhomogeneous anisotropic half-plane, $\Omega = \{x = (x_1, x_2) : x_2 < 0\}$. The mass density $\rho(x)$ varies with the observer position vector. The fourth order elastic tensor $C_{ijkl}(x)$ is symmetric and positively definite

$$C_{ijkl} = C_{jikl} = C_{klij}, \tag{1}$$

$$C_{ijkl}g_{ij}g_{kl} > 0 \quad \text{for each symmetric positive tensor } g_{ij}. \tag{2}$$

Condition (1) is the Green symmetry conditions while (2) corresponds to the requirement that the strain energy density must remain positive since this energy must be minimal in a state of stable equilibrium.

Let us use the compact Voigt form for the material tensor (Su and Sun, 2003). Tensor c_{mn} is obtained from the tensor $C_{ijkl}(x)$ following the rule: $(11) \leftrightarrow 1$, $(22) \leftrightarrow 2$, $(12 = 21) \leftrightarrow 6$. There are altogether six independent constants for two-dimensional problem we consider. In the orthotropic case with elasticity axes parallel to the coordinate axes the independent constants are four (Su and Sun, 2003) because the following condition holds

$$C_{iikl} = 0 \quad \text{for} \quad k \neq l \quad \text{or} \quad c_{16} = c_{26} = 0. \tag{3}$$

Let us assume one more symmetry of the elastic tensor in addition to (1) and (3)

$$C_{iijj} = C_{ijij} \quad \text{for} \quad i \neq j \quad \text{or} \quad c_{12} = c_{66}. \tag{4}$$

Assume that the material parameters depend in the same way as mass density on the position vector $x = (x_1, x_2)$

$$C_{ijkl}(x) = C^0_{ijkl}h(x), \quad \rho(x) = \rho_0 h(x), \tag{5}$$

where

$$h(x) = (ax_2 + 1)^2, \quad a < 0. \tag{6}$$

The analytical solution presented below concerns the special cases of inhomogeneous anisotropic materials satisfying the above discussed restrictions (1)–(6). The considered restrictions for the soil moduli are not as restrictive as it would seem (Su and Sun, 2003; Johnston and Christensen, 1995; Manolis et al., 2007).

The governing frequency dependent equations of motion in the absence of body forces are as follows

$$\sigma_{ij,j}(x,\omega) + \rho(x)\omega^2 u_i(x,\omega) = 0. \tag{7}$$

Here $\sigma_{ij}(x,\omega) = C_{ijkl}(x)u_{k,l}(x,\omega)$ is the stress tensor, u_k is the displacement vector, ω is the frequency of motion, commas indicate spatial derivatives and summation convention under repeating indexes is assumed. Due to the properties (6) of the inhomogeneous function $h(x)$ and condition (2), the system (7) form an strictly elliptic system of partial differential equations in Ω.

The traction free boundary conditions along $\partial\Omega$ are

$$t_j\big|_{x_2=0} = \sigma_{ij}n_i\big|_{x_2=0} = 0. \tag{8}$$

Let us apply a functional transformation for the displacement vector, proposed in Manolis and Shaw (1996) for the isotropic case, see also Manolis et al. (2007)

$$\mathbf{u}(x,\omega) = h^{-1/2}(x)\mathbf{U}(x,\omega). \tag{9}$$

Equation (7) for the transformed displacement U_i reads as follows

$$C_{ijkl}(x)[U_{k,jl} + h^{-1/2}(h^{1/2}_{,j}U_{k,l} - h^{1/2}_{,l}U_{k,j} - h^{1/2}_{,jl}U_k)] + \rho(x)\omega^2 U_i = 0 \tag{10}$$

Using (6) and (4) it follows $C_{ijkl}h^{-1/2}(h^{1/2}_{,j}U_{k,l} - h^{1/2}_{,l}U_{k,j}) = 0$ and $h^{1/2}_{,jl} = 0$ and by reducing the common factor $h(x)$ in both $C_{ijk}(x)$ and $\rho(x)$ we obtain equilibrium equation with constant coefficients for the transformed displacement

$$\Sigma_{ij,j} + \rho_0\omega^2 U_i = 0, \tag{11}$$

where $\Sigma_{ij} = C^0_{ijkl}U_{k,l}$. The traction-free boundary conditions (8) now becomes

$$t_j\big|_{x_2=0} = C^0_{2jkl}(-0.5h^{-3/2}h_{,l}U_k + h^{-1/2}U_{k,l})\big|_{x_2=0} = 0. \tag{12}$$

The radiation condition at infinity has to be satisfied.

3. Half-Plane Solution

First we start with deriving plane wave solutions of equation (11). Using the plane-wave decomposition method (Courant and Hilbert, 1962) for fixed frequency ω and a propagation vector $\xi = (\xi_1, \xi_2)$, $\xi_1 = \sin\theta$, $\xi_2 = \cos\theta$, $\theta \in [0, \pi/2]$ we find solutions to equation (11) in the form

$$\mathbf{U}^m = A^m_0 \mathbf{p}^m \exp ik_m(x_1\xi_1 + x_2\xi_2), \tag{13}$$

where $m = 1$ for P-wave, $m = 2$ for SV wave, $k_m(\xi,\omega)$ are the wave numbers, $\mathbf{p}^m(\xi,\omega)$ are polarization vectors indicating the direction of particle displacement, A^m_0 are wave amplitudes. To find k_m, and $\mathbf{p}^m$ we replace $\mathbf{U}^m$ from (13) in equation (11), reduce the exponential term and obtain a linear system of equations

$$[-N(\xi)k^2_m + \Lambda]\mathbf{p}^m = 0, \tag{14}$$

where $N(\xi) = \{n_{ik}(\xi)\}$, $n_{ik}(\xi) = C^0_{ijkl}\xi_l\xi_j$ and $\Lambda = \{\delta_{ik}\rho\omega^2\}$, $i,k = 1,2$, δ_{ik} is the Kroneker delta. The solutions of (14) are eigenvalues $k^2_m = (\rho^0\omega^2)/\gamma_m$ and polarization vectors $\mathbf{p}^m$, for which $\langle \mathbf{p}^1, \mathbf{p}^2\rangle = 0$, $|\mathbf{p}^m| = 1$, where $\langle\cdot,\cdot\rangle$ is the scalar product in R^2.

As a result we have two types of solutions: P-wave for $m = 1$ and SV wave for $m = 2$. The displacement vector in the original domain is obtained by using the transformation (9)

$$\mathbf{u}^m(x,\xi,\omega) = 1/(ax_2+1)\mathbf{U}^m(x,\xi,\omega), \quad x \in \Omega. \tag{15}$$

The restrictions on the material constants (Payton, 1983) are due to the ellipticity of the differential operator in (7) and positivity properties in (2). Denote $c_{11}^0/c_{66}^0 = \alpha$, $c_{22}^0/c_{66}^0 = \beta$, then all the restrictions are summarized as:

$$\begin{array}{llll} (i) & 0 < \alpha < 1, & 1 < \beta, & \alpha\beta > 1, \\ (ii) & 1 < \alpha, & 0 < \beta < 1, & \alpha\beta > 1, \\ (iii) & 1 < \alpha, & 1 < \beta. & \end{array} \tag{16}$$

Our aim is to find a plane wave solutions of (7) in Ω with traction free boundary conditions (8) on $\partial\Omega$. Following Achenbach (1973) for the isotropic homogeneous half-plane the cases for incident P-wave and for incident SV-wave are described separately below.

3.1. INCIDENT P-WAVE

Let the frequency ω and the inhomogeneity parameter $a < 0$ are fixed and c_{ij}^0 satisfy (1)–(4), i.e., (16). For every incident P-wave with propagation vector $\xi = (\xi_1, \xi_2)$, $\xi_1 \in (0, 1)$ and amplitude A_0 we derive that there exist unique reflected P and SV waves such that the superposition of all three waves is a solution of (7) in Ω with traction-free boundary conditions (8).

The incident P-wave is with displacement vector $\mathbf{u}^0 = 1/(ax_2 + 1)A_0\mathbf{p}^1 \times \exp\{ik_1(x_1\xi_1 + x_2\xi_2)\}$, $\mathbf{p}^1 = (p_1^1, p_2^1)$ and we are asking for two reflected waves: P-wave $\mathbf{u}^1 = 1/(ax_2 + 1)A_1\mathbf{p}\exp\{ik_1(x_1\zeta_1 - x_2\zeta_2)\}$, with wave propagation direction $(\zeta_1, -\zeta_2)$ and polarization vector $\mathbf{p} = (p_1^1, -p_2^1)$ and SV-wave $\mathbf{u}^2 = 1/(ax_2 + 1)A_2\mathbf{q}\exp\{ik_2(x_1\eta_1 - x_2\eta_2)\}$, with wave propagation direction $(\eta_1, -\eta_2)$ and polarization vector $\mathbf{q} = (-p_1^2, p_2^2)$, such that the total wave field is a superposition of incident and reflected waves, i.e., $\mathbf{u} = \mathbf{u}^0 + \mathbf{u}^1 + \mathbf{u}^2$. Also the boundary condition (8) on $x_2 = 0$ is satisfied:

$$\mathbf{t}\big|_{x_2=0} = (\mathbf{t}^0 + \mathbf{t}^1 + \mathbf{t}^2)\big|_{x_2=0} = 0. \tag{17}$$

By means of the transformed displacements, using the expression (12), it is obtained:

$$\begin{cases} t_1^0\big|_{x_2=0} &= c_{66}^0 A_0(ik_1p_1^1\xi_2 - p_1^1 a + ik_1p_2^1\xi_1)e^{ik_1x_1\xi_1}, \\ t_2^0\big|_{x_2=0} &= A_0[ik_1c_{66}^0p_1^1\xi_1 + c_{22}^0(ik_1p_2^1\xi_2 - p_2^1 a)]e^{ik_1x_1\xi_1}, \end{cases}$$

$$\begin{cases} t_1^1\big|_{x_2=0} &= -c_{66}^0 A_1(ik_1p_1^1\zeta_2 + p_1^1 a + ik_1p_2^1\zeta_1)e^{ik_1x_1\zeta_1}, \\ t_2^1\big|_{x_2=0} &= A_1[ik_1c_{66}^0p_1^1\zeta_1 + c_{22}^0(ik_1p_2^1\zeta_2 + p_2^1 a)]e^{ik_1x_1\zeta_1}, \end{cases}$$

$$\begin{cases} t_1^2\big|_{x_2=0} &= c_{66}^0 A_2(ik_2p_1^2\eta_2 + p_1^2 a + ik_2p_2^2\eta_1)e^{ik_2x_1\eta_1}, \\ t_2^2\big|_{x_2=0} &= -A_2[ik_2c_{66}^0p_1^2\eta_1 + c_{22}^0(ik_2p_2^2\eta_2 + p_2^2 a)]e^{ik_2x_1\eta_1}. \end{cases}$$

In order to reduce the exponential factors $e^{ik_1x_1\xi_1}$, $e^{ik_2x_1\zeta_1}$ and $e^{ik_2x_1\eta_1}$ in (17) and to have unique reflected P and SV waves we have to find the unique couple of wave propagation directions of reflected shear and compressional waves η_1, ζ_1 that solves the equation

$$k_1(\xi)\xi_1 = k_1(\zeta)\zeta_1 = k_2(\eta)\eta_1 \tag{18}$$

for fixed wave direction vector ξ of the incident wave. Using the conditions (16) it is proved that for every one direction of the incident wave $\xi_1 \in (0, 1)$ there exist unique solutions of equations (18) for the wave propagation directions of both reflected waves $\eta_1, \zeta_1 = \xi_1$.

After reducing the common exponential multiplier in (17) it is obtained the following system of linear equations with respect to $\chi_1 = A_1/A_0$, $\chi_2 = A_2/A_0$:

$$\begin{cases} d_{11}\chi_1 + d_{12}\chi_2 = d_1, \\ d_{21}\chi_1 + d_{22}\chi_2 = d_2, \end{cases} \tag{19}$$

where

$$\begin{aligned} d_{11} &= -ik_1(p_1^1\xi_2 + p_2^1\xi_1) - p_1^1 a, & d_{21} &= ik_1(c_{66}^0 p_1^1\xi_1 + c_{22}^0 p_2^1\xi_2) + c_{22}^0 p_2^1 a, \\ d_{12} &= ik_2(p_1^2\eta_2 + p_2^2\eta_1) + p_1^2 a, & d_{22} &= -ik_2(c_{66}^0 p_1^2\eta_1 + c_{22}^0 p_2^2\eta_2) - c_{22}^0 p_2^2 a, \\ d_1 &= -ik_1 p_1^1\xi_2 + p_1^1 a - ik_1 p_2^1\xi_1, & d_2 &= -c_{66}^0 ik_1 p_1^1\xi_1 - c_{22}^0(ik_1 p_2^1\xi_2 - p_2^1 a). \end{aligned}$$

System (19) has unique solution since its determinant is non zero. Indeed, $\Delta = d_{11}d_{22} - d_{12}d_{21} = R + iI$, where at least $I \neq 0$ for $a < 0$, because $k_2\eta_2 > 0$, $k_1\xi_2 > 0$ for $\xi_2 \neq 0$ and $p^1(\xi), p^2(\eta)$ are non-collinear.

Using the Kramer's method the unique solution of the system (19) is $\chi_j = \Delta_j/\Delta$, where

$$\Delta_1 = \begin{Vmatrix} d_1 & d_{12} \\ d_2 & d_{22} \end{Vmatrix}, \qquad \Delta_2 = \begin{Vmatrix} d_{11} & d_1 \\ d_{21} & d_2 \end{Vmatrix}.$$

For the case of a normal to the free surface incident P-wave, $k_1 = \sqrt{\rho_0/c_{22}^0}\omega$ and the components of the displacement vector are:

$$u_1 = 0, \quad u_2 = 1/(ax_2 + 1)A_0\left[e^{ik_1x_2} + (a - ik_1)/(a + ik_1)e^{-ik_1x_2}\right].$$

3.2. INCIDENT SV-WAVE

Let the frequency ω and the inhomogeneous parameter $a < 0$ are fixed and c_{ij}^0 satisfy (1)–(4), i.e., (16). We derive that there exists $\eta_1^0 \in [0, 1)$ such that for every incident SV-wave with propagation vector $\eta = (\eta_1, \eta_2)$,

$\eta_2 = \sqrt{1-\eta_1^2}$, $\eta_1 \in (0, \eta_1^0)$ there exist unique reflected P and SV waves such that the superposition of all three waves is a solution of (7) in Ω with traction-free boundary conditions (8). The incident wave field is SV-wave with displacement vector $\mathbf{u}^0 = 1/(ax_2+1)B_0\mathbf{p}^2 \exp\{ik_2(x_1\eta_1 + x_2\eta_2)\}$, $\mathbf{p}^2 = (p_1^2, p_2^2)$. We are asking for two reflected waves: P-wave with displacement vector $\mathbf{u}^1 = 1/(ax_2+1)B_1\mathbf{p}\exp\{ik_1(x_1\xi_1 - x_2\xi_2)\}$ with polarization vector $\mathbf{p} = (p_1^1, -p_2^1)$ and SV-wave with displacement vector $\mathbf{u}^2 = 1/(ax_2+1)B_2\mathbf{q}\exp\{ik_2(x_1\tau_1 - x_2\tau_2)\}$ with polarization vector $\mathbf{q} = (-p_1^2, p_2^2)$, such that the total displacement wave field is $\mathbf{u} = \mathbf{u}^0 + \mathbf{u}^1 + \mathbf{u}^2$ and boundary condition (8) on the free surface boundary $x_2 = 0$ is satisfied.

The displacement free-field motion is obtained in analogous to the case of incident P-wave case.

For the case of a normal to the free surface incident SV-wave, $k_2 = \sqrt{\rho_0/c_{66}^0}\,\omega$ and the components of the displacement vector are:

$$u_1 = 1/(ax_2+1)B_0\left[e^{ik_2x_2} + (a-ik_2)/(a+ik_2)e^{-ik_2x_2}\right], \quad u_2 = 0.$$

4. Numerical Results

The aim of the simulation study is to investigate the dependence of the free-field wave motions on the: (a) incident wave angle; (b) incident wave frequency; (c) type and magnitude of the material inhomogeneity; (d) soil anisotropy. Numerical simulations are performed with reference homogeneous anisotropic soil properties of type *(iii)* in (16) following Johnston and Christensen (1995). The mechanical properties are: $c_{11}^0 = 46.26\times10^6\,\text{N/m}^2$, $c_{22}^0 = 28.78\times10^6\,\text{N/m}^2$, $c_{12}^0 = c_{66}^0 = 10.28\times10^6\,\text{N/m}^2$ for the anisotropic case; $c_{11}^0 = c_{22}^0 = 30.84\times10^6\,\text{N/m}^2$, $c_{12}^0 = c_{66}^0 = 10.28\times10^6\,\text{N/m}^2$ for the isotropic case; $\rho = 2500\,\text{kg/m}^3$. It is specified a frequency range of $0.5\,\text{Hz} \le f \le 2\,\text{Hz}$ values as representative of low frequency seismic motions. Receiver point in Figs. 1 and 2 is the point $R(10,0)$ on the Ox_1 axis. Figure 1 depicts free-field motion at incident longitudinal P-wave, while Fig. 2 is for incident SV-wave. Since displacements are complex valued functions, we plot module of the amplitude corresponding to a unit incoming P (or SV) wave. Four values of the inhomogeneity magnitude are used: $a = -0.00115$; -0.001; -0.02; -0.04. The homogeneous case is recovered with inhomogeneity magnitude of $a = 0$. Figures 1a, b (for incident P-wave) and Figs. 2a, b (for incident SV wave) present amplitude-frequency characteristics at receiver $R(10,0)$, for wave propagating in homogeneous/inhomogeneous isotropic/anisotropic half-plane with fixed incident P-wave angle of 0.6283 (rad) and fixed incident SV-wave angle of 0.4387 (rad). In order to give a sense of magnitude to the

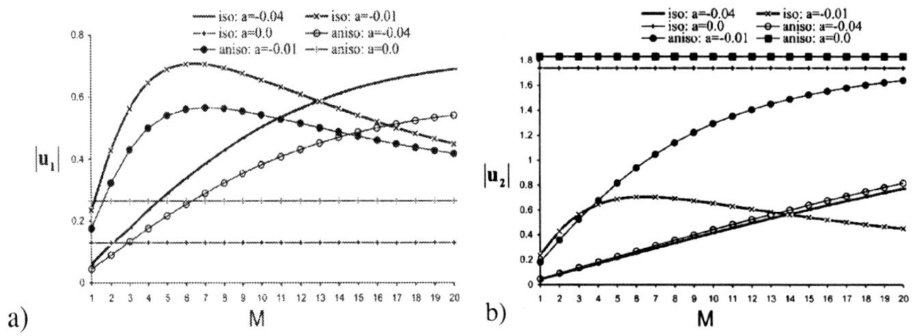

Figure 1. The amplitude of displacement components vs frequency $f = 0.15708M$, $M = 1, \ldots, 20$ of P-wave with incident angle $\theta = 0.6283$ (rad) propagating in isotropic/anisotropic inhomogeneous half-plane: (a) $|u_1|$; (b) $|u_2|$.

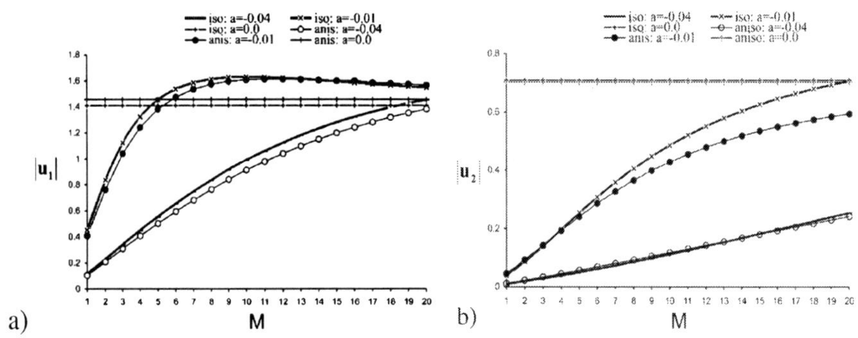

Figure 2. The amplitude of displacement components vs frequency $f = 0.15708M$, $M = 1, \ldots, 20$ of SV-wave with incident angle $\theta = 0.4387$ (rad) propagating in isotropic/anisotropic inhomogeneous half-plane: (a) $|u_1|$; (b) $|u_2|$.

resulting material parameters, a value of $a = -0.005$ yields $h(0, -1000) = 36$, which imply a 360% increase in the soil stiffness coefficients starting from the free surface and moving downwards. At inhomogeneity magnitude of $a = -0.02$, we have that $h(0, -1000) = 441$ and inhomogeneity effect is stronger. Figures 3 and 4 show the amplitude of displacement components vs depth coordinate x_2 at receiver $R(10, x_2)$, for fixed frequency of 1 Hz and fixed incident P-wave angle of 0.6283 (rad) and fixed incident SV-wave angle of 0.4387 (rad) correspondingly in homogeneous/inhomogeneous isotropic/anisotropic half-plane. It can be seen that the effect due to the soil anisotropy and inhomogeneity becomes stronger and increases significantly at depth much greater than the wavelength.

All simulation study results reveal that free-field motions in the continuously inhomogeneous anisotropic half-plane, which consist of incident plus reflected wave contributions, show marked differences with respect to the benchmark isotropic homogeneous case. The influence of the frequency and

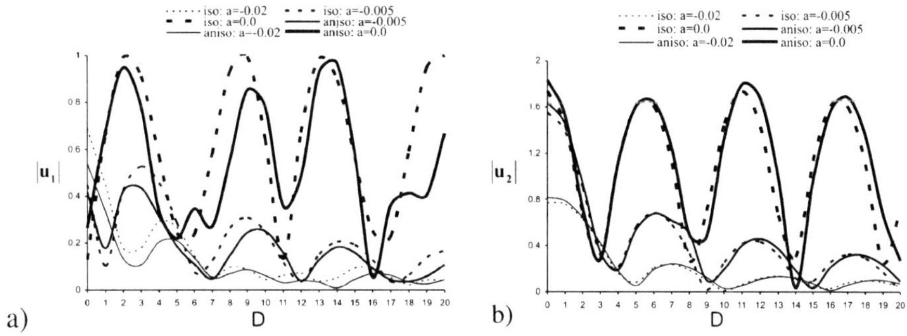

Figure 3. The amplitude of displacement components vs depth at $x = (10, x_2)$, $x_2 = -50D$, $D = 0, \ldots, 20$ for fixed frequency $f = 1$ Hz of P-wave with incident angle $\theta = 0.6283$ (rad) propagating in isotropic/anisotropic inhomogeneous half-plane: (a) $|u_1|$; (b) $|u_2|$.

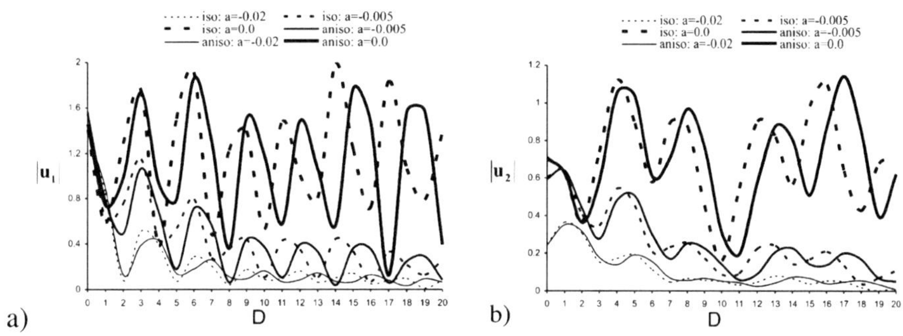

Figure 4. The amplitude of displacement components vs depth at $x = (10, x_2)$, $x_2 = -50D$, $D = 0, \ldots, 20$ for fixed frequency $f = 1$ Hz of SV-wave with incident angle $\theta = 0.4387$ (rad) propagating in isotropic/anisotropic inhomogeneous half-plane: (a) $|u_1|$; (b) $|u_2|$.

wave propagation direction, wave path inhomogeneity and soil anisotropy on the wave fields of waves propagating in half-plane plays an important role and has to be accounted for.

5. Conclusions

An analytical solution of the free-field motion with incident both P and SV-wave is derived for the anisotropic inhomogeneous half-plane. It is supposed that the material parameters vary quadratically with respect to the depth. Numerical examples show the influence of the frequency and propagation direction of the incident seismic signal and soil inhomogeneity and anisotropy on the wave field. The presented solutions can be used as an input data for solving boundary value problems for seismic response of laterally inhomogeneous geological region situated in the half-space as well as for the solution of the inverse problems for identification of mechanical and geometrical properties of the complex soil profiles.

Acknowledgements

The authors wish to acknowledge Prof. G. Manolis, Aristotle University of Thessaloniki, for helpful discussions during the preparation of the paper.

References

Achenbach, J. (1973) *Wave Propagation in Elastic Solids*, North Holland.

Backus, G. E. (1962) Long-wave elastic anisotropy produced by horizontal layering, *Journal of Geophysical Research* **67**, 4427–4440.

Carcione, J. M. (2001) *Wave Fields in Real Media: Wave Propagation in Anisotropic, Anelastic and Porous Media*, Pergamon Press, Amsterdam.

Courant, R. and Hilbert, D. (1962) *Methods of Mathematical Physics,* vol. II, John Wiley, New York.

Daley, P. and Hron, F. (1977) Reflection and transmission coefficients for transversely isotropic solids, *Bulletin of the Seismological Society of America* **67**, 661–675.

Dineva, P., Manolis, G., and Rangelov, T. (2006) Sub-surface crack in an inhomogeneous half-plane, wave scattering phenomena by BEM, *Engineering Analysis with Boundary Elements* **30**, 350–362.

Dineva, P., Manolis, G., and Rangelov, T. (2008) Site effect due to wave path inhomogeneity by BEM, *Engineering Analysis with Boundary Elements* **32**, 1025–1036.

Graebner, M. (1992) Plane-wave reflection and transmission coefficients for a transversely isotropic solid (short note), *Geophysics* **57**, 1512–1519.

Johnston, J. E. and Christensen, N. I. (1995) Seismic anisotropy of shales, *Journal of Geophysical Research* **100**, 5991–6003.

Manolis, G., Rangelov, T., and Dineva, P. (2007) Free-field wave solutions in a half-plane exhibiting a special-type of continuous inhomogeneity, *Wave Motion* **44**, 304–321.

Manolis, G. and Shaw, R. (1996) Green's function for a vector wave equation in a mildly heterogeneous continuum, *Wave Motion* **24**, 185–202.

Payton, R. (1983) *Elastic Wave Propagation in Transversely Isotropic Media*, Martinus Nijhoff, The Hague.

Su, R. and Sun, H. (2003) Numerical solutions of two-dimensional anisotropic crack problems, *International Solids Structures* **40**, 4615–4635.

Thomsen, L. (1986) Weak elastic anisotropy, *Geophysics* **51**, 1954–1966.

Tsvankin, I. (2005) *Seismic Signatures and Analysis of Reflection Data in Anisotropic Media*, Elsevier, Amsterdam.

Wetzel, A. (1987) Sedimentological significance of strain and sonic velocity anisotropy in finegrained turbiditic and hemipelagic deep-sea sediments—An example from the Mississippi fan, *Marine Geology* **74**, 191–207.

Yakhno, V. G. and Akmaz, H. K. (2007) Anisotropic elastodynamics in a half space. An analytic method for polynomial data, *Journal of Computational and Applied Mathematics* **204**, 268–281.

HYBRID SIMULATION OF SEISMIC WAVE PROPAGATION IN LATERALLY INHOMOGENEOUS MEDIA

Tom Schanz (tom.schanz@rub.de)
Laboratory of Foundation Engineering, Soil and Rock Mechanics, Faculty of Civil Engineering, Ruhr-Universität Bochum, Universitätsstraße 150, D-44780 Bochum, Germany

Frank Wuttke
Laboratory of Soil Mechanics, Bauhaus-University Weimar, 99421 Weimar, Germany

Petia Dineva
Department of Continuum Mechanics, Institute of Mechanics, Bulgarian Academy of Sciences, 1113 Sofia, Bulgaria

Abstract. The main aim of this work is to develop, validate and applied in simulation study an efficient hybrid approach to study 2D seismic wave propagation in local multilayered geological region rested on inhomogeneous in depth half-space with a seismic source. Plane strain state is considered. The vertically varying of the soil properties in the half space is modelled by a set of horizontal flat isotropic, elastic and homogeneous layers. The finite local region is with nonparallel layers and with a free surface relief. The hybrid computational tool is based on the analytical wave number integration method (WNIM) and the numerical boundary integral equation method (BIEM). The WNIM is applied considering the bedrock model to compute the input signals for the laterally varying part where the signals are obtained by BIEM at a set of sites. The numerical simulation results reveal that the hybrid method is able to demonstrate the sensitivity of the obtained synthetic signals to the seismic source properties, to the heterogeneous character of the wave path and to the relief peculiarities of the local stratified geological deposit.

Keywords: lateral inhomogeneity, soil stratum, hybrid technique, synthetic seismic signals

1. Introduction

To make use of the advantages of the analytical and numerical approaches, the development of the so-called hybrid technique, see Fah et al. (1993), Gil-Zepeda et al. (2003) and Panza et al. (2000), is recently very actual and of high importance in the computational seismic mechanics. In the hybrid computational schemes, the numerical method is applied merely

T. Schanz and R. Iankov (eds.), *Coupled Site and Soil-Structure Interaction Effects with Application to Seismic Risk Mitigation,* NATO Science for Peace and Security Series C: Environmental Security,

in the lateral heterogeneous part of the medium, which is a small part of the modeled geology. The laterally homogeneous layers of the geological model are treated analytically. Although the advantages of the boundary integral equation method (BIEM) for solving the seismic wave propagation problems are well known, there is a lack of hybrid computational tools based on this method, that allow to take into consideration the properties of the all three important components—seismic source, wave path and finite laterally inhomogeneous multilayered region. The aim of the present paper is to combine the facilities of both methods the analytical Wave Number Integration Method (WNIM) and the numerical BIEM in order to develop an accurate and efficient hybrid approach for synthesis of seismic signals in a laterally varying seismic region accounting for: (a) the seismic source characteristics; (b) the wave path inhomogeneity from the source to the area of interest; (c) the local geological media with its complex mechanical and geometrical properties; (d) the existence of free surface relief peculiarities. The proposed hybrid technique is based on the WNIM for investigating wave propagation in depth inhomogeneous half-space, while BIE method is used for synthesis of theoretical seismograms on the free surface of the local finite soil stratum.

The paper is organized in the next way. First, the problem statement for seismic wave propagation in a laterally varying geological media is described in Section 2. Next, the hybrid tool is presented in Section 3. Finally, numerical examples for different seismic scenarios are solved in Section 4, followed by conclusions in Section 5.

2. Problem Statement

It is considered 2D in-plane seismic wave propagation problem in finite local multilayered geological region Ω_{LGR} disposed in inhomogeneous in depth half-space Ω_0 with seismic source in it, as it is shown in Fig. 1. Plane strain state holds. Non-parallel isotropic, elastic and homogeneous layers Ω_i, $i = 1, 2, \ldots, N$, are situated in the local soil stratum. The vertical varying of the mechanical properties in the wave path Ω_0 is modelled with series of M elastic isotropic homogeneous flat layers, parallel to the free surface, overlaying the homogeneous half-space where is the seismic bed. The aim is to obtain the synthetic seismograms in some receiver points along the free surface of the local region. The governing wave equations for the l-th layer are the following partial differential equations:

$$(\alpha_l^2 - \beta_l^2)u_{j,ji}(x,z,t) + \beta_l^2 u_{i,jj}(x,z,t) = \ddot{u}_i(x,z,t) \quad \text{in} \quad Q_B = \Omega_B \times (0,T). \quad (1)$$

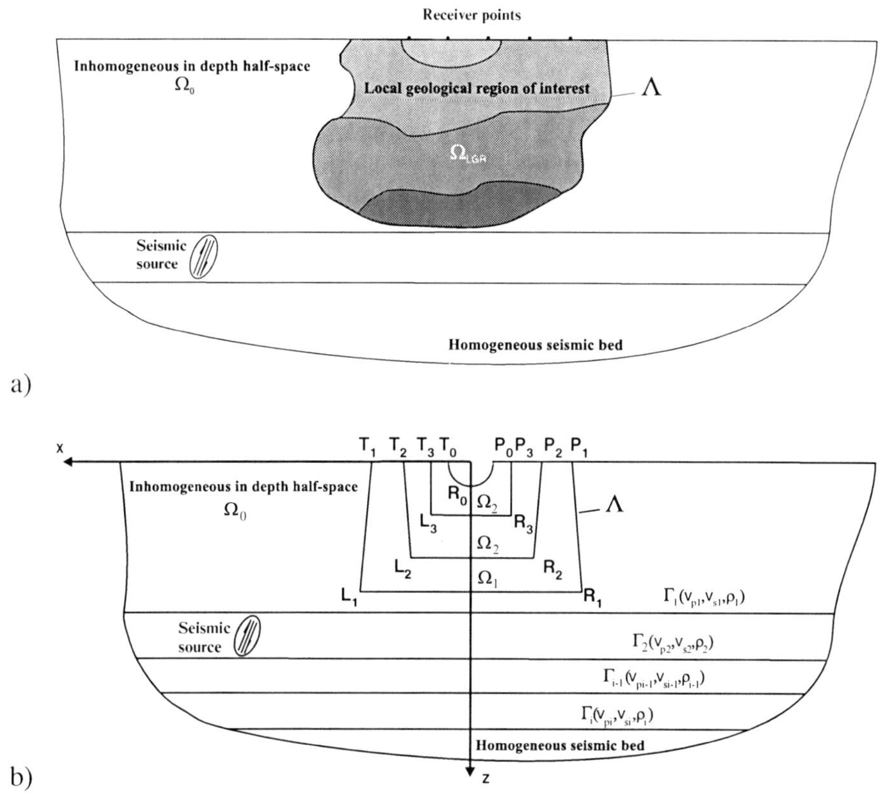

Figure 1. The geometry of the problem (a) and geometry of geological region (b).

Here: $\Omega_B = \Omega_{LRG} \cup \Omega_0$, longitudinal and shear wave velocities, α_l, β_l, are different for different soil layers in Ω_{LRG}, where $l = 1, 2, \ldots, N$, and in Ω_0, $l = 1, 2, \ldots, M$, T is the duration of the seismic load, u_i is the displacement, $\ddot{u}_i$ is the acceleration, $i = x, z$. In order to exclude the time variable and to solve the boundary-value problem in the frequency domain, Fourier transform is applied to the time variable. The governing partial differential equation for each l-th layer becomes of elliptic type. The boundary-value-problem consists of the governing Eq. (1) and the following boundary conditions: (a) the traction at the free surface is zero; (b) the displacement compatibility and traction equilibrium conditions on the interface between each two layers is satisfied; (c) Sommerfeld radiation condition at infinite holds. The BVP is solved for a sufficient number of values of frequency ω and a numerical inverse fast Fourier transformation (FFT) is applied in order to obtain the time-dependent solution.

3. Hybrid Computational Tool

Efficient hybrid wave number integration-boundary integral equation method is proposed. Each of the two techniques is applied in that part of the geological model where it works most efficiently. The BIEM is used in the local stratified geological area, while WNIM is applied for simulation the wave propagation from the source position to the local region of interest. The following governing boundary integral equations are used in each soil layer separately:

$$C_{ij}u_i(x,z,\omega) = \int_{\Omega_m} U^*_{ij}(x,z,x_0,z_0,\omega)p_j(x_0,z_0,\omega)\,d\Gamma - \int_{\Omega_m} P^*_{ij}(x,z,x_0,z_0,\omega)u_j(x_0,z_0,\omega)\,d\Gamma, \quad m = 1,2,3,\ldots,N. \quad (2)$$

where C_{ij} are constants depending on the geometry at the collocation point (x,z); (x_0,z_0) denotes the position vector of the source point; Γ_{Ω_m} is the boundary of the Ω_m layer; u_i and p_j are the unknown total displacements and total tractions on the boundaries Γ_{Ω_m}; U^*_{ij} and P^*_{ij} are the displacement and traction frequency dependent fundamental solutions of Eq. (1). The following governing boundary integral equations over the common boundary Λ between the finite soil region Ω_{LGR} and the infinite medium Ω_0 are used for the scattered wave field $u_i^{sc} = u_i - u_i^{fr}$ and $p_i^{sc} = p_i - p_i^{fr}$, see Dominguez (1993)

$$C_{ij}\left(u_i(x,z,\omega - u_i^{fr}(x,z,\omega))\right) = \int_{\Lambda} U^*_{ij}(x,z,x_0,z_0,\omega)\left(p_j(x_0,z_0,\omega) - p_j^{fr}(x_0,z_0,\omega)\right)d\Gamma - \int_{\Lambda} P^*_{ij}(x,z,x_0,z_0,\omega)\left(u_j(x_0,z_0,\omega) - u_j^{fr}(x_0,z_0,\omega)\right)d\Gamma. \quad (3)$$

In order to solve the system of BIEs (2)–(3), the wave free-field motion u_i^{fr}, p_i^{fr} has to be known. Here the solution of the wave propagation problem for the vertically inhomogeneous half-space with a buried seismic source represents the free-field motion and it is determined by the analytical wave number integration method.

4. Numerical Simulations

The proposed hybrid tool has been tested by solution of a benchmark example that can be solved by both the pure WNIM and the proposed hybrid approach. The geometry of the geological model is a valley with flat free surface that is placed in the first layer of a layered half-space. The mechanical properties of the valley are the same as ones of the first layer in the horizontally layered half-space. In this case it is possible to solve the problem by the pure WNIM and by the proposed here hybrid tool. The comparison between studied hybrid solutions against analytical results shows that the computational error is not more than 7%. More details about validation study of the proposed hybrid technique can be seen in (Dineva et al., 2006). The conclusion is that it is necessary to establish the limits and possibilities of the proposed hybrid technique and to this aim one-dimensional simulation experiment allows us to consider the WNIM result as a reference benchmark. The validation study has to be done for each new seismic scenario, because the comparison between solutions of the pure analytical method and of the proposed hybrid method gives an efficient control over the accuracy of the BIEM calculations depending on the mesh discretization.

In order to illustrate the efficiency of the proposed hybrid method the response of a multilayered region with geometry given in Fig. 1b is analyzed. The finite local soil stratum has three layers with non-parallel interfaces. Three types of the free-surface geometry relief are considered: (a) flat model; (b) semi-circle canyon with radius $r = a = 30$ m; (c) semi-circle hill with radius $r = a = 30$ m. The coordinates of the points indicating the geometrical boundaries of the layers are: $T_0(30, 0)$; $T_3(60, 0)$; $T_2(90, 0)$; $T_1(100, 0)$; P_0 $(-30, 0)$; $P_3(-60, 0)$; $P_2(-90, 0)$; $P_1(-100, 0)$; $L_1(110, 270)$; $R_1(-110, 270)$; $L_2(90, 180)$; $R_2(-90, 180)$; $L_3(60, 90)$; $R_3(-60, 90)$. The mechanical properties of the finite geological region are given in Table II. The aim of the simulation study is to demonstrate that the obtained through the proposed hybrid method synthetic signals depend on: (a) wave path inhomogeneity (b) relief existence on the free surface; (c) seismic source characteristics.

4.1. SENSITIVITY OF THE OBTAINED SYNTHETIC SEISMIC SIGNALS TO THE WAVE PATH PROPERTIES

Canyon relief on the free surface of the local geological region is considered. Synthetic seismograms are obtained at two different wave paths 1 and 2 with properties given in Table Ia, b. The mechanical properties of the finite local stratum are given in Table II. The vertical line seismic source located at $x = 2$ km and at depth of 6 km is used. The source term was assumed as a Ricker wavelet of the second order with time duration of 1.28 s. Two receiver points are considered: receiver that is at the edge of the canyon and receiver $(0, 30)$ that is at the bottom of the canyon. Figure 2a, b present the normalized displacement amplitudes vs. frequency at both receivers in the case of wave path 1. Normalization is made with respect to the maximal amplitude of the corresponding displacement component, synthesized for the bedrock

TABLE Ia. The properties of the layered half-space (wave path 1)

Number of Soil Layer Γ_i	Thickness (km)	Depth (km)	Density ρ_1 (kg/m^3)	Shear Wave Velocity β_i (m/s)	Longitudinal Wave Velocity α_i (m/s)
1	5.0	5.0	2750	3500	6100
2	8.0	13.0	2900	3600	6200
3	4.0	17.0	3200	4100	7650
4	2.0	19.0	3200	4200	7500
5	2.0	21.0	3200	4200	7650
6	2.0	23.0	3200	4300	7800
7	5.0	28.0	3300	4350	8000
8	22.0	50.0	2900	3800	6800
Seismic bed	∞	∞	3350	4600	8200

TABLE Ib. The properties of the layered half-space (wave path 2)

Number of Soil Layer Γ_i	Thickness (km)	Depth (km)	Density ρ_1 (kg/m^3)	Shear Wave Velocity β_i (m/s)	Longitudinal Wave Velocity α_i (m/s)
1	0.35	0.35	2100	1400	2400
2	0.65	1.0	2200	1400	2400
3	1.5	2.5	2300	1400	2400
4	1.0	3.5	2400	1400	2400
5	1.5	5.0	2500	2200	3800
6	2.0	7.0	2600	2550	4500
7	5.0	12.0	2650	3100	5400
8	13.0	25	2750	3500	6200
9	10.0	35.0	2900	4200	7500
Seismic bed	∞	∞	3350	4600	8200

TABLE II. The properties of the local geological region

Number of Soil Layer	Density ρ_1 (kg/m^3)	Shear Wave Velocity β_i (m/s)	Longitudinal Wave Velocity α_i (m/s)
3	1800	800	1500
2	2000	1000	1700
1	2000	1400	2400

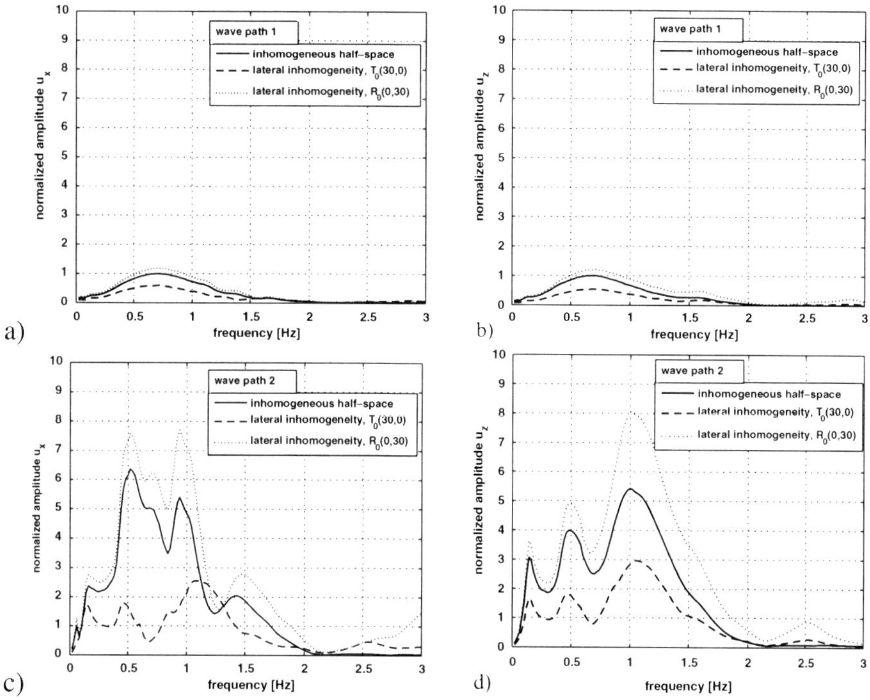

Figure 2. Sensitivity of the obtained synthetic seismic signals on the wave path properties.

reference model at wave path 1. Figure 2c, d concern the same curves but in the case of wave path 2. Comparison is done with solutions on the free surface of inhomogeneous in depth half-space with seismic source without lateral heterogeneous finite soil stratum. Figure 2 demonstrates very clear the existence of site effects and also their dependence on the type of the wave path inhomogeneity. The effect of site amplification is stronger in the case of wave path 2. One of the strong advantages of the proposed hybrid technique is that the seismic signals obtained by this tool depend on the inhomogeneous properties of the wave path.

4.2. SENSITIVITY OF THE OBTAINED SYNTHETIC SEISMIC SIGNALS TO THE FREE SURFACE RELIEF OF THE LOCAL GEOLOGICAL REGION

Figure 3a, b show the sensitivity of the seismic signals on the free surface relief. The mechanical properties of the half-space are presented by wave path 2 and the line vertical seismic source located at $x = 2$ km and at depth of 6 km is assumed. Three types of the free surface geometry are considered: (i) flat surface; (ii) canyon relief; (iii) hill-like relief, correspondingly. Figure 3a, b

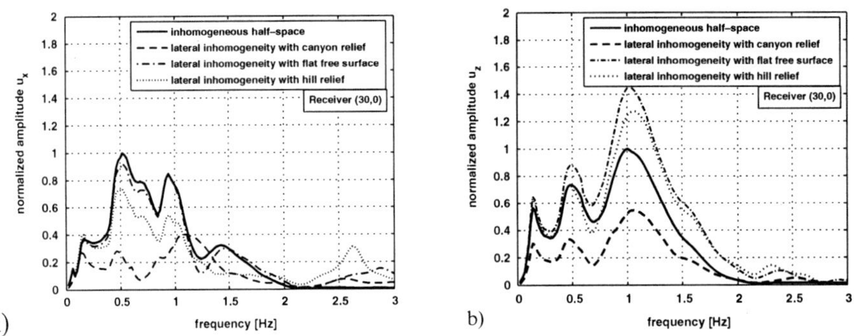

Figure 3. Sensitivity of the obtained synthetic seismic signals on the free surface relief.

show the normalized horizontal and vertical displacement amplitude in the case of local multilayered region with and without relief. Normalization in figures is made with respect to the max amplitude of the corresponding displacement component, synthesized for the bedrock reference model at wave path 2. The discussed results reveal that the site effects are much stronger in the case of relief on the free surface, while in the case of flat free surface there is no clear presence of the site amplification.

4.3. SENSITIVITY OF THE OBTAINED SYNTHETIC SEISMIC SIGNALS TO THE SEISMIC SOURCE PROPERTIES

The synthetic seismograms are obtained for double coupled seismic source at two different depths of 2 and 6 km. The source parameters are: the scalar seismic moment $M_0 = 5.98e14$ Nm, strike angle $\varphi = 151°$, dip angle $\delta = 83°$, rake angle $\theta = 7°$. The mechanical properties of the inhomogeneous in depth half-space are as that of the wave path 2, the canyon relief on the free surface is assumed and the mechanical properties of the local soil region are as those given in Table II. The seismograms for two different receivers and for receiver on the free-surface of the inhomogeneous in depth half-space without the finite soil stratum are shown in Figs. 4 and 5. These figures demonstrate the sensitivity of the synthetic signals to the depth of the earthquake source.

The discussed numerical results in this item show that the synthetic signals depend on all essential components of the seismic wave path: seismic source, inhomogeneous wave path plus the lateral inhomogeneous local soil stratum with its complex mechanical and geometrical peculiarities.

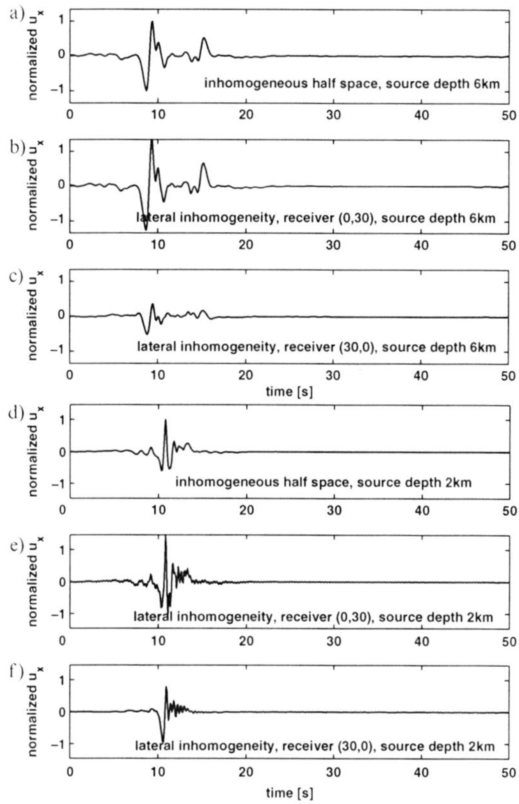

Figure 4. Synthetic seismic signals for horizontal displacement component.

5. Conclusion

Efficient hybrid wave number integration-boundary integral equation method is proposed. Each of the two techniques is applied in that part of the geological model where it works most efficiently. The BIEM is used in the local stratified geological area, while WNIM is applied for simulation of wave propagation from the source position to the local region of interest. The main advantage of this hybrid tool, is that it can take into consideration the basic properties of the seismic source, wave path and local geological deposit. The obtained by the WNI-BIE method seismic signal can be used as a necessary base at solution of the following engineering problems: (a) seismic wave propagation with accounting for the more complex mechanical behaviour of the soil as poroelasticity, anisotropy, nonelasticity, nonhomogeneity, etc.; (b) soil-structure interaction and its effects on the dynamics of structures during earthquake; (c) solution of inverse problems for dynamic site characterization and identifying the soil profiles.

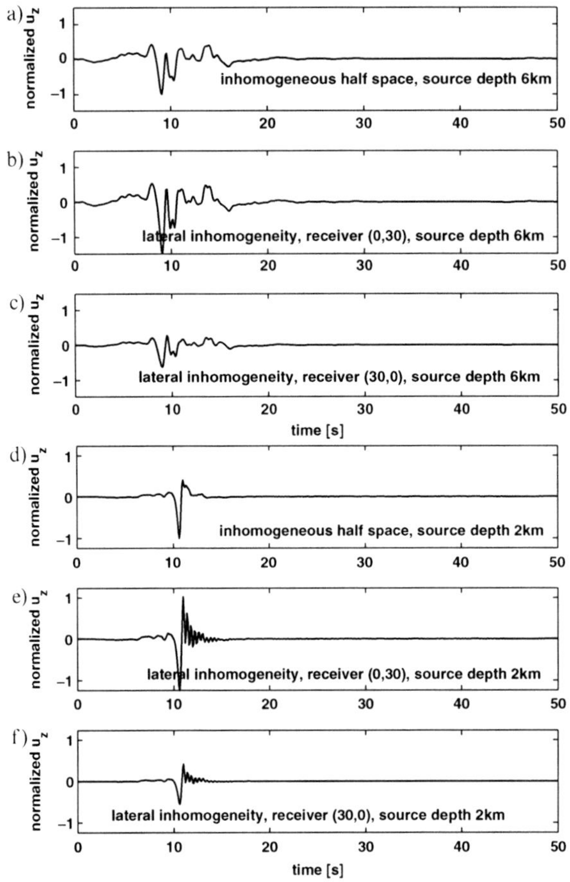

Figure 5. Synthetic seismic signals for vertical displacement component.

Acknowledgements

The authors acknowledge the support of the NATO under the grant number CLG982064.

References

Dineva, P., Wuttke, F., and Schanz, T. (2006) Validation study of the wave number integration-boundary integral equation method for seismic wave propagation problems, *Comptes rendus de l'Academie bulgare des Sciences* **59**, 939–944.

Dominguez, J. (1993) *Boundary Elements in Dynamics*, Computational Mechanics, Southampton/Amsterdam, Elsevier.

Fah, D., Suhadolc, P., and Panza, G. (1993) Variability of seismic ground motion in complex media: The Fruili area (Italy), *J. Appl. Geophys.* **30**, 131–148.

Gil-Zepeda, S., Montalvo-Arrieta, J., Vai, R., and Sanchez-Sesma, F. (2003) A hybrid indirect boundary element-discrete wave number method applied to simulate the seismic response of stratified alluvial valleys, *Soil Dynamics Earthquake Eng.* **23**, 77–86.

Panza, G., Romanelli, F., and Vaccai, F. (2000) Seismic wave propagation in laterally heterogeneous anelastic media: Theory and applications to seismic zonation, *Adv. Geophys.* **43**, 1–95.

ON THE INFLUENCE OF SATURATION AND FREQUENCY ON MONOCHROMATIC PLANE WAVES IN UNSATURATED SOILS

Bettina Albers (albers@grundbau.tu-berlin.de)
Institute for Soil Mechanics and Geotechnical Engineering
Technische Universität Berlin, Germany

Abstract. The propagation of sound waves in partially saturated sandstone is investigated by use of a macroscopic linear model which is based on the two-component model of Biot and on the Simple Mixture Model by Wilmanski. The considered porous medium consists of a deformable skeleton and two compressible pore fluids (water and air).

The wave analysis of the poroelastic model reveals the existence of four body waves: three longitudinal waves, $P1$, $P2$, $P3$, and one shear wave, S. The dependence of their phase velocities and attenuations on the saturation and on the frequency is studied and compared to experimental observations and the sound velocity of suspensions.

Keywords: monochromatic waves, porous media, partially saturated soils

1. Introduction

By means of a model (Albers, 2009) which is an extension of both the Biot Model (Tolstoy, 1992) for two-component saturated media and the Simple Mixture Model by Wilmański (Wilmanski, 1999) the propagation of sound waves in partially saturated poroelastic media is studied. Both these models follow as particular cases of the three-component model which describes linear processes in unsaturated poroelastic materials. Biot's model incorporates a coupling, Q, between the components through an additional contribution to the partial stresses . Due to the presence of one more component in the model for unsaturated porous media three of such couplings (solid–fluid Q^F, solid–gas Q^G and fluid–gas Q^{FG}) appear. The characteristic of the Simple Mixture Model which does not include such couplings is that the porosity belongs to the set of fields and satisfies an own balance equation. This is also the case for the three-component model.

It is well known that the number of components of a medium increases the number of longitudinal waves. While in a one-component medium besides the transversal wave S (secondary or shear) only one longitudinal wave P (primary) occurs in a saturated porous medium one more longitudinal wave (the

T. Schanz and R. Iankov (eds.), *Coupled Site and Soil-Structure Interaction Effects with Application to Seismic Risk Mitigation,* NATO Science for Peace and Security Series C: Environmental Security,

$P2$-wave or Biot-wave) is observed. Due to the existence of the second pore fluid (the gas) in the unsaturated medium a third compressional wave ($P3$) emerges. In this work the phase velocities and attenuations of the four waves are shown in dependence on the frequency ω and on the initial saturation S_0.

2. Linear Model

The linear field equations

$$\begin{aligned}
\rho_0^S \frac{\partial \mathbf{v}^S}{\partial t} &= \operatorname{div}\left\{\lambda^S e \mathbf{1} + 2\mu^S \mathbf{e}^S + Q^F \varepsilon^F \mathbf{1} + Q^G \varepsilon^G \mathbf{1}\right\} \\
&\quad + \pi^{FS}\left(\mathbf{v}^F - \mathbf{v}^S\right) + \pi^{GS}\left(\mathbf{v}^G - \mathbf{v}^S\right), \\
\rho_0^F \frac{\partial \mathbf{v}^F}{\partial t} &= \operatorname{grad}\left\{\rho_0^F \kappa^F \varepsilon^F + Q^F e + Q^{FG} \varepsilon^G\right\} - \pi^{FS}\left(\mathbf{v}^F - \mathbf{v}^S\right), \\
\rho_0^G \frac{\partial \mathbf{v}^G}{\partial t} &= \operatorname{grad}\left\{\rho_0^G \kappa^G \varepsilon^G + Q^G e + Q^{FG} \varepsilon^F\right\} - \pi^{GS}\left(\mathbf{v}^G - \mathbf{v}^S\right), \\
\frac{\partial \mathbf{e}^S}{\partial t} &= \operatorname{sym}\operatorname{grad} \mathbf{v}^S, \quad \frac{\partial \varepsilon^F}{\partial t} = \operatorname{div} \mathbf{v}^F, \quad \frac{\partial \varepsilon^G}{\partial t} = \operatorname{div} \mathbf{v}^G, \quad e \equiv \operatorname{tr} \mathbf{e}^S,
\end{aligned} \tag{1}$$

are satisfied by the essential fields $\left\{\mathbf{v}^S, \mathbf{v}^F, \mathbf{v}^G, \mathbf{e}^S, \varepsilon^F, \varepsilon^G\right\}$. This set of equations coincides with the Biot model if the third component, i.e., the gas, is neglected.

The quantities $\mathbf{v}^S$, $\mathbf{v}^F$, $\mathbf{v}^G$ are the velocity fields of the components, $\mathbf{e}^S$ is the deformation tensor. Quantities with subindex zero are initial values of the corresponding current quantity.

The model is formulated using the volume changes of the components e, ε^F, ε^G instead of the partial mass densities of the components, ρ^S, ρ^F, ρ^G. They are related by

$$e = \frac{\rho_0^S - \rho^S}{\rho_0^S}, \quad \varepsilon^F = \frac{\rho_0^F - \rho^F}{\rho_0^F}, \quad \varepsilon^G = \frac{\rho_0^G - \rho^G}{\rho_0^G}. \tag{2}$$

The porosity, n, also belongs to the fields and satisfies an own balance equation. However, if we neglect memory effects, the balance equation can be solved. On the other hand, the current saturation of the fluid, S, i.e., the fraction of fluid in the voids, is not a field and a constitutive law will be formulated for this quantity.

3. Material Parameters

The diffusion velocities $\mathbf{v}^F - \mathbf{v}^S$ and $\mathbf{v}^G - \mathbf{v}^S$ in (1) appear multiplied by the resistance parameters π^{FS} and π^{GS}. Both in the Biot model and in the Simple Mixture Model only one of them appears (called π) because only the resistance of the flow of the fluid through the channels of the skeleton is reflected. It is easy to show that this parameter is connected to the classical permeability parameter $\overline{K}$ in the following way

$$\overline{K} = \frac{K}{\rho g} \sim \frac{1}{\pi}, \tag{3}$$

where the hydraulic conductivity is denoted by K, the mass density of the fluid by ρ and the earth acceleration by g. If there appear two pore fluids both $\overline{K}$ and π are split into parts for the fluid and the gas: $\overline{K}_f = k_f \overline{K}$ and $\overline{K}_g = k_g \overline{K}$ and thus

$$\pi^{FS} = \frac{\pi^F}{k_f}, \qquad \pi^{GS} = \frac{\pi^G}{k_g}. \tag{4}$$

The parameters π^F and π^S depend on the viscosity of the material and thus are extremely different for water and air. As shown in Fig. 1 (left) the relative permeabilities k_f and k_g for water and air in sand have been measured in dependence on the liquid saturation (Wyckoff and Botset, 1936). Theoretical relationships for these permeabilities have been proposed by van Genuchten (1980):

$$k_f = S^{\frac{1}{2}} \left[1 - \left(1 - S^{\frac{1}{m}} \right)^m \right]^2, \qquad k_g = (1 - S)^{\frac{1}{3}} \left(1 - S^{\frac{1}{m}} \right)^{2m}, \tag{5}$$

where S is the degree of saturation and m a material parameter.

Also the material parameters $\left\{ \lambda^S + \frac{2}{3}\mu^S, \kappa^F, \kappa^G, Q^F, Q^G, Q^{FG} \right\}$ appearing in the Cauchy stress tensors have to be specified according to the material. This is done by applying a transition from the micro- to the macro-scale.

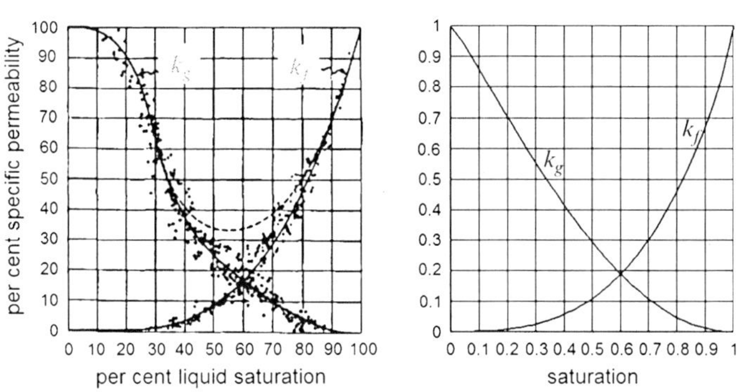

Figure 1. Experimentally obtained relative permeabilities of a water–air mixture in sand in dependence on the liquid saturation (Wyckoff and Botset, 1936), right: calculated relative permeabilities according to formulae (5) with $m = 0.85$.

Therein the van Genuchten formula for the dependence of the capillary pressure on the saturation is used (van Genuchten, 1980)

$$p_c(S) = \frac{1}{\alpha_{vG}}\left[S^{(-1/m_{vG})} - 1\right]^{1/n_{vG}}, \tag{6}$$

where α_{vG}, m_{vG}, n_{vG} are parameters which depend on the properties of the soil. The corresponding capillary pressure curve for an air–water mixture in sandstone is shown in Fig. 2 (left). The right hand side of this figure illustrates the set of material parameters in dependence on the initial saturation which follows for those materials from the micro–macro-transition (for details see Albers, 2009). The numerical values of microscopic parameters are given in Table I. The initial porosity is denoted by n_0, K_θ is the microscopic compressibility of the θ-component ($\theta = S, F, G$) and $\rho^{\theta R}$ is its true mass density. The expression for the compressibility modulus of the empty matrix, K_d, has been proposed in this form by Geertsma (White, 1983). The true initial pressures of skeleton and fluid are negligible in comparison to K_s and K_f. Therefore they are chosen to be zero. For the corresponding pressure in the gas this is not the case. It is of the same order as K_g. It is determined by the capillary pressure $p_c(S_0)$.

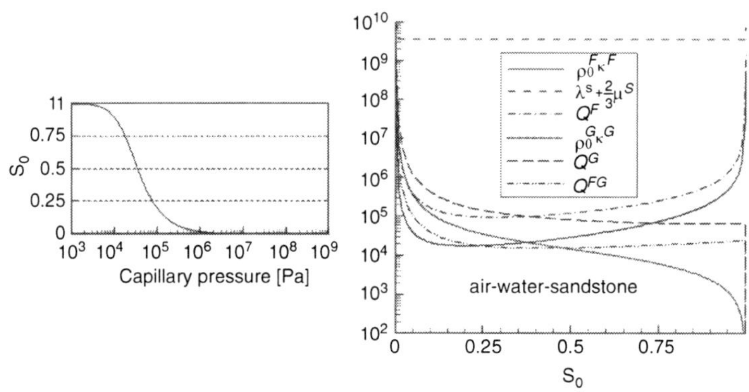

Figure 2. Left: capillary pressure curve for an air–water mixture in sandstone calculated using the formula of van Genuchten, right: material parameters $\left\{\lambda^S + \frac{2}{3}\mu^S, \rho_0^F\kappa^F, \rho_0^G\kappa^G, Q^F, Q^G, Q^{FG}\right\}$ in log-scale in dependence on the initial saturation S_0.

TABLE I. Some material parameters for an air–water mixture in sandstone

$n_0 = 0.25,$	$K_s = 48 \cdot 10^9$ Pa,	$\rho_0^{SR} = 2650$ kg/m^3,	$p_0^{SR} = 0,$	$m_{vG} = 0.5,$
$K_d = \dfrac{K_s}{1 + 50n_0},$	$K_f = 2.25 \cdot 10^9$ Pa,	$\rho_0^{FR} = 1000$ kg/m^3,	$p_0^{FR} = 0,$	$n_{vG} = 2,$
	$K_g = 1.01 \cdot 10^5$ Pa,	$\rho_0^{GR} = 1.2$ kg/m^3,	$p_0^{GR} = p_c,$	$\alpha_{vG} = 5 \cdot 10^{-5}.$

4. General Propagation Condition of Monochromatic Waves

It is assumed that the fields satisfy the following relations

$$\begin{aligned} \varepsilon^F &= E^F \mathcal{E}, \quad \varepsilon^G = E^G \mathcal{E}, \quad \mathbf{e}^S = \mathbf{E}^S \mathcal{E}, \quad n - n_0 = D\mathcal{E}, \\ \mathbf{v}^F &= \mathbf{V}^F \mathcal{E}, \quad \mathbf{v}^G = \mathbf{V}^G \mathcal{E}, \quad \mathbf{v}^S = \mathbf{V}^S \mathcal{E}, \quad \mathcal{E} := \exp i \left(\mathbf{k} \cdot \mathbf{x} - \omega t\right), \end{aligned} \tag{7}$$

where $\mathbf{E}^S$, E^F, E^G, $\mathbf{V}^S$, $\mathbf{V}^F$, $\mathbf{V}^G$, D are constant amplitudes, ω is a given frequency, $\mathbf{k} = k\mathbf{n}$ is the, possibly complex, wave vector. Therein k is the complex wave number and $\mathbf{n}$ a unit vector in the direction of propagation. A solution of this form describes the propagation of plane monochromatic waves in an infinite medium whose fronts are perpendicular to $\mathbf{n}$.

Substitution of relations (7) into the field equations $(1)_4$ yields the compatibility relations

$$\begin{aligned} E^F &= -\tfrac{1}{\omega} k\mathbf{n} \cdot \mathbf{V}^F, \quad E^G = -\tfrac{1}{\omega} k\mathbf{n} \cdot \mathbf{V}^G, \\ \mathbf{E}^S &= -\tfrac{1}{2\omega} k \left(\mathbf{n} \otimes \mathbf{V}^S + \mathbf{V}^S \otimes \mathbf{n}\right), \quad \text{i.e.,} \quad e = -\tfrac{1}{\omega} k\mathbf{n} \cdot \mathbf{V}^S \mathcal{E}. \end{aligned} \tag{8}$$

Insertion of (7) into the remaining field equations leads to further relations:

$$\begin{aligned} \omega^2 \mathbf{V}^S &= \frac{\lambda^S}{\rho_0^S} k^2 \left(\mathbf{V}^S \cdot \mathbf{n}\right) \mathbf{n} + \frac{\mu^S}{\rho_0^S} k^2 \left(\left(\mathbf{V}^S \cdot \mathbf{n}\right) \mathbf{n} + \mathbf{V}^S\right) \\ &\quad + \frac{Q^F}{\rho_0^S} k^2 \left(\mathbf{V}^F \cdot \mathbf{n}\right) \mathbf{n} + \frac{Q^G}{\rho_0^S} k^2 \left(\mathbf{V}^G \cdot \mathbf{n}\right) \mathbf{n} + \\ &\quad + i \frac{\pi^{FS} \omega}{\rho_0^S} \left(\mathbf{V}^F - \mathbf{V}^S\right) + i \frac{\pi^{GS} \omega}{\rho_0^S} \left(\mathbf{V}^G - \mathbf{V}^S\right) = 0, \\ \omega^2 \mathbf{V}^F &= \kappa^F k^2 \left(\mathbf{V}^F \cdot \mathbf{n}\right) \mathbf{n} + \frac{Q^F}{\rho_0^F} k^2 \left(\mathbf{V}^S \cdot \mathbf{n}\right) \mathbf{n} \\ &\quad + \frac{Q^{FG}}{\rho_0^F} k^2 \left(\mathbf{V}^G \cdot \mathbf{n}\right) \mathbf{n} - i \frac{\pi^{FS} \omega}{\rho_0^F} \left(\mathbf{V}^F - \mathbf{V}^S\right) = 0, \\ \omega^2 \mathbf{V}^G &= \kappa^G k^2 \left(\mathbf{V}^G \cdot \mathbf{n}\right) \mathbf{n} + \frac{Q^G}{\rho_0^G} k^2 \left(\mathbf{V}^S \cdot \mathbf{n}\right) \mathbf{n} + \\ &\quad + \frac{Q^{FG}}{\rho_0^G} k^2 \left(\mathbf{V}^F \cdot \mathbf{n}\right) \mathbf{n} - i \frac{\pi^{GS} \omega}{\rho_0^G} \left(\mathbf{V}^G - \mathbf{V}^S\right) = 0. \end{aligned} \tag{9}$$

In order to investigate transversal and longitudinal waves (oscillations perpendicular and parallel to the propagation direction of the wave) the contributions of the normal and transversal components of the wave vector $k\mathbf{n}$ are separated.

Presenting the dispersion relation of the transversal wave is an easy task:

$$\omega^2\left[1-\frac{\mu^S}{\rho_0^S}\left(\frac{k}{\omega}\right)^2\right]-\pi^{FS}\pi^{GS}\frac{\rho_0^S+\rho_0^F+\rho_0^G}{\rho_0^S\rho_0^F\rho_0^G}\left[1-\frac{\mu^S}{\rho_0^S+\rho_0^F+\rho_0^G}\left(\frac{k}{\omega}\right)^2\right]$$
$$+i\omega\left(\frac{\pi^{FS}+\pi^{GS}}{\rho_0^S}+\frac{\pi^{FS}}{\rho_0^F}+\frac{\pi^{GS}}{\rho_0^G}\right)\left[1-\frac{\frac{\pi^{FS}}{\rho_0^F}+\frac{\pi^{GS}}{\rho_0^G}}{\frac{\pi^{FS}+\pi^{GS}}{\rho_0^S}+\frac{\pi^{FS}}{\rho_0^F}+\frac{\pi^{GS}}{\rho_0^G}}\frac{\mu^S}{\rho_0^S}\left(\frac{k}{\omega}\right)^2\right]=0. \tag{10}$$

However, the dispersion relation for the longitudinal waves is very lengthy. Instead of presenting it here, the results of a numerical example will be shown in the next section.

5. Numerical Analysis of the Wave Propagation

The dispersion relations are solved for the complex wave number k. The solution gives rise to the phase velocities $\omega/(\mathrm{Re}\,k)$ and the attenuations $\mathrm{Im}\,k$ of the four waves. Numerical data of Table I and the additional values for the resistances which incorporate also the different viscosities of the two pore fluids are used

$$\pi^F = 10^7\ \mathrm{kg/m^3s}, \quad \pi^G = 1.82\cdot 10^5\ \mathrm{kg/m^3s}, \tag{11}$$

to investigate the wave propagation in sandstone filled with an air–water mixture. For results on further soil textures or different pore fillings it is referred to Albers (2009).

Figures 3 and 4 show the numerical results of the wave analysis. The velocities of the shear wave, S, and the three longitudinal waves, $P1$, $P2$ and $P3$, are shown on the left hand side of the figures, the attenuations on the right hand side.

In Figure 3 the phase velocities and attenuations of the four waves are given in dependence on the frequency ω. The different curves represent different values of the initial saturation, namely $S_0 = 0.2, 0.4, 0.6$ and 0.99999. The latter corresponds to a nearly water-saturated sandstone.

The plots in the frequency domain of the phase speeds of the transversal wave and of the $P1$-wave show a result which is already known from the wave analysis in saturated porous media. The speeds of both waves expose a low and a high limit value. For each value of the initial saturation the phase speed of both S- and $P1$-waves grows a little from its initial value to the asymptotic speed for $\omega \to \infty$. In the low frequency range which appears in geotechnical applications the speeds of both waves are nearly constant. The increase occurs in the region of ultrasonic frequencies which may be excited in laboratory experiments. The $P1$-wave is the fastest of the appearing waves. This is related

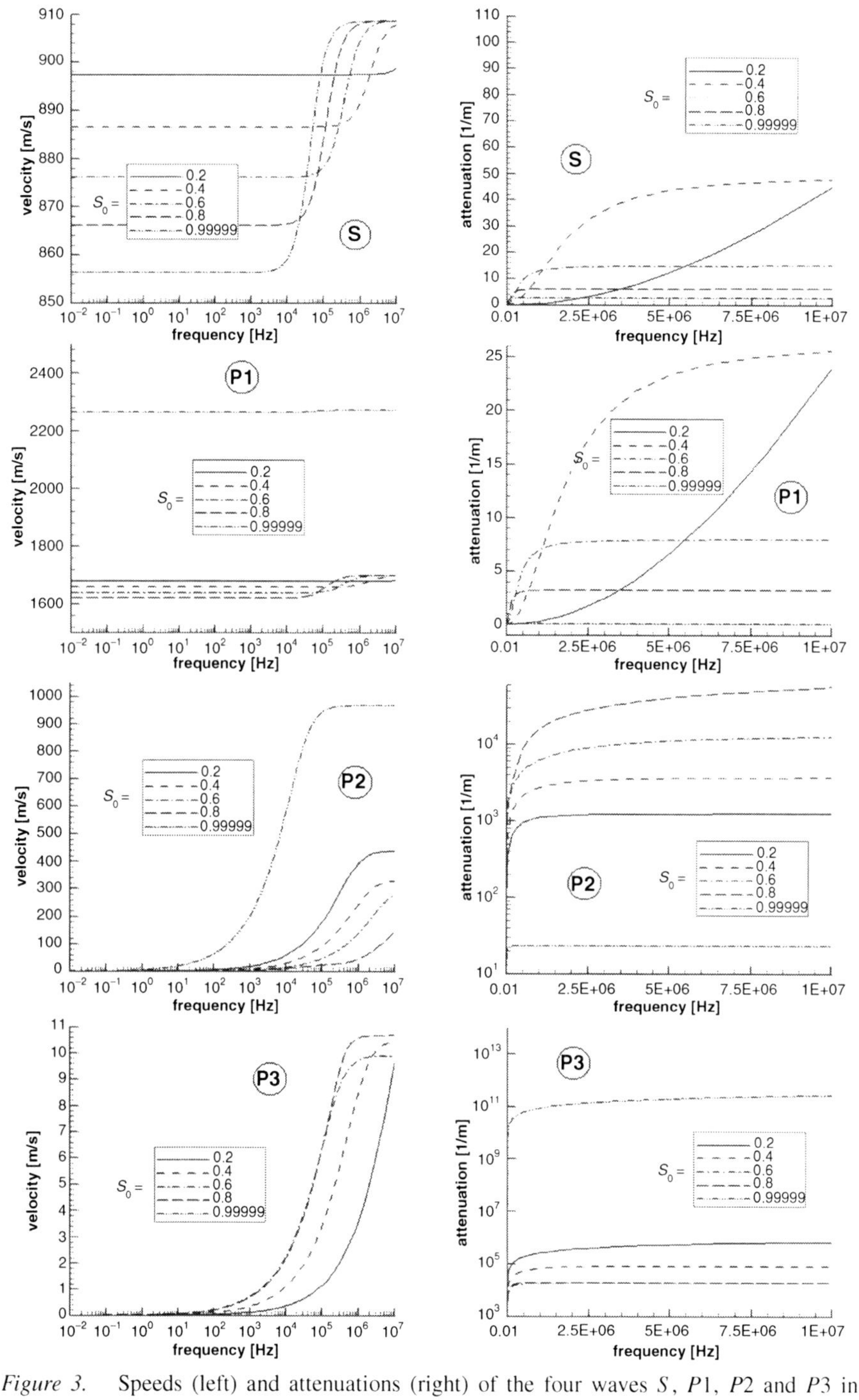

Figure 3. Speeds (left) and attenuations (right) of the four waves S, $P1$, $P2$ and $P3$ in dependence on the frequency ω.

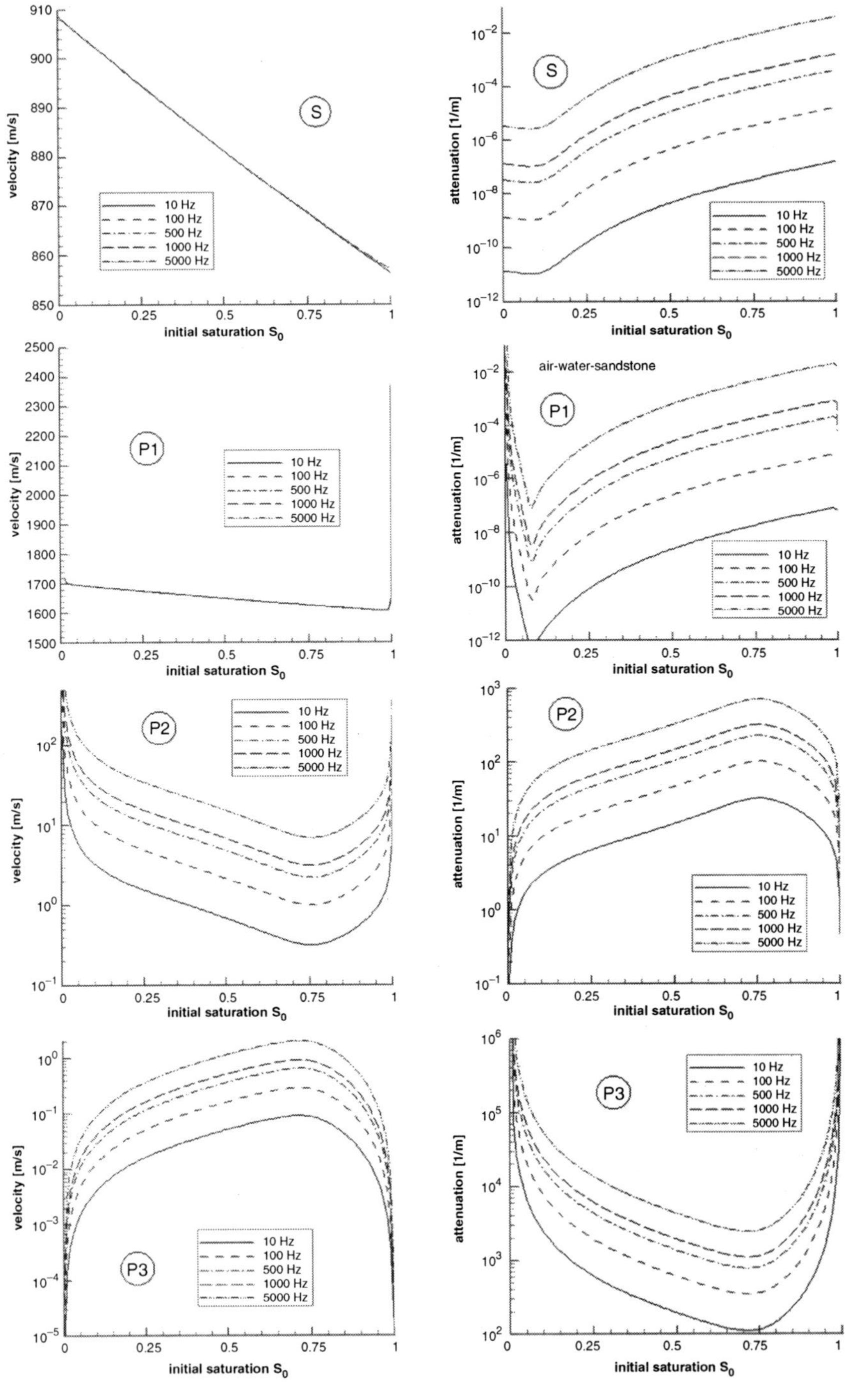

Figure 4. Speeds (left) and attenuations (right) of the four waves S, $P1$, $P2$ and $P3$ in dependence on the initial saturation S_0.

to the fact that it propagates mainly in the skeleton. The shear wave is slower and also the difference of the limit values is smaller. For very large values of the frequency the speeds of both waves are, again, nearly constant.

Both the $P2$-wave and the $P3$-wave start from zero velocity for zero frequency. Otherwise the frequency dependence is similar to this of the $P1$- and S-wave. At relative high values of the frequency a strong increase of the velocity takes place before an asymptotic value is reached. However, at least near water saturation, the increase is stronger than for the other waves. In the next figure the dependence of the waves on the saturation is shown and it will be obvious that the $P3$-wave only exists if a second pore fluid is present. But here it can be seen already that the velocities for each degree of saturation are much lower than for all other waves. For the highest value of the saturation the velocity of this wave is even so low that it does not appear in the figure anymore.

This is consistent with the attenuation of the $P3$-wave for this degree of saturation. It is some orders of magnitude larger than any other value of the attenuation. Both $P1$- and S-wave have a low attenuation. In the region of high frequencies shown in Fig. 3 (normal scale of the frequency axis) it is especially low for high degrees of saturation. From the analysis of wave propagation in saturated poroelastic media it is already known that the second longitudinal wave is highly damped and therefore hard to observe especially in non-artificial materials. This problem is even more pronounced for the $P3$-wave because its attenuation is even higher than this of the $P2$-wave. Thus the observation of this wave will be nearly impossible.

Figure 4 shows the dependence of the four waves on the initial saturation. Again, on the left hand side the speeds are given, on the right hand side the corresponding attenuations are shown. Now, the various lines represent different values of the frequency. Because there result some astonishing curves in the region of very high frequencies (especially if the second pore fluid is oil and not air, see Albers, 2009) which have no clear interpretation up to now, velocities and attenuations are given for the relative low frequencies $\omega = 10$, 100, 500, 1000 and 5000 Hz. For geophysical applications these are anyway frequencies which have a practical bearing.

The saturation axis covers the whole region from $0 \leq S_0 \leq 1$, however, observation of the curves shows that the model is not applicable for very low degrees of saturation. That was to be expected because in this region presumptions of the model are not fulfilled anymore. Instead of a continuous fluid with air inclusions in this region of saturation a frothy structure of the the pore fluids in encountered. A second reason for acceptance of the limitation is that smaller values of the saturation do not appear in reality (a residual amount of fluid is trapped in the channels).

The shear modulus, μ^S, is for a given Poisson number constant and the wave velocity depends only on the mass density of the skeleton. Thus it decreases linearly with increasing saturation. In nearly the whole range of saturations also the other elastic constant of the skeleton, λ^S, changes only marginally. However, for a degree of saturation which is closely to the state of water saturation it increases abruptly and reaches approximately the double of the value before. Thus, the $P1$-wave proceeds nearly constant for almost all values of the saturation and increases only for very large values of S_0. Again, it is obvious that the changes in attenuation of both waves are small.

For both other waves, $P2$ and $P3$, the influence of the degree of saturation is higher. These waves are effected by the existence of the fluid and the gas. The velocity of the $P2$-wave decreases as the degree of saturation increases up to approximately 75% and then increases rapidly until saturation is reached. It is opposite for the $P3$-wave: it increases with increasing S_0 up to 75% and then decreases. Both for gas saturation and for water saturation this wave disappears. This shows that it is driven by the capillary pressure and thus only exists if a second pore fluid is existent (see the work (Tuncay and Corapcioglu, 1996) whose authors were most likely the first in predicting this form of $P3$-waves in such media). Its velocity is much smaller than this of the other waves, however, its attenuation is much higher. Due to the high values of attenuation it is not astonishing that the author is not aware of any attempt to measure this third longitudinal wave, either in artificial media or even less in non-artificial ones.

6. Comparison with Suspensions and Experiments

The behavior of the wave velocities in the air–water mixture in sandstone, of course, shows similarities to the propagation of sound in suspensions, especially in an air–water mixture. It is well known that the existence of air bubbles in water reveals a minimum in the sonic velocity. Figure 5 (left) is taken from the book (Brennen, 1995) and shows results of the velocity of sound in an air–water mixture for a frequency of 1000 Hz in dependence on the air volume fraction. As shown in right panel of the figure such a behavior also appears in the partially saturated porous medium since the air–water mixture is only jacketed by the solid particles.

Besides the similarity to the propagation of sound in suspensions there exists also qualitative agreement of the calculated wave velocities with experimental results. The reproduction of one figure of (Murphy, 1982), Fig. 6 (left), shows the measured wave velocities of the S- and $P1$-waves in Massilon sandstone (porosity $n_0 = 0.23$). Due to the high attenuation of the $P2$- and $P3$-wave these waves could not be observed. However, in spite of somewhat

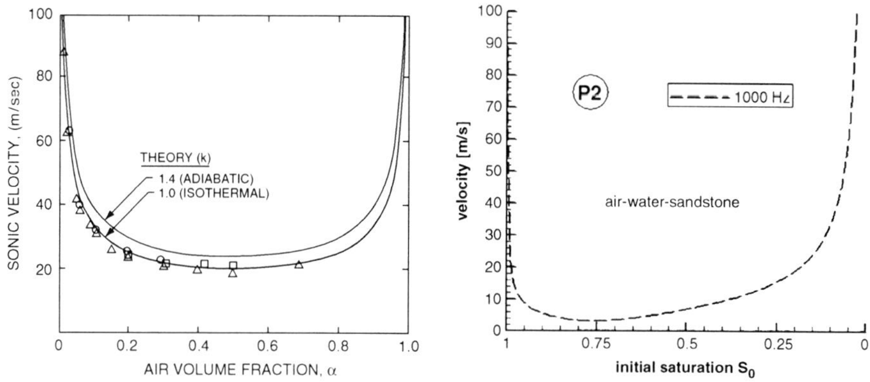

Figure 5. Left: velocity of sound in air–water mixtures (Brennen, 1995) in dependence on the air volume fraction, right: speed of the *P*2-wave in dependence on the air fraction (initial saturation in reverse axis).

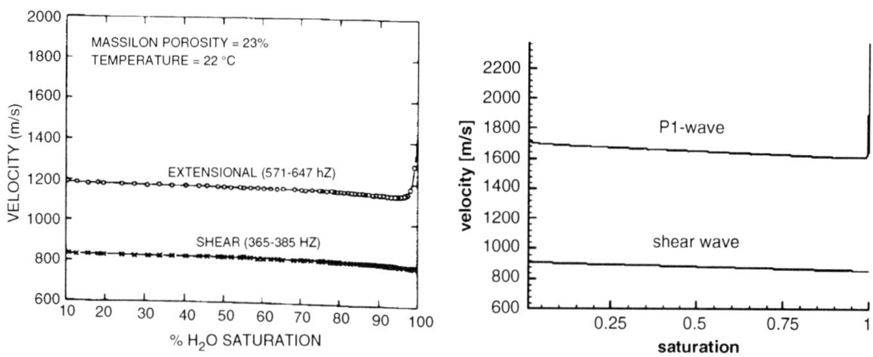

Figure 6. Comparison of experimental results (left) by Murphy and numerical results for the velocities of the fast longitudinal and the shear wave.

different conditions (porosity, frequency, temperature, etc.) the accordance of the experimentally observed *S*- and *P*1-wave velocities and their calculated counterparts shown in Fig. 6 (right) is quite well. Both *P*1-velocities show the strong increase of this velocity for high degrees of saturation and both *S*-wave velocities behave linearly and nearly constant in the whole range of saturations. Also the order of magnitude of the wave velocities matches closely.

7. Concluding Remarks

The wave analysis of a model for partially saturated soils containing three elastic constants and three coupling constants has been accomplished. For the special case of an air–water mixture in sandstones four body waves are predicted: three longitudinal waves, *P*1, *P*2, *P*3, and one shear wave, *S*.

The fastest longitudinal wave, $P1$, which propagates mainly in the skeleton, and the shear wave, S, are relatively little affected by the saturation except for a small region near water saturation. In this region the velocity of the $P1$-wave increases abruptly to nearly the double of its value for other saturations. Experimental results found in the literature support the occurrence of this effect. This may be an important feature for applications in geotechnics. It provides the hope for the development of a non-destructive testing method to warn against land slides. The latter occur if the degree of saturation exceeds a certain value. It is favourable for such a method that this feature occurs for the $P1$-wave which is the first arrival on an oscillogram. However, it is disadvantageous that it appears in a very narrow range of saturation. Further applications are imaginable as for example in piping, oil industry, landfilled waste management or for the remediation of compacted residual soils.

The $P2$- and $P3$-waves, which are effected by the existence of the fluid and the gas, are much more affected by the degree of saturation. They behave in the opposite way. While the velocity of the $P2$-wave has a strong minimum for medium values of saturation (as it is the case for suspensions) the $P3$-wave only exists in this region of saturation. Both for the water- and for the gas-saturated medium it does not emerge since it is evoked by the capillary pressure between the pore fluids. Both these waves are strongly damped, for the $P3$-wave the attenuation is so high that an observation of this wave in the field is nearly impossible.

References

Albers, B. (2009) *Modeling and Numerical Analysis of Wave Propagation in Saturated and Partially Saturated Porous Media*, Shaker, Habilitation thesis, Veroeffentlichungen des Grundbauinstitutes der TU Berlin.

Brennen, C. E. (1995) *Cavitation and Bubble Dynamics*, New York, Oxford University Press.

Murphy, W. F. (1982) Effects of partial water saturation on attenuation in Massilon sandstone and Vycor porous glass, *J. Acoust. Soc. Am.* **71**, 1458–1468.

Tolstoy, I. (ed.) (1992) *Acoustics, Elasticity and Thermodynamics of Porous Media: Twenty-One Papers by M. A. Biot*, Acoustical Society of America, Knockvennie, Castle Douglas, Scotland.

Tuncay, K. and Corapcioglu, M. Y. (1996) Body waves in poroelastic media saturated by two immiscible fluids, *J. Geophys. Res.* **101**, 25, 149–159.

van Genuchten, M. T. (1980) A closed-form equation for predicting the hydraulic conductivity of unsaturated soils, *Soil Sci. Soc. Am. J.* **44**, 892–898.

White, J. E. (1983) *Underground Sound. Application of Seismic Waves, Methods in Geochemistry and Geophysics*, Vol. 18, Amsterdam, New York, Elsevier.

Wilmanski, K. (1999) Waves in porous and granular materials. In K. Hutter and K. Wilmanski (eds.), *Kinetic and Continuum Theories of Granular and Porous Media*, No. 400 in CISM Courses and Lectures, Wien New York, Springer, pp. 131–186.

Wyckoff, R. and Botset, H. (1936) The flow of gas–liquid mixtures through unconsolidated sands, *Physics* **7**, 325–345.

LOCAL SITE EFFECTS AND SEISMIC RESPONSE OF BRIDGES

Prodromos N. Psarropoulos (prod@central.ntua.gr)
Department of Infrastructure Engineering, Hellenic Air-force Academy, Themistocleous St. 43, 16674 Athens, Greece

Abstract. The term "local site effects" is used to describe not only the potential effects of the soil stratigraphy, but the effects of the topographic irregularities and of the geomorphic conditions as well. The latter, usually characterizing an alluvial valley, tend to modify the amplitude, the frequency content, the duration, and the spatial variability of seismic ground shaking. The current study aims at shedding some light on these important issues by analyzing numerically the effects of the sub-surface geomorphic conditions of a valley on its ground surface seismic motion. Two-dimensional linear ground response analyses are performed to study an alluvial valley in Japan, the behavior of which has been monitored during many small earthquakes in the past. Additionally, equivalent-linear ground response analyses for the valley show that the potential soil nonlinearity (due to hypothetical strong shaking or highly nonlinear soil behavior) may reduce substantially the observed valley effects. Finally, as a road bridge founded on the valley has also been extensively monitored, special emphasis is given on the numerical simulation of the distress of its pile foundation and of the dynamic response of the superstructure.

Keywords: local site effects, valley effects, amplification, aggravation, bridge, pile foundation, soil nonlinearity, finite-element analyses

1. Introduction

In the seismic analysis of important or/and sensitive structures (like bridges, dams, etc.) ground response analysis is regarded as an essential initial step. In the case of a bridge (which is usually founded in a valley) the success in calculating the distress of its foundation depends on the ability of the geotechnical earthquake engineer to estimate realistically the level of stresses and strains developed on the surrounding soil under free-field conditions.

The dynamic stress field developed in the soil is a function of the characteristics of excitation at the base of the soil deposit and the local site conditions. In valleys or basins the term "local site conditions" is being used to describe both material and geomorphic conditions. Records and analyses (e.g., Aki, 1988; Finn, 1991; Gatmiri and Arson, 2008) have shown that, apart from the soil material conditions, the geomorphic conditions tend to alter the amplitude, frequency content, duration, and spatial variability of

T. Schanz and R. Iankov (eds.), *Coupled Site and Soil-Structure Interaction Effects with Application to Seismic Risk Mitigation,* NATO Science for Peace and Security Series C: Environmental Security,

ground shaking. Hence, their importance in seismic design of sensitive long structures, such as bridges, is substantial.

In geotechnical earthquake engineering it is a common practice to estimate the ground seismic response performing 1D analysis, assuming parallel soil layers of infinite extent, and neglecting thereby the potential impact of geomorphic conditions. On the other hand, objective difficulties in classifying the large variety of geomorphic features makes it a formidable task to account for these effects in simplistic, code-type prescriptions. To cope with this, 2D site-specific ground response analyses become essential. Such 2D *linear* analyses, as well as records from microtremors and small-magnitude earthquakes, have usually shown very substantial "valley effects" (Bard, 1994; Yegian et al., 1994). A small number of published *nonlinear* analyses have seriously questioned such an "aggravation". For instance Zhang and Papageorgiou (1996), simulating the seismic response of the Marina District Basin during the 1989 Loma Prieta Earthquake and utilising an equivalent-linear method (to account for soil nonlinearity), showed that in the case of strong-motion excitation "*three-dimensional (3D) focusing and lateral interferences, while still present, are not as prominent as in the weak-motion excitation case*". Additionally, the above mentioned study underlined the fact that the energy dissipation during strong-motion excitation dampens substantially the surface waves, and thus, the response of the valley is dominated by the nearly vertically propagating waves.

The present work has been motivated by the need to interpret the seismic response of a bridge and its foundation. The bridge, known as Ohba Ohashi, and the extremely soft alluvial valley it traverses, are in Japan and constitute a natural experimental site, as the motion of the ground (at the base and the surface of the valley), the response of a bridge pier, and the bending strains in the piles of this pier have been instrumentally recorded during a number of earthquakes (Tazoh et al., 1984, 1988). The available records are used for verification of the methods for computing free-field motion, loading of piles, and structural response.

One-dimensional analyses had proved inadequate to explain the level of shaking at the ground surface of the valley (Fan, 1992). Thus, 2D linear and equivalent-linear ground response analyses were performed to estimate the amplitude and the spatial variability of the free-field motion. Verification of the linear model has been demonstrated with a reasonable good reproduction of recorded ground motions, while the nonlinear model revealed the influence of soil nonlinearity on valley effects.

The finite-element model is then extended to incorporate, in a simple but realistic way, firstly the pile foundation, and then, the pile-superstructure

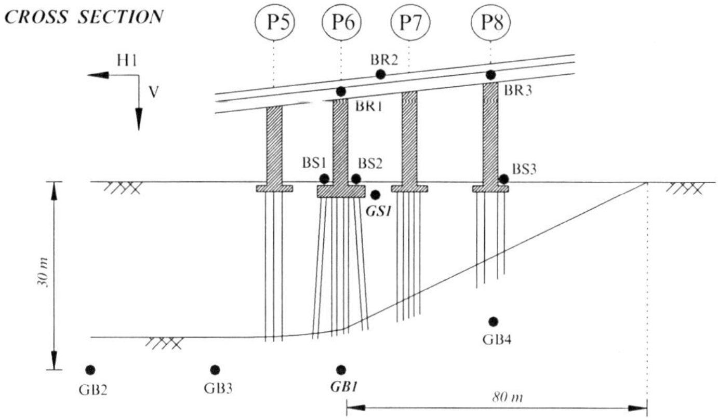

Figure 1. Cross section of the bridge between pier P5 and P8 (vertical scale is exaggerated).

system. The model developed seems able to reproduce satisfactorily the recorded strains on the piles and the acceleration time histories on the superstructure.

2. The Valley and the Bridge

The 600 m-long bridge is supported by seventeen piers. Figure 1 sketches the cross section of the bridge between pier P5 and pier P8, and the arrangement of the accelerographs. Of special interest is pier P6, which is supported by a pile group consisting of sixty-four steel pipe piles, thirty-two of which are battered.

The strain gauges had been installed by the Institute of Technology of Shimizu Corporation, Japan (Tazoh et al., 1984, 1988), along one vertical and one batter pile, at four depths.

The soil profile near pier P6 is shown in Fig. 2. The top layers consist of extremely soft layers of humus and clay. Despite the soil improvement, the N_{SPT} were almost null, while V_S ranged between 40 and 65 m/s. The underlying substratum consists of stiff clay and fine sand with V_S around 400 m/s and N_{SPT} over 50. Ground water table is almost at the ground surface, while the water content of the top layers exceeds 100%. The top layers are characterized by large plasticity index *PI* (in excess of 200). According to Vucetic and Dobry (1991), linear behavior is expected under dynamic loading.

3. Recorded Motions and Pile Strains

Earthquake observations had also been carried out by the Institute of Technology of Shimizu Corporation, Japan (Tazoh et al., 1984, 1988). Five

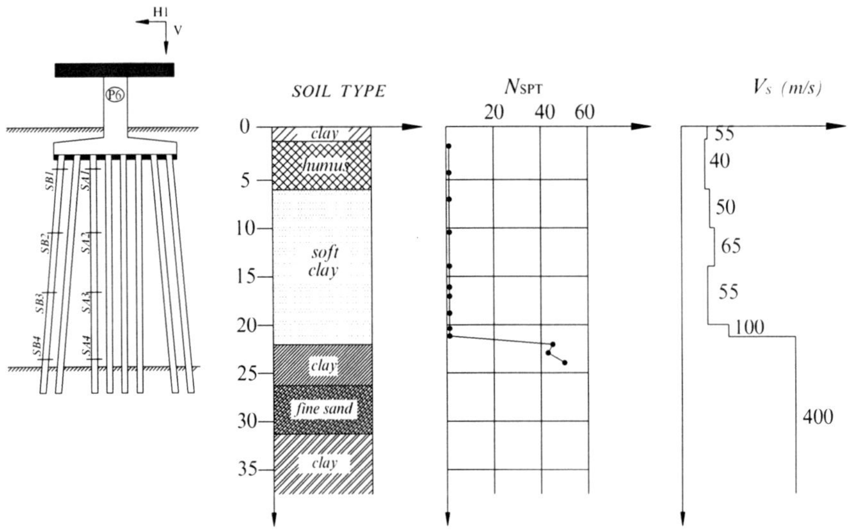

Figure 2. Foundation of pier P6 and soil profile characteristics.

accelerometers had been installed on the valley, six on the bridge superstructure, while eight strain gauges were installed on its piled foundation. Among fourteen earthquakes, recorded from 1981 to 1985, the event with the highest ground surface acceleration (0.114 g) was the Kanagawa-Yamanashi-Kenzakai earthquake (KYK), with magnitude $M_{JMA} = 6$, and epicentral distance 42 km.

Three of the records have been used for the linear ground response analyses, while only the earthquake event with the highest acceleration levels (earthquake KYK) has been used for the non-linear analyses and the estimation of the bending strains and the response of the superstructure. The free-field motion has been adequately recorded with accelerometers installed at the ground surface (e.g., GS1 near pier P6), and at the base of the superficial deposits (e.g., GB1, GB2, GB3, and GB4). The recorded acceleration time-histories at the base of the profile (GB1) and at the ground surface (GS1) and the corresponding elastic response spectra during earthquake KYK are shown in Fig. 3.

Three accelerometers (BS1, BS2, and BS3) have been installed on the pile caps, two (BR1, BR3) are on the bridge piers P6 and P8, respectively, while one (BR2) is located on the girder, between the piers P6 and P7.

The pile distress was traced by strain gauges that recorded the bending strains at two directions. Four instruments (SA1, SA2, SA3, SA4) were installed along a vertical pile of pier P6, while four instruments (SB1, SB2, SB3, SB4) were placed along one of the batter piles (see Fig. 2).

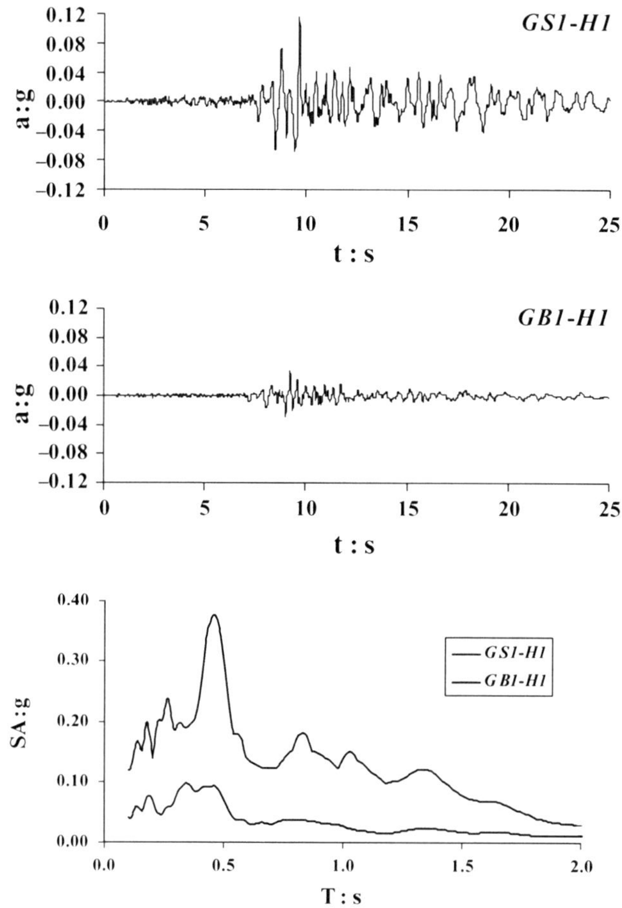

Figure 3. Acceleration time-histories and response spectra at the base (GB1) and the surface (GS1) (earthquake KYK).

4. Linear 2D Ground Response Analyses

As 1D analyses had been proved insufficient to estimate the recorded free-field motion of the valley (Fan, 1992), 2D finite-element and spectral-element analyses were performed by Psarropoulos et al. (1999). As shown in Fig. 4, the geometry and the soil properties of the valley were simplified, assuming a trapezoidal shape and mean low-strain V_S of the soil stratum equal to 60 m/s. In the same figure, the point of interest M is shown, located on the surface (identical with location of receiver GS1).

The analyses were initially based on the assumption of linear visco-elastic behavior of the soil, which is quite acceptable for earthquakes with relatively low values of accelerations and for clayey deposits with very high plasticity index (as is the case here).

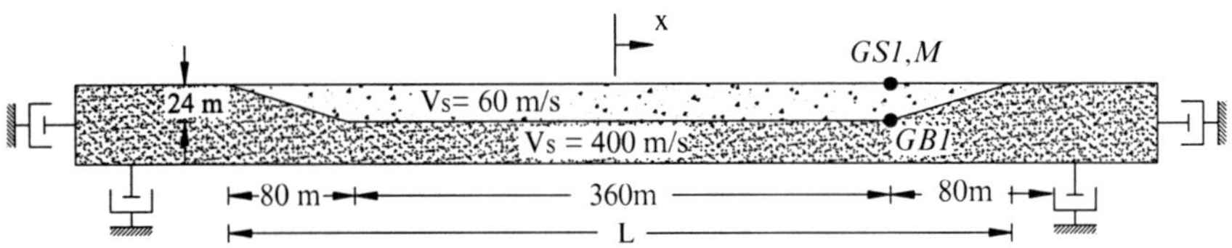

Figure 4. Idealized geometry of the valley.

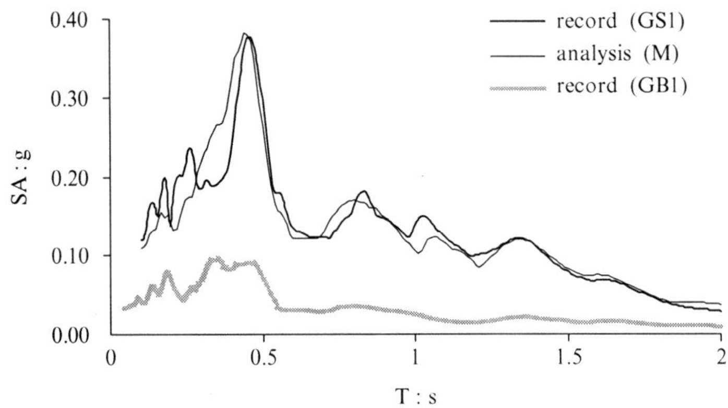

Figure 5. Records vs. linear analysis: response spectra for earthquake KYK.

The approach for the linear finite-element analysis utilizing ABAQUS code (2004) was based on the "effective seismic excitation" technique (Bielak and Christiano, 1984). With this approach, the problem of seismic response of a 2D valley is transformed into an equivalent one in which the source is located in the interior of the domain of computation. The advantages of the technique are discussed in detail in Loukakis (1988).

To verify the linear models, each of the recorded ground base (GB1) acceleration time histories of the three different earthquakes was applied as input excitation, and the recorded acceleration time-histories at the ground surface (GS1) were obtained. As there were no additional records available from the valley surface, other than GS1, the satisfactory comparison between records and analyses at this location offers only a first validation for the linear model and the method of analysis. Figure 5 depicts the results obtained for point *M* in the case of earthquake KYK.

5. Equivalent-Linear 2D Ground Response Analyses

A significant issue relating to "microzonation" research is to quantify the degree and extent of reduction due to soil nonlinearity of the huge 2D amplifications often computed and, especially, recorded during weak shaking (where soil behavior is linear). To account for hypothetically nonlinear soil, a parametric investigation is conducted, where 2D equivalent-linear analyses are performed using QUAD4M code (Hudson et al. 1994). Initially, earthquake KYK is used as base excitation, scaled from 0.034 up to 0.34 g, while the effect of *PI* (from 200 to 50) is also examined.

The distribution along the surface of the amplification *A* of the peak ground acceleration (defined as the ratio $\alpha_{S\,\max}/\alpha_{B\max}$ of peak accelerations at surface and base) is shown in Fig. 6. Several trends are noted:

- Increasing the intensity of base excitation (from 0.034 to 0.34 g) for a deposit with a moderately high *PI* (= 50) reduces *A* almost everywhere on the surface, except near the edges of the valley. Apparently the increased nonlinearity, with the large hysteretic damping it generates, leads to reduced accelerations. In addition, waves generated at the edges and propagating laterally across the valley attenuate substantially; thus the undulatory (in space) nature of surface acceleration is diminished.
- With decreasing *PI* the soil becomes less linear, and the trends are essentially the same as with increasing intensity of motion.

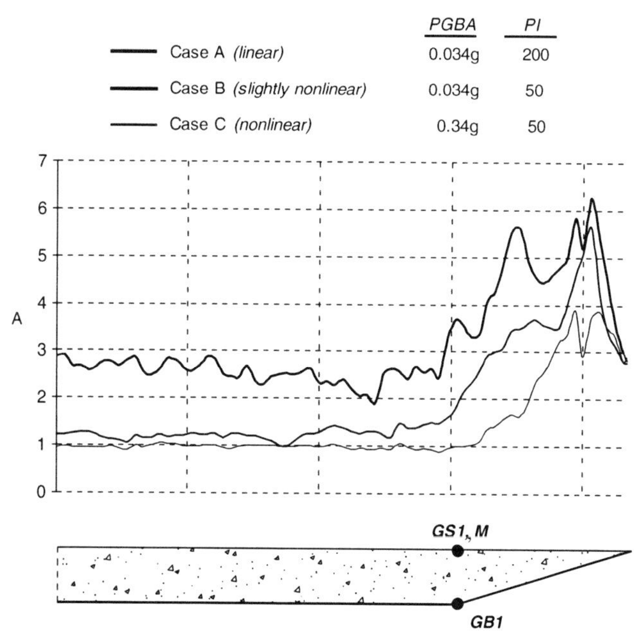

Figure 6. Equivalent linear analysis: Distribution along the right half of the valley surface of the amplification (*A*), for two different values of base acceleration (0.034 and 0.34 g), and two different values of plasticity index (PI = 200 and 50). Notice the progressive reduction of the amplification as the degree of nonlinearity increases (from Case A to Case C).

- Spatial variability of ground shaking may be substantial even for relatively short distances, but is restricted only near the inclined boundaries of the valley if nonlinearity is strong. The phenomenon may be of extreme importance for bridges, and cannot be evaluated with 1D analysis, unless very strong nonlinear response takes place.

Finally, in order to make the wave propagation in the valley more comprehensive, a simple Ricker pulse is used as base excitation. The pulse has a central frequency f_o of 2 Hz, close to the fundamental frequency of KYK. Figure 7 shows the acceleration time history of the applied Ricker pulse and the corresponding response spectrum (scaled to PGA = 0.34 g), while Fig. 8 shows the wavefields of acceleration calculated along the surface for the three cases of nonlinearity. The effect of nonlinearity is evident. While shifting progressively from the linear case to the nonlinear case, both energy dissipation (due to the hysteretic nature of soil behavior) and soil stiffness degradation take place. It is obvious that the energy dissipation dampens all the waveforms. In this way the interferences of vertical or inclined body waves from multiple reflections at the soil-base interface and the free surface are substantially affected, while on the other hand, the laterally propagating surface Rayleigh waves are so strongly attenuated that cannot even be distinguished in the nonlinear case at distances beyond the first point from the edge.

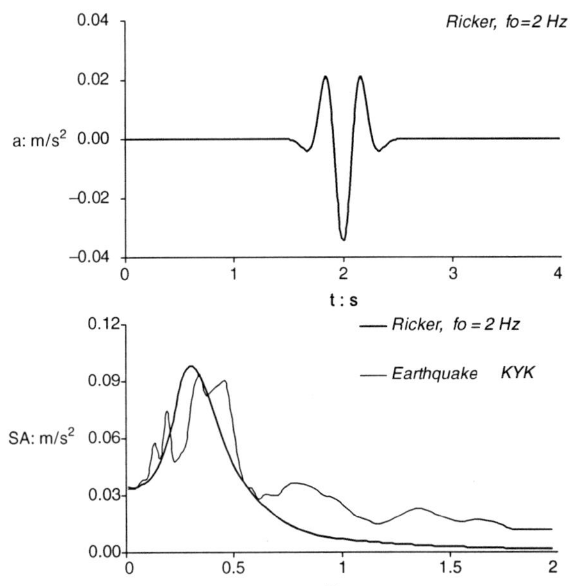

Figure 7. Acceleration time history and the corresponding response spectrum of the Ricker pulse applied as excitation (scaled to PGA = 0.34 g). Spectrum of KYK is also included.

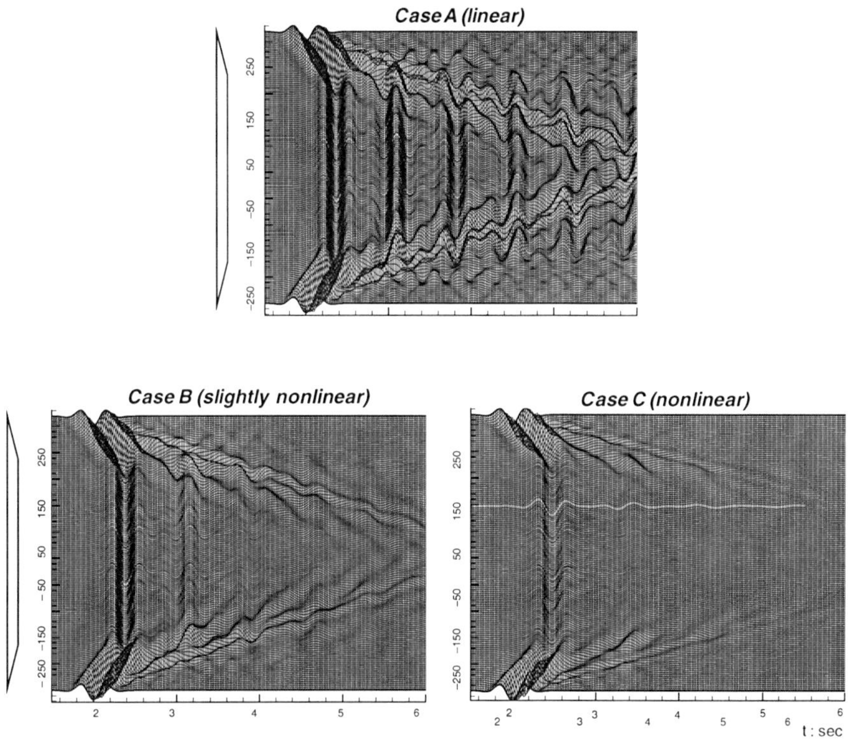

Figure 8. Wavefields of acceleration calculated along the surface of the valley for the three cases of nonlinearity examined (after Psarropoulos et al., 2007).

6. Soil–Pile–Structure Interaction

To estimate numerically the imposed pile bending, the finite-element model was extended by incorporating, in a simple but realistic way, the pile foundation. Properties of the soil are kept exactly the same. As a 3D modeling (that could possibly take into account the entire pile group) was a formidable task, the new model was based on the following simplistic assumption: plane-strain conditions were considered, and to this end an "equivalent diaphragm" was used. The diaphragm is characterized by longitudinal stiffness $E_P I_P$, equal to the one that characterizes the piles per current meter (in the transversal direction). The 4-noded quads used for the modeling of the "equivalent diaphragm" are equipped with incompatible modes to improve bending behavior. The maximum kinematic bending strains developed close to the tip are consistent with the recorded strains at depth.

The finite-element model is then extended a step further, as the bridge pier and the corresponding mass of the girder were incorporated as an additional single degree of freedom system. The entire soil–pile–structure system is then

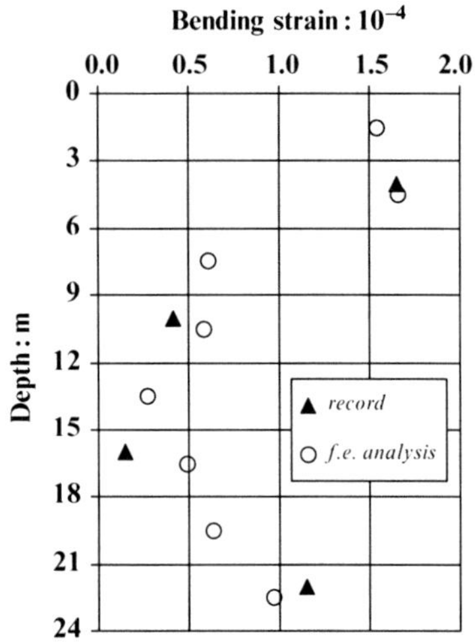

Figure 9. Maximum bending strains on piles (earthquake KYK).

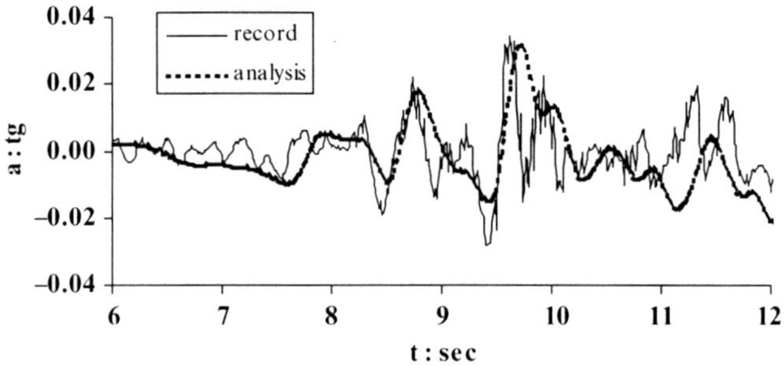

Figure 10. Acceleration time history predicted for the superstructure in comparison with the corresponding record (earthquake KYK).

analyzed. In Fig. 9 the bending strains predicted from the simulation are being compared with the recorded bending strains. In Fig. 10 the acceleration time histories predicted for the superstructure are successfully compared with the recorded time histories (BR2).

7. Conclusions

The present work has been motivated by the need to interpret the seismic response of a bridge and its foundation. The bridge and the extremely soft alluvial valley it traverses constitute a natural experimental site, as the motion of the ground, the response of a pier, and the bending strains in the piles of this pier have been instrumentally recorded during a number of earthquakes.

The available records are used for verification of the methods for computing free-field motion, loading of piles, and structural response.

Trying to capture any significant 2D valley effects on the amplitude and the variability of ground shaking, it was found out that the *linear* 2D numerical analyses can successfully explain the observed ground shaking, the pile distress, and the structural response of the superstructure. This consistency may be attributed to the low acceleration levels of excitation and the high plasticity index of the soil.

On the other hand, *equivalent linear* 2D ground response analyses have shown that an increase of the intensity of base shaking or/and a decrease of the plasticity index of the soil may lead to substantially lower valley effects, The significant energy dissipation that takes place in such a case dampens substantially the laterally propagating Rayleigh waves generated at the valley edges, while the changing with time soil modulus renders any wave resonance of vertically propagating body waves less important of vertical or inclined body waves from multiple reflections at the interfaces.

Finally, the numerical example and the state of the art in the reviewed literature make clear that there exists a continued need for more research on nonlinear multi-dimensional analyses and observations.

Acknowledgements

The current study was partially funded by the Greek Organization for Earthquake Planning and Protection in the framework of the project "Experimental and theoretical support of microzonation methods". The author is grateful to Prof. G. Gazetas for his valuable contribution to the project, and Dr. T. Tazoh for his support of the seismic observations.

References

ABAQUS (2004) *Analysis User Manual*, v. 6.4, Abaqus Inc., USA.

Aki, K. (1988) Local site effects on strong ground motion, *Earthquake Engineering and Soil Dynamics II*, New York, pp. 103–155, ASCE.

Bard, P. Y. (1994) Effects of surface geology on ground motion: some results and remaining issues. In *Proc. 10th European Conference on Earthquake Engineering*, Vienna, Vol. 1, pp. 305–323.

Bielak, J. and Christiano, P. (1984) On the effective seismic input for non-linear soil-structure interaction systems, *Earthquake Eng. Structural Dynamics* **12**, 107–119.

Fan, K. (1992) Seismic response of pile foundations evaluated through case histories, Ph.D. Thesis, S.U.N.Y. at Buffalo.

Finn, W. D. L. (1991) Geotechnical engineering aspects of seismic microzonation. In *Proc. 4th Int. Conf. on Seismic Zonation*, Vol. 1, Stanford, CA, pp. 199–250.

Gatmiri, B. and Arson, C. (2008) Seismic site effects by an optimized 2D BE/FE method II. Quantification of site effects in two-dimensional sedimentary valleys, *Soil Dynamics Earthquake Eng.* **28**, 646–661.

Hudson, M., Idriss, I. M., and Beikae, M. (1994) *User's Manual for QUAD4M*, Center for Geotechnical Modeling, Department of Civil and Environmental Engineering, University of California, Davis, USA.

Loukakis, K. (1988) Transient response of shallow layered valleys for inclined SV waves calculated by the finite-element method, M.S. Thesis, Carnegie Mellon University.

Psarropoulos, P. N., Gazetas, G., and Tazoh, T. (1999) Seismic response analysis of alluvial valley at bridge site. In *Proc. 2nd Int. Conf. on Geotechnical Earthquake Eng.*, Lisbon, pp. 41–46.

Psarropoulos, P. N., Tazoh, T., Gazetas, G., and Apostolou, M. (2007) Linear and non-linear valley amplification effects on seismic ground motion, *Soils Foundations* **47**(5), 857–872.

Tazoh, T., Dewa, K., Shimizu, K., and Shimada, M. (1984) Observations of earthquake response behavior of foundation piles for road bridge. In *Proc. 8th World Conf. on Earthquake Eng.*, Vol. 3, pp. 577–584.

Tazoh, T., Shimizu, K., and Wakahara, T. (1988) Seismic observations and analysis of grouped piles, Technical Research Bulletin, No. 7, pp. 17–32, Shimizu Corp.

Vucetic, M. and Dobry, R. (1991) Effect of soil plasticity on cyclic response, *J. Geotechnical Eng.*, ASCE **117**, 89–107.

Yegian, M. K., Ghahraman, V. G., and Gazetas, G. (1994) Seismological, soil and valley effects in Kirovakan, 1988 Armenia earthquake, *J. Geotechnical Eng.*, ASCE **120**(2), 349–365.

Zhang, B. and Papageorgiou, A. S. (1996) Simulation of the response of the Marina District Basin, San Francisco, to the 1989 Loma Prieta earthquake, *BSSA* **86**(5), 1382–1400.

LOCAL SITE EFFECT EVALUATION IN SEISMIC RISK MITIGATION

Behrooz Gatmiri (`gatmiri@enpc.fr`)
Université Paris-Est, Ins. Navier, Ecole Nationale des Ponts et Chaussées, Paris, France

Abstract. The response of a site to a seismic solicitation depends on local topographical and geomechanical characteristics. This paper deals with the evaluation of seismic site effects due to the local topographical and geomechanical characteristics and tries to contribute to establishment of a simple method to evaluate site effects. The amplification of surface motions is calculated by a numerical tool, HYBRID software, combining finite elements in the near field and boundary elements in the far field (FEM/BEM). In this paper, horizontal ground movements in various points of 2D empty valleys with irregular configurations subjected to synthetic SV waves of vertical incidence are calculated. Parametric studies are done on slopes, ridges and different shapes of canyons to characterize topographical site effects. The second part deals with sedimentary valleys. The complexity of the combination of geometrical and sedimentary effects is underlined. Extensive parametrical studies are achieved to discriminate the topographical and geotechnical effects on seismic ground movement amplifications in two-dimensional irregular configurations. The results are also shown in the form of pseudo-acceleration response spectra. For the empty valleys, we can classify the spectral response according to a unique geometric criterion: the "surface/angle" ratio, where surface is the area of the valley opening, and angle denotes the angle between the slope and horizontal line in the above coin. To assess the influence of the two dimensional effect on the spectral response of filled valleys, the response of alluvial basins are compared with the response of one-dimensional columns of soil. Finally, an offset criterion is proposed to choose a relevant computation method for the spectral acceleration at the surface of alluvial basins. The accuracy of this quantitative evaluation technique is tested and discussed.

Keywords: hybrid numerical technique, time truncation, poroelastodynamics, seismic amplification, 2D site effects, topographical irregularities, seismic risk

1. Introduction

Unfortunately, the complete responses of movement taking into account the influence of topographical and geological conditions on seismic movement are not considered in normal engineering practice. This work aims at characterizing and quantifying the combined site effects in the bidimensional configurations, in the spectral domain.

T. Schanz and R. Iankov (eds.), *Coupled Site and Soil-Structure Interaction Effects with Application to Seismic Risk Mitigation,* NATO Science for Peace and Security Series C: Environmental Security,

This work considers combined influence of topography and geology on the seismic response of alluvial valleys. The two-dimensional wave scattering is studied with the aid of HYBRID program, combining finite elements in the near field and boundary elements in the far field (FEM/BEM). This program has been developed by Gatmiri and his co-workers (Gatmiri and Kamalian, 2002a; Gatmiri and Kamalian, 2002b; Gatmiri and Nguyen, 2005; Kamalian et al., 2006). The time integration process is optimized in this study by a new time truncation method (Gatmiri and Dehghan, 2005; Gatmiri and Nguyen, 2007; Gatmiri et al., 2007). Seismic solicitation is a vertically incident SV Ricker wave. In this paper, acceleration response spectra of different empty valleys will be studied. This will help better understanding of the problems encountered in the modelling of topographical effects. Curves will be collected on a unique figure, which will characterize topographical effects in a quantitative and qualitative way in the spectral domain. The influence of bi-dimensionality on the response of a sedimentary site is studied also. For this purpose, acceleration responses of filled valleys are compared to the responses of an equivalent one-dimensional column of soil. A civil engineer needs to know how to identify the preponderant site effect in a point of the surface of a given shape. That is why in the latest part of this work, a preponderance criterion is presented.

2. Formulation of Problems Combining FE and BE Methods

The finite-element method (FEM) is particularly adapted to work with anelastic or non-linear soils. The boundary-element method (BEM) reduces the problem by one dimension and is relevant for half-plane problems. The study of site effects requires the resolution of mechanical wave radiation equations in irregular configurations, defined by specific topographical and geotechnical conditions. That is why hybrid models combing both methods are often used. In our study, sediments are modelled by finite elements. Substratum is represented by boundary elements, which is adapted to the study in the far field. The region of interest is a half-space and must be enclosed with fictitious boundary elements known as "enclosing elements". All materials are supposed to be elastic.

In the finite-element domain, application of the modified Newton–Raphson iterative method leads to:

$$M \bullet \ddot{U}^{t+\Delta t(k)} + K^t \bullet \left(U^{t+\Delta t(k)} - U^{t+\Delta t(k-1)}\right) = R^{t+\Delta t} - F^{t(k)}, \qquad (1)$$

where M is the mass matrix and K^t is the rigidity matrix at instant t. $U^{t+\Delta t(k)}$ is the displacement vector for the kth iteration done to reach the load increment $R^{t+\Delta t}$ imposed at $t + \Delta t$. $F^{t(k)}$ is the force calculated by the behaviour law of the material at the kth iteration.

Using the Newmark method, in which:

$$U^{t+\Delta t} = U^t + \frac{\Delta t}{2} \cdot \left(\dot{U}^t + \dot{U}^{t+\Delta t}\right) \tag{2}$$

Eq. (1) becomes:

$$\overline{K}^t \bullet \Delta U^{(k)} = R^{t+\Delta t} - F^{t(k)} - \left(\frac{4}{(\Delta t)^2} \cdot M\right) \bullet \left(U^{t+\Delta t(k-1)} - U^t\right) + \left(\frac{4}{\Delta t} \cdot M\right) \bullet \dot{U}^t + M \bullet \ddot{U}^t \tag{3}$$

where:

$$\overline{K}^t = K^t + \frac{4}{(\Delta t)^2} \cdot M. \tag{4}$$

Adding $\overline{K}^t \bullet U^{t+\Delta t(k-1)}$ at both sides of Eq. (3) and assuming that zone 1 is modelled with finite elements:

$$\overline{K}^t \bullet U^{t+\Delta t(k)} = R_1^{t+\Delta t} - Z_1^{t+\Delta t(k)}, \tag{5}$$

where:

$$Z_1^{t+\Delta t(k)} = F_1^{t(k)} - K_1^t \bullet U_1^{t+\Delta t(k-1)} - \left(\frac{4}{(\Delta t)^2} \cdot U_1^t + \frac{4}{\Delta t} \cdot \dot{U}_1^t + \ddot{U}_1^t\right) \bullet M_1. \tag{6}$$

The boundary integral equation of elastodynamics in time-domain for a homogeneous isotropic elastic medium, occupying a volume Ω, bounded by a surface Γ, and subjected to an incident plane wave is:

$$c_{ij}(\xi) \cdot u_j(\xi) = \int_\Gamma \left[G_{ij}(\xi, x, t)^* t_j(x, t) - F_{ij}(\xi, x, t)^* u_j(x, t)\right] \cdot d\Gamma + u_i^{\mathrm{eq}}(\xi, t) \tag{7}$$

if the contributions of initial conditions and body forces are neglected. ξ is the source point, x is the field point; u_i and t_i are the amplitudes of the ith component of displacement and traction vectors respectively, at the boundary; u_i^{eq} represents the incident wave; the symbol * indicates a Riemann convolution integral; c_{ij} is the discontinuity term depending on the local geometry of the boundary at ξ and on the Poisson's ratio; G_{ij} and F_{ij} are the fundamental solutions representing the displacement and traction at x in direction i due to a unit point force applied at ξ in the j direction.

The numerical implementation of Eq. (7) requires a discretization in both time and space. For this purpose, the boundary Γ is discretized into a defined number of elements, and time axis is divided into N equal intervals so that $t = N \cdot \Delta t$.

In this work, quadratic spatial variation of the field variables (displacement and traction) is assumed over each element. Both constant and linear temporal variations can be used for each field variable. Space discretization gives the following matricial expression at instant $t = N \cdot \Delta t$:

$$F^1 \bullet U^N = G^1 \bullet T^N + \sum_{n=1}^{N-1} (G^{N+1-n} \bullet T^n - F^{N+1-n} \bullet U^n). \tag{8}$$

The equations obtained from the FEM are expressed in force and displacement whereas in the BEM, stresses replace forces. Therefore, equations need to be adapted. In the last term of (8), stresses are transformed into forces as follows:

$$Z^{t(k)} = N \bullet (G^1)^{-1} \bullet \sum_{n=1}^{N-1} (F^{N+1-n} \bullet U^n - G^{N+1-n} \bullet T^n). \tag{9}$$

Consider that zone 2 modelled by boundary elements and has a common frontier with zone 1, which is modelled by finite elements (Gatmiri and Kamalian, 2002b; Gatmiri and Nguyen, 2005). The governing matricial equation of zone 2 can be written in the same way as (5):

$$\overline{K}_2^t \bullet U_2^{t+\Delta t(k)} = R_2^{t+\Delta t(k)} - Z_2'^{\,t+\Delta t(k)}. \tag{10}$$

The presented formulation is integrated in HYBRID code and validated in Gatmiri and Kamalian (2002a, 2002b). Two-dimensional fundamental solutions were also found and implemented in HYBRID (Gatmiri and Kamalian, 2002b; Gatmiri and Nguyen, 2005; Gatmiri and Jabbari, 2005a, 2005b).

2.1. OPTIMIZED INTEGRATION TECHNIQUE BY TIME TRUNCATION

An optimized time integration process has been integrated in HYBRID code (Gatmiri and Dehghan, 2005). The technique lies on time truncation. Integration is limited to a number of time steps (m) chosen at the beginning of the calculation. Consider the case $m = 2$. The two first integral equations are the same as in the Wrobel method (Eqs. (2) and (4)), whereas the third one is expressed as:

$$c.u_3 = \int_\Gamma \int_{t_1}^{t_3} G_{ij}^* t_j(x,t) \cdot d\Gamma - \int_\Gamma \int_{t_1}^{t_3} F_{ij}^* u_j(x,t) \cdot d\Gamma + \int_\Omega G_{ij}^2 \cdot F^1 \cdot d\Omega. \tag{11}$$

At each step of the calculation, the domain integral can be approximated by generalized mean value theorem applied to the average value of the body force. An iterative process yields to the general formulation of the optimised integration technique. The integration process stops by a convergence criterion based on a small tolerance L_m. Precision is thus, controlled by two parameters: the number of time steps which gives the backtracking limit (m), and a tolerance coefficient that cuts the calculation when the terms become negligible (L_m). The numerical studies reported in Gatmiri and Dehghan (2005) show that the time-truncation process used in HYBRID program is fast and accurate. The minimal meshing size is less than the size required in a traditional integral computation frame. Moreover, artificial waves generated at the truncation points of the model vanish easier if the optimized method is used.

3. Topographic Site Effects in 2D Configurations

3.1. PROBLEM PARAMETERS

In order to give some salient features of topographic effects, various examples that cover different 2D geometries are considered. The configurations of the studied rocky valleys are triangle, trapezium, rectangle, ellipse and truncated ellipse. Valleys are characterized by their half width at the surface and at the base and their depth (L, L_1 and H).

For triangular, trapezoidal and rectangular valleys, the angle formed by the slope of the relief relatively to the horizontal line in the above corner is considered, for the ellipsoidal and truncated ellipsoidal configurations, α is the angle between the tangent at the top corner of the valley and the horizontal line.

3.2. STUDY OF THE TOPOGRAPHICAL EFFECTS IN THE VARIOUS VALLEYS

For a given empty valley, the curves of the acceleration response spectrum are obtained for each observation point and for the reference site that is situated on the outcrop. The spectral ratios are also obtained. A triplet of curves is thus obtained for every chosen observation point and for every type of valley. In order to find the more critical point, the curves of the spectral ratio are represented as a function of the dimensionless offset variable X/L for the various valleys (for the detail see Gatmiri et al., 2008).

3.2.1. *Quantification of topographical effects in empty valleys*

For the non-curved geometries (triangle, trapezium and rectangle), amplification ratio curves are presented in Fig. 1. Table I establishes a clear relation between the geometrical parameters of the remaining shapes and the classification of the calculated seismic amplifications.

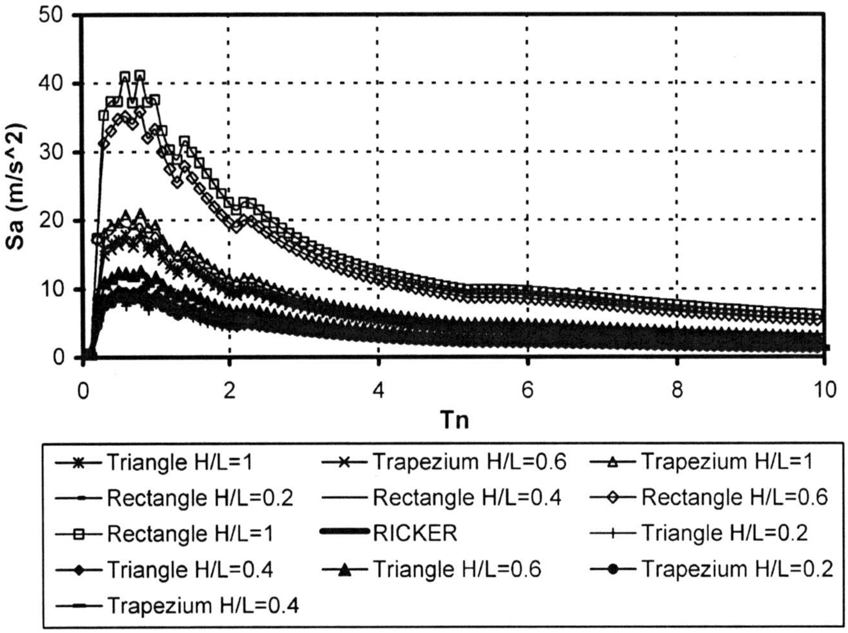

Figure 1. Acceleration response spectra for non-curved valleys at $X/L = 1$.

TABLE I.

Classification	Figure	H/L	Surface (S) m^2	Angle (A)	S/A
1	Rectangle	1	20,000	90°	222.2
2	Rectangle	0.6	12,000	90°	133.3
3	Trapezium	1	14,000	120°	116.7
4	Rectangle	0.4	8,000	90°	88.8
5	Triangle	1	10,000	135°	74.07
6	Trapezium	0.6	8,400	135°	62.2
7	Triangle	0.6	6,000	150°	40
8	Trapezium	0.4	5,600	146°	38.4
9	Triangle	0.4	4,000	158°	44.4
10	Rectangle	0.2	4,000	90°	25.3
11	Triangle	0.2	2,000	169°	17.3
12	Trapezium	0.2	2,800	162°	11.8

There is a very clear correlation between the parameter S/A (surface/angle) and the classification. It is thus possible to model topographical site effects in the various configurations only by means of this new parameter (S/A).

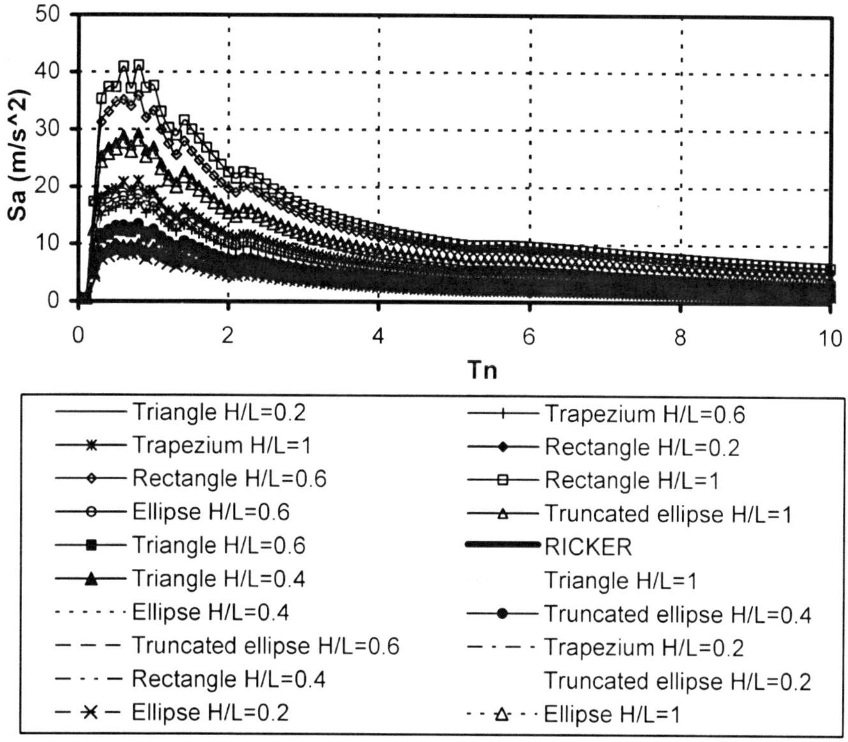

Figure 2. Acceleration response spectra for various form of valleys at $X/L = 1$.

For the curved geometries (ellipse and truncated ellipse), we study all geometric shapes with a fixed ratio of H/L (Fig. 2). For every value of H/L, the spectral response increases with the parameter of S/A. It is noted that the behaviour of curved forms (ellipse and truncated ellipse) is intermediate: the spectral curves are always located between those of the rectangular valleys and the trapezoidal valleys.

4. Two-Dimensional Analysis of the Response of the Filled Valleys

The aim of this section is to study the influence of 2D effects on the seismic response of filled valleys. Acceleration response of filled valleys will be compared to the responses of one-dimensional columns of soil. The height of the 1D reference column is chosen equal to the thickness of the sedimentary layer underlying the observation point considered in the filled valley. The same geometrical characteristics of valleys are chosen as empty valleys. It is assumed that valleys are completely filled by a homogeneous sedimentary layer. The impedance contrast β between rock and sediment is equal to 0.31 (for more details see Gatmiri et al., 2008).

4.1. STUDY OF COMBINED EFFECTS (2D) IN THE VARIOUS VALLEYS

For a given filled valley, some of the points at surface are chosen as observation stations to study the combined topographical-sedimentary effects. The curves of acceleration response spectra are drawn for each observation point and for the reference site. The spectral ratio from previous spectra is also obtained. In order to study the response of each valley to a given seismic solicitation, the curves of the spectral ratio versus dimensionless offset variable X/L are shown for the various configurations (Gatmiri et al., 2008). It has been concluded that:

All curves have two parts. A decreasing part from the central point ($X/L = 0$) to a point whose abscissa is between $X/L = 0.5$ and $X/L = 1$, and an increasing part between the intermediate point and the top of the slope $X/L = 1$. For the first part, it is obvious that as we move away from the central point, amplitude decreases, due a decreasing influence of the sedimentary effect (Gatmiri and Arson, 2008). The increasing part of the curve shows the predominance of topographical effects on the slopes covered by sediment. In the central part of the valley, one-dimensional sedimentary effect controls the local response of the site. On the slopes of the sedimentary basin, the presence of alluvium attenuates the predominant topographical amplification. Practically, the maximal amplification is reached at the central point of the valley ($X/L = 0$). This point seems to be the most critical. This is why afterwards; combined effects are modelled at the centre of valleys.

5. Comparison of 2D Combined and 1D Effects

In Table II, the curve denoted by "n° 1" corresponds to the strongest amplification of spectral acceleration response. In this table, a height/surface parameter has been used. It is clear that if the height/surface parameter reduces, amplification increases.

It is important to note that:

$$\frac{H}{S} \propto \frac{1}{L}. \tag{12}$$

On the other hand,

$$\frac{H}{S} \downarrow \Rightarrow \frac{1}{L \uparrow} \Rightarrow L \uparrow \Rightarrow \text{amplification} \uparrow . \tag{13}$$

It means that, at a given point $X/L = 0$, when L increases, amplification increases. On the other hand, at a given point $X/L = 0$, if the value of height is fixed, the influence of topographical effects in a filled valley disappears, as the surface width increases. Therefore, combined effects are transformed into 1D

TABLE II.

Classification	Figure	H/L	Surface (S) m^2	Height (H) m	H/S
1	Rectangle	1	20,000	100	0.005
2	Truncated ellipse	1	15,700	100	0.006
3	Rectangle	0.6	12,000	60	0.005
4	Ellipse	1	15,708	100	0.006
5	Truncated ellipse	0.6	9,055	60	0.006
6	Trapezium	0.6	8,400	60	0.007
7	Triangle	0.4	4,000	40	0.01

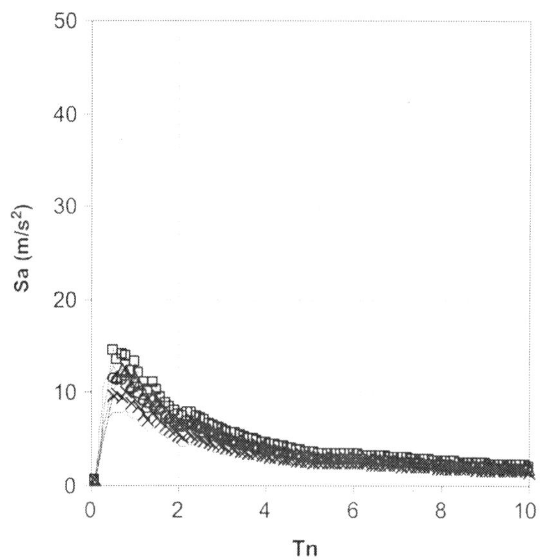

Figure 3. Acceleration spectra of the various valleys at $X/L = 0$: (a) $H = 20$ and (b) $H = 100$ m.

geological effect. To illustrate this conclusion, the valleys characterized by the same depth, but at the different surfaces S are compared. In this case, varying S with a fixed height changes the values of L. The curves are shown with H equal to 20 and 100 m, which are the minimal and maximal values of height in the present parametric study (Fig. 3). The same result will be achieved; if S increases, the amplification of the spectral response in acceleration at the centre of the valley increases. It can be concluded that on a point far from topographical effect, if the value of L tends to ∞, the curves of combined effect tend to the characteristic curve of 1D effect; and in this point, 1D effect is predominant.

6. Conclusion

6.1. DEFINITION OF A CLEAR CRITERION FOR TOPOGRAPHICAL EFFECTS IN AN EMPTY VALLEY

As presented in the sections dealing with topographical effects, the spectral acceleration responses are classified according to a unique geometrical criterion: the "surface/angle" ratio. By using Fig. 1, and making a regression between the minimal and maximal values of the surface/angle parameter, we can find the spectral responses in acceleration for any valley.

6.2. DEFINITION OF A CLEAR CRITERION FOR THE COMBINED EFFECTS AT THE VARIOUS POINTS ON THE SURFACE OF A FILLED VALLEY

A criterion is proposed to determine the predominant site effect at the surface of a filled valley. According to the sections dealing with topographical effects, the spectral acceleration response is attenuated for all points whose ordinate is lower than mid-slope point. On the other hand, in the section concerning the combined site effects, it was observed that in the central zone (first part), the seismic response was amplified. Therefore, in the central zone, sedimentary effects dominate topographical effects. This result is illustrated in Fig. 4.

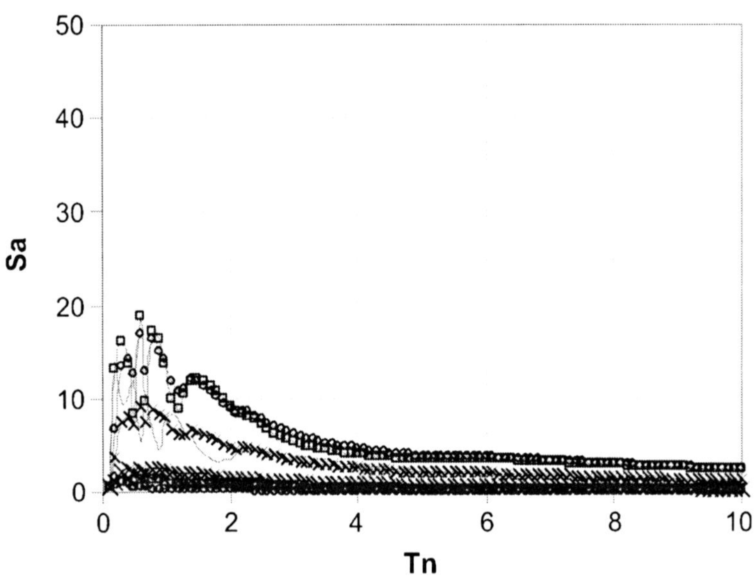

Figure 4. Comparison between the topographical effect in the empty valleys and the combined effect in the filled valleys at corner and at $X/L = 0$: (a) $H = 60$ and (b) $H = 100$ m.

It has already been noticed that at the centre, the spectral responses related to 2D combined effects tend to the one-dimensional analysis results. In central zone, sedimentary effects are always predominant in the centre of the alluvial valleys.

In conclusion, in the central zone (from $X/L = 0$ to mid-slope), results provided by one-dimensional analyses can be used to estimate the spectral acceleration response of a filled valley (similar to actual seismic codes), and in the lateral zone, the spectral response of the sedimentary valleys can be deduced from the characteristic spectra of topographical effects, shown in the figures of the section concerning the topographical site effects.

References

Gatmiri, B. and Arson, C. (2008) Seismic site effects by an optimized 2D BE/FE method. II. Quantification of site effects in two dimensional sedimentary valleys, *Int. J. Soil Dynamics Earthquake Eng.* **28**, 646–661.

Gatmiri, B., Arson, C., and Nguyen, K. V. (2008) Seismic site effects by an optimized 2D BE/FE method. I. Theory, numerical optimization and application to topographical irregularities, *Int. J. Soil Dynamics Earthquake Eng.* **28**, 632–645.

Gatmiri, B. and Dehghan, K. (2005) Applying a new fast numerical method to elasto-dynamic transient kernels in HYBRID wave propagation analysis. In *Proc. 6th Conf. on Structural Dynamics (EURODYN 2005)*, Paris, France, pp. 1879–1884, Rotterdam, Millpress.

Gatmiri, B. and Jabbari, E. (2005a) Time-domain Green's functions for unsaturated soils. Part I: Two dimensional solution, *Int. J. Solid Structure* **42**, 5971–5990.

Gatmiri, B. and Jabbari, E. (2005b) Time-domain Green's functions for unsaturated soils. Part II: Three dimensional solution, *Int. J. Solid Structure* **42**, 5991–6002.

Gatmiri, B. and Kamalian, M. (2002a) On the fundamental solution of dynamic poroelastic boundary integral equations in the time domain, *Int. J. Geomechanics* **2**, 381–398.

Gatmiri, B. and Kamalian, M. (2002b) Two-dimensional transient wave propagation in anelastic saturated porous media by a Hybrid FE/BE method. In *Proc. 5th European Conf. on Numerical Methods in Geotechnical Engineering*, Paris, France, pp. 947–956.

Gatmiri, B. and Nguyen, K. V. (2005) Time 2D fundamental solution for saturated porous media with incompressible fluid, *Int. J. Comm. Num. Meth. Eng.* **21**, 119–132.

Gatmiri, B. and Nguyen, K. V. (2007) Evaluation of seismic ground motion induced by topographic irregularity, *Int. J. Soil Dynamics Earthquake Eng.* **27**, 183–188.

Gatmiri, B., Nguyen, K.-V., and Dehghan, K. (2007) Seismic response of slopes subjected to incident SV wave by an improved boundary element approach, *Int. J. Num. Anal. Meth. Geomechanics* **31**, 1183–1195.

Kamalian, M., Jafari, M. K., Sohrabi-Bidar, A., Razmkhah, A., and Gatmiri, B. (2006) Time-domain two-dimensional site response analysis of non-homogeneous topographic structures by a Hybrid FE/BE Method, *Int. J. Soil Dynamics Earthquake Eng.* **26**, 753–765.

SEISMIC SITE EFFECT MODELLING BASED ON *IN SITU* BOREHOLE MEASUREMENTS IN BUCHAREST, ROMANIA

Andrei Bala (bala@infp.ro)* and S. Florin Balan
National Institute for Earth Physics, Bucharest-Magurele, Romania

Joachim Ritter and Dieter Hannich
Universität Karlsruhe (TH), Germany

Abstract. Within the NATO Science for Peace Project 981882 "*Site-effect analyses for the earthquake-endangered metropolis Bucharest, Romania*" we determined a complete and homogeneous dataset of seismic, soil-mechanic and elasto-dynamic parameters. Ten 50 m deep boreholes were drilled in the metropolitan area of Bucharest in order to recover cores for dynamic tests and to measure vertical seismic profiles. These are used for an updated microzonation map related to earthquake wave amplification. The boreholes are placed near former or existing seismic station sites to allow a direct comparison and calibration of the borehole data with actual seismological measurements. A database is assembled which contains P- and S-wave velocity, density, geotechnical parameters measured at rock samples and geological characteristics for each sedimentary layer. Using SHAKE2000 we compute spectral acceleration response and transfer functions obtained from the in situ measurements. The acceleration response spectra correspond to the shear-wave amplifications excited in the sedimentary layers from 50 m depth (maximum depth) up to the surface. We present the acceleration response results from four sites.

Keywords: seismic site effects; in situ measurements; equivalent linear modelling

1. Introduction

Bucharest, the capital of Romania, with more than two million inhabitants, is considered, after Istanbul, the second-most earthquake-endangered metropolis in Europe. It is identified as a natural disaster hotspot by a recent global study of the World Bank and the Columbia University (Dilley et al., 2005). Four major earthquakes with moment magnitudes between 6.9 and 7.7 hit Bucharest in the last 68 years. The most recent destructive earthquake of 4 March 1977, with a moment magnitude of 7.4, caused about 1500 casualties in the capital alone. All disastrous earthquakes are generated within a small epicentral area – the Vrancea region – about 150 km northeast of

*Corresponding author.

T. Schanz and R. Iankov (eds.), *Coupled Site and Soil-Structure Interaction Effects with Application to Seismic Risk Mitigation*, NATO Science for Peace and Security Series C: Environmental Security,

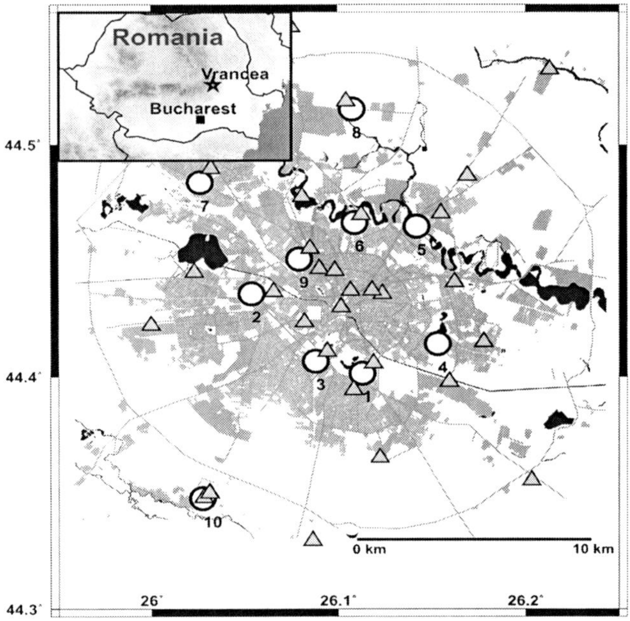

Figure 1. Map with area under investigation. The metropolitan region of Bucharest, Romania, is mainly inside the characteristic ring road with a diameter of about 20 km. Residential and industrial areas are indicated in grey; lakes, channels and rivers in black. The ten borehole sites are shown as circles. Sites with broadband instruments during the URS experiment are indicated with triangles.

Bucharest (Fig. 1). Thick unconsolidated sedimentary layers below Bucharest amplify the arriving seismic waves causing severe destruction. Thus, disaster prevention and mitigation of earthquake effects is an issue of highest priority.

There are only a few sites which were investigated coincidentally with geophysical and geotechnical techniques to relate the local geology with seismic wave propagation properties in Bucharest City (especially amplitude-amplification properties). Therefore, the main purpose of the NATO SfP Project 981882 (Ritter et al., 2006) is to obtain a complete and homogeneous dataset of soil-mechanic and elasto-dynamic parameters of the subsurface of Bucharest from ten new boreholes to model the so-called seismic site responses. Here we present the seismic measurements and modelling results from four selected sites.

2. Description of the NATO Science for Peace Project 981882

The microzonation of the metropolitan area of Bucharest for seismic amplification pattern has been pursued with great effort since the 1977 disastrous event. Geophysical groups at the National Institute for Earth Physics (NIEP) and civil engineers at the National Institute for Building Research worked on this problem, as well as foreign institutions like the Universität Karlsruhe (TH), the University of Trieste and the Japanese International Cooperation Agency. Their work resulted in an improved seismic database obtained from modern seismic observation networks as well as several borehole analyses. Based on these observations recent microzonation studies were done by, e.g., Aldea et al. (2004), Cioflan et al. (2004), Kienzle et al. (2004), Moldoveanu et al. (2004), Mandrescu et al. (2007) and Wirth et al. (2003). However, all of these studies could cover only fraction of the microzonation problem, because either seismic data alone (Wirth et al., 2003) or numerical modelling based on the assumed geological layering (Cioflan et al., 2004) was done. Sokolov et al. (2004) used spectral amplification factors and a probabilistic method to determine ground motion site effects in Bucharest. A major drawback of all studies is missing geophysical and geotechnical information from well-distributed boreholes in the Bucharest City area.

New high-quality seismic waveforms were measured during the *URS* (*URban Seismology*) Project from October 2003 until August 2004 (Fig. 1). Within this project 32 state-of-the-art broadband stations were continuously recording in the metropolitan area of Bucharest (Ritter et al., 2005). This unique dataset provides important information on the seismic amplitude variation across the area. Additionally, there is a modern ground acceleration observation network (K2-network) which has been upgraded in the last years by the Universität Karlsruhe (KA) and NIEP and which is run by NIEP. From this network a database with valuable strong motion recordings emerged.

To complement the seismic data with a coherent set of borehole measurements and dynamic core analysis, we received funding from NATO to drill ten boreholes in the city, to recover cores for dynamic geotechnical testing and to conduct seismic borehole measurements. The main objective of the project is earthquake risk mitigation and better seismic safety of Bucharest. The boreholes are placed near URS stations (Ritter et al., 2005) or K2 stations (a strong-motion recording network) of the NIEP to allow a direct comparison and calibration of borehole data with actual seismic measurements. The determined dynamic material parameters and the structural information will be used as input for linear and non-linear waveform modelling to estimate the seismic amplitude amplification at specific sites in Bucharest. These modelled waveforms will be compared and calibrated with observations from seismic stations in the city. The results from the site-effect analysis will be gathered in an updated seismic microzonation map of Bucharest which will be disseminated to the public and especially to the end-users who will introduce our results in future city planning.

3. Results of the Down-Hole Seismic Measurements

Most P-wave seismic velocities (V_P) values recorded in Bucharest City are in a narrow range (Table I). The velocities recorded at the Ecologic University site are a bit larger. The seismic shear wave velocities (V_S) are in a very close range: between 120 and 160 m/s at the surface and 405 and 450 m/s at 50 m depth. Results obtained by the down-hole method in four boreholes drilled in Bucharest City are presented in Table I. They were used as input data in the program SHAKE2000 as described below.

The mean weighted seismic velocities for the first six (of seven types) of Quaternary layers are computed and given also in Table I for the four sites, in order to be compared with seismic velocity values obtained from previous seismic measurements. Weighted mean values for V_S are computed according to the following equation:

$$\overline{V_S} = \frac{\sum_{i=1}^{n} h_i}{\sum_{i=1}^{n} \frac{h_i}{V_{S_i}}}, \tag{1}$$

TABLE I. Mean weighted seismic velocities for the first six (of seven types) of Quaternary layers in four boreholes in Bucharest City (Fig. 1). For a complete description of the layers see Ciugudean-Toma and Stefanescu (2006), Bala et al. (2006)

Geological layer	1		2		3		4		5		6		
Mean weighted velocities	V_P m/s	V_S m/s	V_P m/s	V_S m/s	V_P m/s	V_S m/s	V_P m/s	V_S m/s	V_P m/s	V_S m/s	V_P m/s	V_S m/s	V_{S30} m/s
Tineret Park (site 1)	180	140	570	220	856	299	–	–	1666	398	–	–	263
Ecology Univ. (site 2)	300	120	1180	220	1250	241	1610	354	1850	390	2042	405	286
Astronomy Inst. (site 3)	200	120	914	260	1200	330	1440	350	1900	390	2124	433	283
Titan2 Park (site 4)	290	160	800	250	800	250	980	350	1576	381	1850	450	308
Mean weighted *V*	233	138	854	251	972	290	1273	351	1687	390	1991	423	

where h_i and V_{S_i} denote the thickness (in meters) and the shear-wave velocity (in m/s) of the ith layer, in a total of n layers for the same type of layer (Romanian Code, 2006). According to the same code, the weighted mean values $\overline{V}_S$, computed for at least 30 m depth, determine four classes of the soil conditions:

$$\begin{array}{ll} \text{1. Class A, rock type:} & \overline{V}_S \geq 760\,\text{m/s}; \\ \text{2. Class B, hard soil:} & 360 < \overline{V}_S < 760\,\text{m/s}; \\ \text{3. Class C, intermediate soil:} & 180 < \overline{V}_S < 360\,\text{m/s}; \\ \text{4. Class D, soft soil:} & \overline{V}_S \leq 180\,\text{m/s}. \end{array} \tag{2}$$

All the $V_{S_{30}}$ values in Table I belong to type C (intermediate soil) after this classification (Romanian Code, 2006; EUROCODE-8, 2001). Even the $V_{S_{50}}$ values fall in the type C of the classification. According to this code, the elastic response spectra characterising the four classes of the soil conditions will be determined using the methodologies in the international practice.

4. One-Dimensional Ground Response Analysis at Four Sites in the Bucharest City Area

A ground response analysis consists of studying the behaviour of a soil/rock deposit subjected to an acceleration time history applied to a layer of the profile. When dealing with earthquake ground motions, the acceleration time history is usually specified at the bedrock. Examples of response quantities that can be obtained are the acceleration, velocity, displacement, stress, and strain time histories for any layer. Some of the applications of these analyses are the liquefaction potential or the seismic risk assessment in earthquake-prone regions.

Different methods of ground response analysis have been developed including one dimensional, two dimensional, and three dimensional approaches. Various modelling techniques like the finite element method were implemented for linear and non-linear analysis. Extended information on these analyses is given by Kramer (1996). Here we apply an equivalent linear one-dimensional analysis, as implemented in the computer program SHAKE2000 (Ordónez, 2003). The term one-dimensional refers to the assumption that the soil profile extends to infinity in all the horizontal directions and the bottom layer is considered a half space. Only the vertical propagation of seismic waves is considered, usually shear waves. The equivalent linear one-dimensional analysis is often an approximate linear approach. The non-linear behavior of the soil is accounted for by means of an iterative process in which the soil damping ratio and shear modulus are changed such that they are consistent with a certain level of strain calculated with linear assumptions. The non-linearities of the soil are not implicitly considered as in fully non-linear methods; rather at each iteration cycle the equations of motion solved are those of an equivalent linear model.

The equivalent linear method was proved to obtain good approximations of the response of a soil deposit subjected to an earthquake. and it had been successfully compared with finite element methods and fully non-linear analysis. A recent comparison made with finite element non-linear codes was performed in a seismic amplification study in Lotung, Taiwan (Borja et al., 1999).

4.1. INPUT DATA

The *static soil properties* required in the 1D ground response analysis with SHAKE2000 are: maximum shear wave velocity or maximum shear strength and unit weight. Since the analysis accounts for the non-linear behaviour of the soils using an iterative procedure, *dynamic soil properties* play an important role. The shear modulus reduction curves and damping curves are usually obtained from laboratory test data (cyclical triaxial soil tests). The variation in geotechnical properties of the individual soil layers are mostly impossible to model because of the lack of appropriate data. Therefore, these poorly constraint properties should be assumed constant for each defined soil layer. In built shear modulus reduction curves and damping curves for specific types of layers are used in SHAKE2000 based on published geotechnical tests (Ordónez, 2003).

As input data the interval seismic velocities V_S (in m/s), as well as the natural unit weight (in kN/m^3) and thickness of each layer (in m) are used. All these data are stored in a database for four new borehole sites in Bucharest area (Table I, after Bala et al., 2007). The velocity values from Table I are close to other measured values by Hannich et al. (2006). The recorded motion of the 27.10.2004 earthquake (Mw = 5.8) at K2 accelerometer station BBI in Bucharest is used as seismic input motion. All three components (one vertical and two horizontal components) were available. This station is placed in the borehole at INCERC site at 100 m depth. The strong motions BBI_E (east-west component) and BBI_N (north-south component) were chosen as being a representative acceleration recorded in a borehole in Bucharest from a moderate Vrancea earthquake.

4.2. ACCELERATION RESPONSE SPECTRA OBTAINED AT FOUR SITES IN THE BUCHAREST CITY CENTRAL AREA

Acceleration response spectra computed at different depth levels with the programme SHAKE2000 show the amplification due to the package of

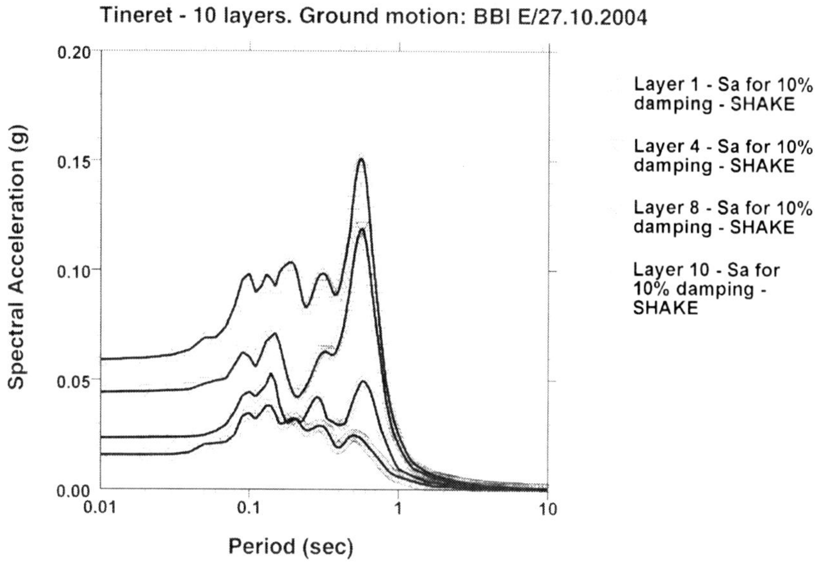

Figure 2. Calculated spectral acceleration response computed for the site Tineretului Park (site 1 in Fig. 1). As input wavelet the strong motion from BBI_E is taken (see text). Results for four layers are shown, layer 1 is the surface layer.

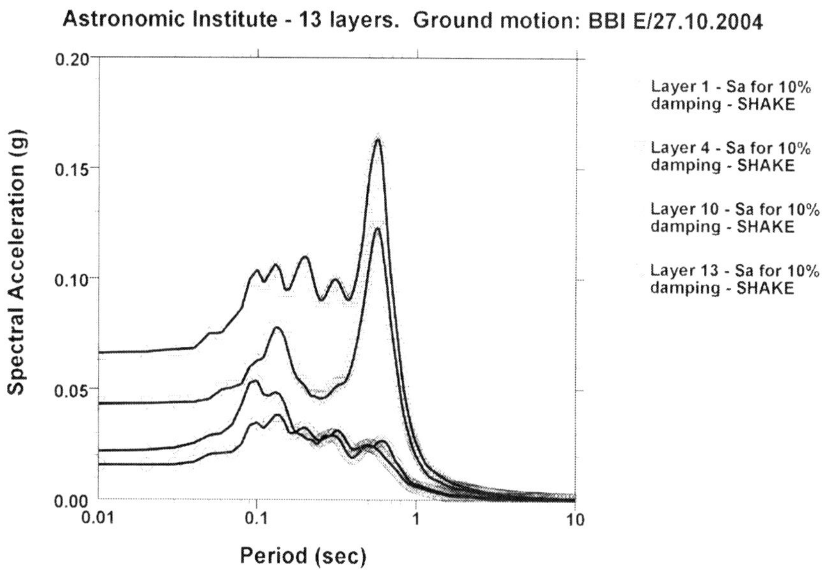

Figure 3. Calculated spectral acceleration response computed for the site Astronomic Institute (site 2 in Fig. 1). Results for four layers are shown.

sedimentary layers from 50 m depth to the surface (Figs. 2–6). These response spectra are thought to represent the case of a moderate to strong earthquake motion (see above).

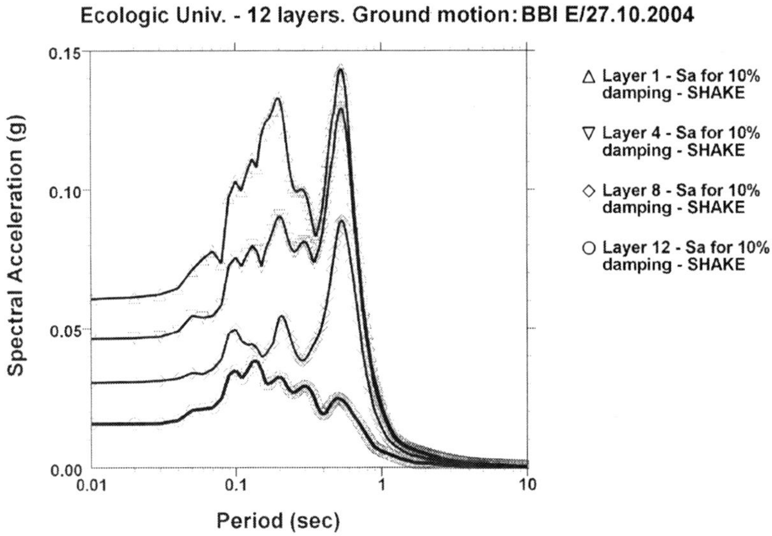

Figure 4. Calculated spectral acceleration response computed for the site Ecologic University.

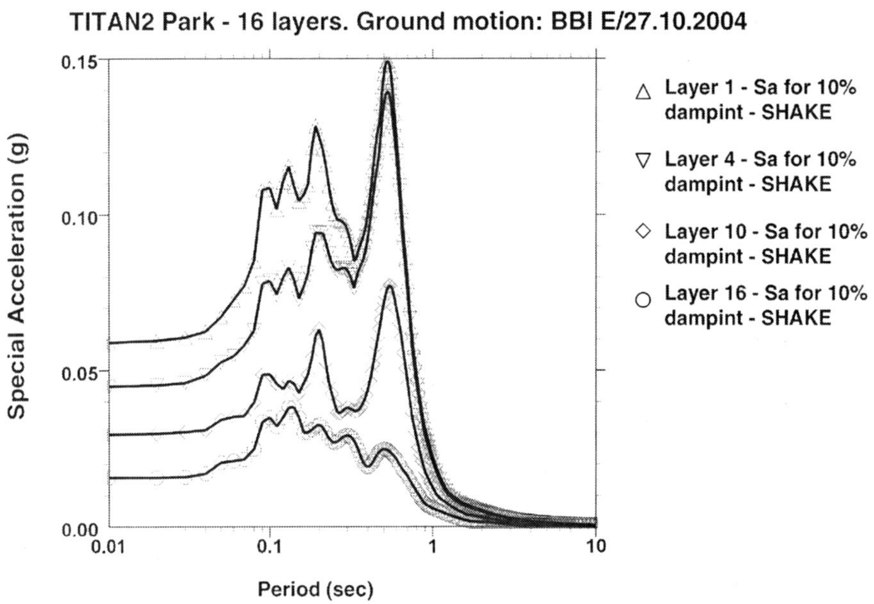

Figure 5. Calculated spectral acceleration response computed for the site Titan 2 Park.

The response diagram for Tineretului Park (Fig. 2) is very similar to the diagram for Astronomy Institute (Fig. 3), with a peak of 0.16 g acceleration at 0.55 s period.

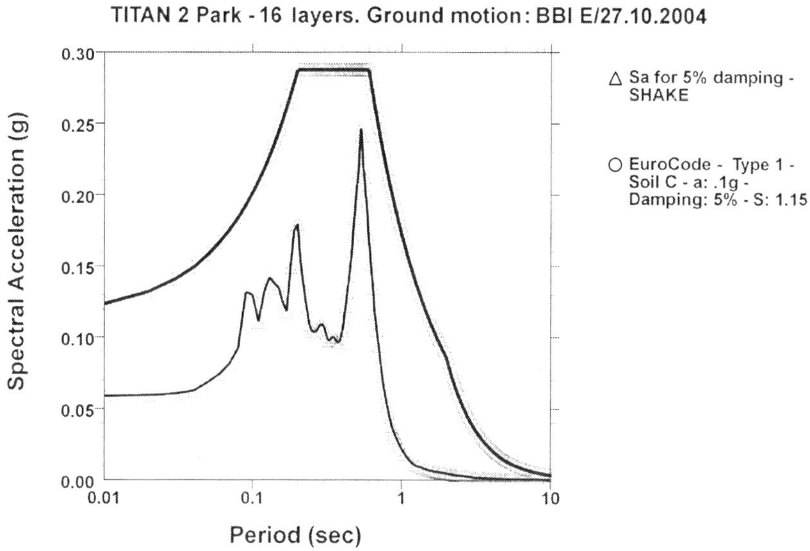

Figure 6. Calculated spectral acceleration response (5% damping) computed for the site Titan 2 Park and the corresponding design spectra from EUROCODE 8.

For the site Ecology University (Fig. 4), near Dambovita river, we find two acceleration peaks, one in the range 0.1–0.4 s (0.14 g), especially in the first layer, and another again at 0.55 s (0.15 g). For the site Titan 2 Park (Fig. 5), the amplification occurs between 0.1 and 0.2 s (0.12 g) and also at 0.5 s (0.15 g).

The EUROCODE 8 currently proposes two standard shapes for the design response spectra. Type 1 spectra are enriched in long periods and are suggested for high seismicity regions and magnitude $M_S > 5.5$. Type 2 spectra are proposed for moderate seismicity areas and exhibit both a larger amplification at short periods, and a much smaller amplification at long period contents, with respect to Type 1 spectra ($M_S < 5.5$). In Fig. 6 the type 1 spectral acceleration was found from EUROCODE 8, type 1, and comparison is presented with the spectral acceleration at the site Titan 2 Park.

Finally, for a pertinent soil site analysis, the SHAKE2000 program needs specific geotechnical inputs such as: input signal (scenario earthquakes), shear wave velocity and soil thickness. Some real borehole profiles are available now with shear wave velocity and soil thickness values, including some measurements on the core samples. As both the numerical or experimental seismic microzonation technique needs a critical evaluation of the results, this should be done in future work preferably by quantitatively analysis.

The proper input signal, either a real earthquake or an artificial strong motion created with SHAKE2000, remain to be chosen and tested in future work. If a synthetic signal will be applied, this signal should include the main frequency characteristics of the earthquakes recorded in the area.

The level in the geologic profile considered as "geotechnical bedrock", where this input signal should be applied, should be properly documented from field measurement.

5. Conclusions

1. An international research project was initiated in 2006 – NATO SfP Project 981882: *Site-effect analyses for the earthquake-endangered metropolis Bucharest, Romania*. This project, conducted by the *National Institute for Earth Physics*, Bucharest, Romania and *Universität Karlsruhe (TH)*, Germany, has the target to fill the gap in the knowledge concerning seismic and geotechnical parameters in the shallow ($h < 50$ m) layers in Bucharest, especially in the Quaternary layers 1–6, as they are described by Ciugudean-Toma and Stefanescu (2006).
2. The computed values for seismic velocities are in the same range as others obtained by *in situ* seismic measurements of different types. They are added to the database at NIEP, which is a valuable collection of elastic and dynamic parameters of the sedimentary rocks obtained by direct measurements. The values will be used for further studies on the seismic microzonation of Bucharest City using linear and non-linear approaches.
3. The velocity values obtained in the first three layers (Table I) are very important and among the first results measured in these sedimentary layers and reported for Bucharest City. They are well correlated with the values obtained in Bucharest by Hannich et al. (2006) using the SCPT method.

Acknowledgements

This paper is a result of the NATO SfP Project 981882 *Site-effect analyses for the earthquake-endangered metropolis Bucharest, Romania*. This research is sponsored by NATO's Scientific Affairs Division in the framework of the Science for Peace Programme.

References

Aldea, A., Lungu, D., and Arion, C. (2004) GIS mapping of seismic microzonation and site effects in Bucharest based on existing seismic and geophysical evidence. In D. Lungu, F. Wenzel, P. Mouroux, and I. Tojo (eds.), *Proc. Int. Conf. on Earthquake Loss Estimation and Risk Reduction*, Vol. 1, 237–249.

Bala, A., Raileanu, V., Zihan, I., Ciugudean, V., and Grecu, B. (2006) Physical and dynamic properties of the shallow sedimentary rocks in the Bucharest Metropolitan Area, *Romanian Reports Phys.* **58**(2), 221–250.

Bala, A., Ritter, J. R. R., Hannich, D., Balan, S. F., and Arion, C. (2007) Local site effects based on in situ measurements in Bucharest City, Romania. In *Proc. Int. Symp. on Seismic Risk Reduction, ISSRR-2007*, Bucharest, Romania, pp. 367–374.

Borja, R. I., Chao, H. Y., Montáns, F. J., and Lin, C. H. (1999) Nonlinear Ground Response At Lotung LSST Site, *J. Goetechnical Geoenvironmental Eng., ASCE* **125**(3), 187–197.

Cioflan, C. O., Apostol, B. F., Moldoveanu, C. L., Panza, G. F., and Marmureanu, G. (2004) Deterministic approach for the microzonation of Bucharest, *Pure Appl. Geophys.* **161**, 1–16.

Ciugudean-Toma, V., and Stefanescu, I. (2006) Engineering geology of the Bucharest city area, Romania. In *IAEG-2006 Proceedings, Engineering Geology for Tomorrow's Cities*, paper no. 235.

Dilley, M., Chen, R. S., Deichmann, U., Lerner-Lam, A. L., and Arnold, M., with Agwe, J., Buys, P., Kjekstad, O., Lyon, B., and Yerman, G. (2005) *Natural Disaster Hotspots: A Global Risk Analysis*, Synthesis report, pp. 29.

EUROCODE-8 – prEN1998-1-3 (2001) Design provisions for earthquake resistance of structures, European Committee for Standardisation.

Hannich, D., Huber, G., Ehret, D., Hoetzl, H., Balan, S., Bala, A., Bretotean, M., and Ciugudean, V. (2006) SCPTU techniques used for shallow geologic/hydrogeologic site characterization in Bucharest, Romania, *Third Int. Symp. on the Effects of Surface Geology on Seismic Motion*, Grenoble, France, paper 71.

Kienzle, A., Hannich, D., Wirth, W., Ciugudean, V., Rohn, J., and Czurda, K. (2004) Seismic zonation of Bucharest. In D. Lungu, F. Wenzel, P. Mouroux, and I. Tojo (eds.), *Earthquake Loss Estimation and Risk Reduction* **1**, 251–259.

Kramer, S. L. (1996) *Geotechnical Earthquake Engineering*, Prentice Hall.

Moldoveanu, C. L., Radulian, M., Marmureanu, Gh., and Panza, G. F. (2004) Microzonation of Bucharest: state of the art, *Pure Appl. Geophys.* **161**, 1125–1147.

Mandrescu, N., Radulian, M., and Marmureanu, Gh. (2007) Geological, geophysical and seismological criteria for local response evaluation in Bucharest urban area, *Soil Dynamics Earthquake Eng.* **27**, 367–393.

Ordónez G. A. (2003) SHAKE2000: A Computer Program for the 1-D Analysis of Geotechnical Earthquake Engineering Problem, User's Manual, www.shake2000.com.

Ritter, J. R. R., Balan, S., Bonjer, K.-P., Diehl, T., Forbriger, T., Marmureanu, G., Wenzel, F., and Wirth, W. (2005) Broadband urban seismology in the Bucharest metropolitan area, *Seism. Res. Lett.* **76**, 573–579.

Ritter, J. R. R., Balan, S., Bala, A., and Rohn, J. (2006) Annual Technical Report for the NATO SfP Project 981882, Bucharest and Karlsruhe.

Romanian Code for the seismic design for buildings P100-1/2006 (2006).

Sokolov, V. Y., Bonjer, K.-P., and Wenzel, F. (2004) Accounting for site effect in probabilistic assessment of seismic hazard for Romania and Bucharest: a case study of deep seismicity in Vrancea zone, *Soil Dynamics Earthquake Eng.* **24**, 927–947.

Wirth, W., Wenzel., F., Sokolov, V. Y., and Bonjer, K. (2003) A uniform approach to seismic site effect analysis in Bucharest, Romania, *Soil Dynamics Earthquake Eng.* **23**, 737–758.

Part II

Soil-Structure Interaction

ISSUES RELATED TO THE DYNAMIC INTERACTION OF RETAINING WALLS AND RETAINED SOIL LAYER

Yiannis Tsompanakis (jt@science.tuc.gr)
Division of Mechanics, Department of Applied Sciences, Technical University of Crete, Chania GR-73100, Greece

Abstract. The present work aims to examine how and to what extent potential soil nonlinearity may affect: (a) the dynamic distress of a rigid fixed-base retaining wall and (b) the seismic response of the retained soil layer. For this purpose, a parametric study is conducted which is based on 2-D dynamic finite element analyses. Soil nonlinearity is realistically taken into account via the commonly used equivalent-linear procedure. In order to examine more thoroughly the influence of material nonlinearity, the developed numerical model is studied under idealized seismic excitations and several intensity levels of the imposed ground acceleration. The results justify the perception that the nonlinear soil behavior has a considerable impact on the dynamic earth pressures developed on the wall and the amplification of the acceleration developed on the backfill as well.

Keywords: retaining walls, dynamic wall–soil interaction, soil nonlinearity, amplification, earthquake-induced pressures

1. Introduction

Retaining walls have many applications in geotechnical engineering practice: harbor quay-walls, bridge abutments, deep excavations are some characteristic cases. Despite their structural simplicity, the seismic response of retaining systems is a rather complicated problem. This can be attributed to the dynamic interaction between the wall and the retained soil, especially when material and/or geometry nonlinearities are considered (Kramer, 1996; Wu and Finn, 1999). The seismic response of various types of walls supporting a single soil layer has been examined by a number of researchers in the past either experimentally, analytically, or numerically (Veletsos and Younan, 1997; Iai, 1998; Psarropoulos et al., 2005). Depending on the assumed material behavior of the retained soil and the assumptions regarding the development of the displacements of the wall, there exist two main categories of analytical methods used in the design of retaining walls: (a) *pseudo-static methods* incorporating the limiting-equilibrium concept (Mononobe-Okabe

T. Schanz and R. Iankov (eds.), *Coupled Site and Soil-Structure Interaction Effects with Application to Seismic Risk Mitigation,* NATO Science for Peace and Security Series C: Environmental Security,

type solutions), which assume yielding walls and rigid-perfect plastic behavior of the soil (Okabe, 1926; Mononobe and Matsuo, 1929; Seed and Whitman, 1970) and (b) *elasticity-based solutions* that regard the soil as a linear (visco-) elastic continuum (Scott, 1973; Wood, 1975; Veletsos and Younan, 1997).

According to an efficient simplification of the Mononobe-Okabe method, developed by Seed and Whitman (1970), the (maximum) normalized dynamic active earth force imposed on the wall is:

$$\overline{\Delta P}_{AE} = \frac{\Delta P_{AE}}{A\rho H^2} \approx 0.4 \tag{1}$$

where A is the peak base acceleration, ρ is the soil mass density, and H is the wall height. During the same period, the elastic solutions developed by Scott (1973) or Wood (1975) suggest that for low-frequency (quasi-static) motions the normalized dynamic active earth force developed on a rigid fixed-base wall is:

$$\overline{\Delta P}_{AE} \approx 1 \tag{2}$$

It is evident that this quantity is almost 2.5 times bigger than the proposal made by Seed and Whitman. This discrepancy may be even more intense when the fundamental frequency of the base excitation approaches that of the retained soil layer under 1-D conditions, i.e.:

$$f_o = \frac{V_S}{4H} = \frac{\sqrt{G_{\max}/\rho}}{4H} \tag{3}$$

where V_S is the shear-wave velocity of the soil, and $G_{\max}$ is the corresponding small-strain shear modulus.

The two aforementioned modes of wall–soil system behavior are rather extreme and in many cases are unrealistic. The limiting-equilibrium solutions imply the capability of the system to develop relatively large displacements (*geometric nonlinearity*) together with the formation of plastic zones (*material nonlinearity*). In contrast, the elasticity-based solutions include only the potential geometric nonlinearity by taking into account the wall flexibility and/or the wall foundation compliance (Veletsos and Younan, 1997). In some cases, such as bridge abutments, braced excavations, or basement walls, the existence of kinematic constraints on the wall motion is incompatible with the limiting-equilibrium concept, while on the other hand, the elasticity-based solutions overlook the potential nonlinear behavior of the soil, leading thus to over-conservative designs.

The objective of the present study is to examine more thoroughly the influence of material nonlinearity on the dynamic distress of a wall retaining a single soil layer. Apart from the dynamic earth pressures developed on the wall, emphasis is given on the soil amplification of the base acceleration. Note that seismic norms (such as the Eurocode 8 (EC8) (2003), or the Greek Seismic Code (EAK) (2000)), being based on the limit-equilibrium methods, underestimate the role of the potential soil amplification. Results provide a clear indication of the impact of potential soil nonlinearity on the dynamic wall–soil interaction. Nonlinearity increases the complexity of the wall–soil interaction, being either beneficial or detrimental for the wall distress. Therefore, seismic design of retaining walls should incorporate these interrelated phenomena.

2. Numerical Modeling

In order to examine the nonlinear dynamic wall–soil interaction, 2-D numerical simulations of the retaining system depicted in Fig. 1 were conducted. The simulations were performed utilizing the popular QUAD4M finite element code developed by Hudson et al. (1994), which performs dynamic nonlinear analyses incorporating the well-known iterative equivalent-linear procedure. This approach is considered accurate enough provided that the magnitude of maximum shear strain is of the order of 1% (Kramer, 1996), and as it was observed in the present study, the strain levels are within the acceptable range. The $G/G_{\max} - \gamma$ and $\xi - \gamma$ curves used are characteristic of sandy soil material (Seed and Idriss, 1970; Idriss, 1990).

As the wall flexibility is examined in relation to soil stiffness and the earth pressures are normalized with respect to ρ and H, the soil material properties and the wall height values do not affect the dynamic pressures on the wall (Veletsos and Younan, 1997). This was also verified in a recent numerical study by Psarropoulos et al. (2005) where linear analyses of the examined model were performed utilizing ABAQUS software. Thus, the parameters used in that study have also been used in the current investigation. All equivalent-linear analyses in QUAD4M were performed considering an

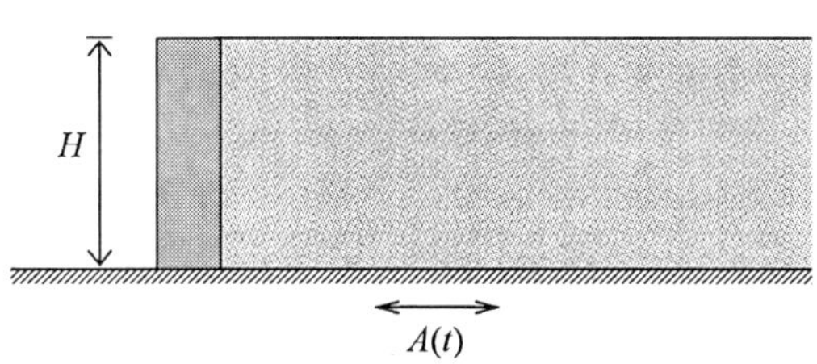

Figure 1. The retaining system examined in the present study: a rigid fixed-base wall retaining a single soil layer with strain-dependent material behavior, both excited by an acceleration time history $A(t)$.

8 m-high wall and the retained soil layer is characterized by a relatively low small-strain shear-wave velocity V_S equal to 100 m/s and a mass density of 1.8 Mg/m^3.

The discretization of the retained soil was performed by four-node plane-strain quadrilateral elements. The model was adequately elongated so as to reproduce adequately the free-field conditions at its right-hand side (see Fig. 1). The rigid wall was simulated by an extremely stiff column with linear elastic behavior. The simplifying assumption of no de-bonding or relative slip at the wall–soil interface was used.

The base of both the wall and the soil stratum were considered to be excited by a horizontal motion, assuming an equivalent force-excited system. Dynamic response of any system depends on the seismic excitation characteristics (both in the time and in the frequency domain). However, in the present numerical study the excitations were limited on purpose to simple pulses in order to examine more thoroughly this complex phenomenon. Thus, the model was subjected to harmonic and Ricker pulses which allow for a better understanding and interpretation of the results. Furthermore, the results of harmonic excitations can be easily generalized for any real earthquake excitations via Fourier transformations. Moreover, various values of peak base acceleration were used, aiming at the development of different levels of material nonlinearity.

3. Linear Harmonic Response

The dynamic linear response of a single horizontal soil layer under 1-D conditions has been studied by many researchers, and analytical solutions for harmonic excitation can be found in the literature (e.g., Kramer, 1996). In the case of harmonic excitation the response is controlled by the ratio f/f_o, where f is the dominant period of the excitation, and f_o the fundamental period of the soil layer. Thus, the employed harmonic excitations had two characteristic frequencies: the first was set equal to the low-strain fundamental eigenfrequency of the soil layer ($f = f_o$), while the second had much lower value ($f = f_o/6$), approximating a quasi-static excitation. In the examined model the fundamental frequency of the soil layer f_o is almost 3.1 Hz (the fundamental period of the soil layer T_o is 0.32 s). The duration of the sinusoidal pulse was such that steady state conditions were reached. The maximum amplification factor (AF) for linear response is given by:

$$AF \cong \frac{2}{\pi\xi}\frac{1}{2n+1} \tag{4}$$

where ξ is the critical damping ratio and n the eigen-mode number. For the first mode ($n = 0$) and $\xi = 5\%$, AF is approximately equal to 12.7.

The aforementioned amplification was verified numerically in this study by developing the 2-D finite element model for the soil layer under "free-field" conditions, which actually resembles 1-D conditions due to the horizontal soil stratigraphy. The presence of a rigid retaining wall imposes a vertical boundary condition, leading thus to a real 2-D model in the vicinity of the wall. In addition, this model has a fundamental low-strain eigenfrequency slightly lower than the corresponding one of the 1-D model, since the existence of the rigid wall makes the model stiffer. However, the difference of the two eigenfrequencies is negligible, as it is lower than 0.03 Hz for the examined test case.

The response of the horizontal soil layer under "free-field" conditions is compared with the corresponding 2-D model when the wall is included. The distribution of the amplification factor (AF) on the surface of the backfill in the case of harmonic excitation at resonance ($f = f_o$) is plotted in Fig. 2. For the examined rigid fixed-base wall, the motion in the vicinity of the wall is practically induced by the wall itself, and therefore, no amplification is observed ($AF = 1$). The amplification factor converges to its maximum value ($AF \approx 12.7$) at a distance of $10H$ from the wall, since at that distance 1-D conditions are present (*free-field motion*). Note that this distance was also calculated by Wood as the minimum distance required to eliminate the effects of the wall on the retained soil.

Figure 3 depicts the height-wise distribution of the normalized induced dynamic earth pressures for the two harmonic excitations examined. It can be observed that when the fundamental frequency of the input motion approaches that of the retained soil layer, the normalized dynamic earth

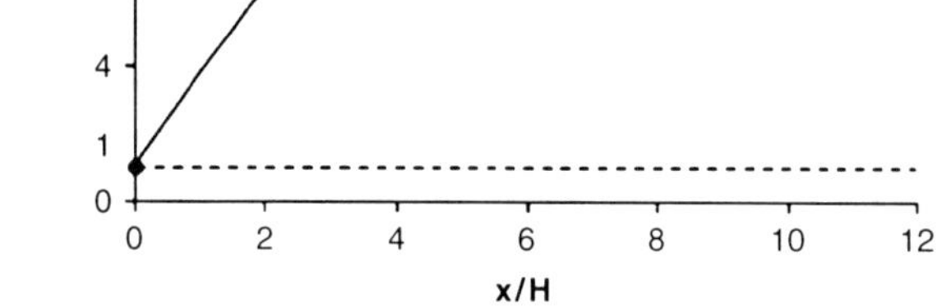

Figure 2. Distribution of the soil amplification factor (AF) along the free surface of the backfill in the case of harmonic excitation at resonance ($f = f_o$). Note that AF is equal to unity just behind the rigid wall examined, and along the backfill–base interface.

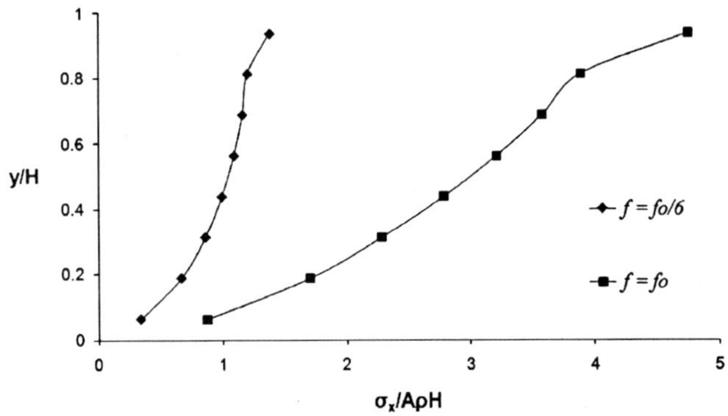

Figure 3. Height-wise distribution of the normalized induced dynamic earth pressures for the two harmonic excitations examined (f_o is the soil layer eigenfrequency for 1-D conditions).

force is almost three times greater in the case of resonance, compared to the corresponding value in the case of quasi-static excitation, which according to equation (2) is almost equal to unity.

4. Nonlinear Harmonic Response

The aforementioned results referring to linear soil behavior are valid for very low levels of base acceleration, when the induced strains remain small ($\gamma <$ 0.005%). However, when the maximum acceleration acting on the soil mass takes more realistic values the induced strains are substantially greater, and thus, the impact of material nonlinearity (expressed by the $G/G_{\max} - \gamma$ and $\xi - \gamma$ curves) is more evident. The wall–soil system behavior depends heavily, not only on the level of the applied acceleration, but also on the f/f_o ratio, as it is justified by the subsequent results.

The distribution of the amplification factor on the surface of the backfill in the case of the harmonic excitation at resonance ($f = f_o$) is plotted in Fig. 4, for five levels of peak base acceleration: 0.0001 g (corresponding to linear soil behavior), 0.12, 0.24, 0.36, and 0.50 g, covering a broad range of induced dynamic strains. Note that in the range of small shear strains the critical damping ratio, ξ, was set equal to 5%, instead of the much lower values of the curves proposed by Seed and Idriss (1970), in order to ensure that the theoretical amplification ($AF \approx 12.7$) for linear conditions is also numerically achieved for the lowest peak base acceleration case (0.0001 g). As it was expected, increasing the degree of material nonlinearity makes the system more flexible, thus decreases its fundamental frequency and leads to the avoidance of resonance. This phenomenon can be easily observed by examining the substantially reduced values of AF, for all levels (0.12 to 0.50 g) of nonlinear behavior, as shown in Fig. 4.

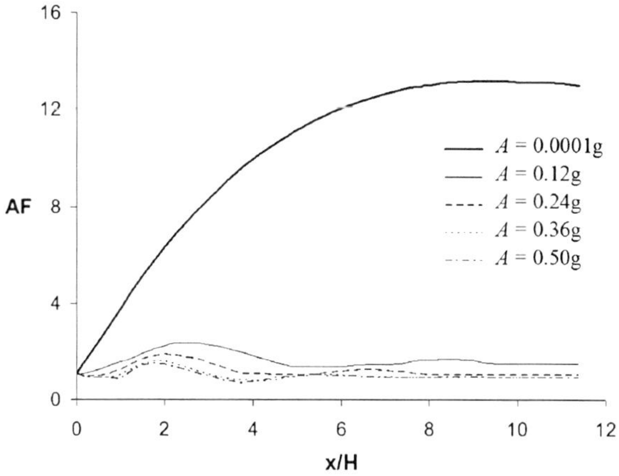

Figure 4. istribution of the soil amplification factor (AF) along the surface of the backfill for the harmonic excitation with frequency f equal to f_o.

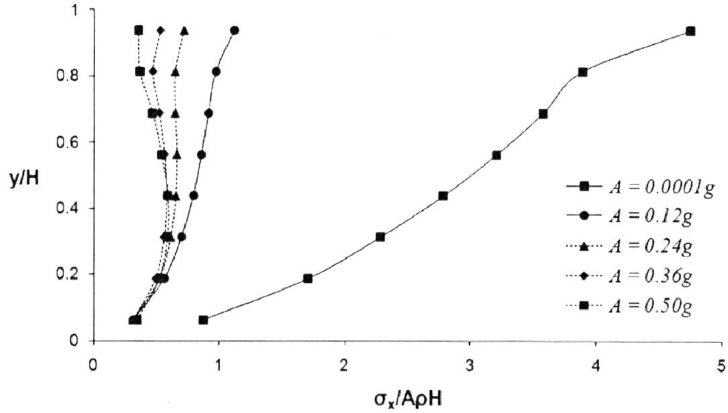

Figure 5. Height-wise distribution of the normalized induced dynamic earth pressures for the harmonic excitation with frequency f equal to f_o.

Figure 5 depicts the height-wise distribution of the normalized induced dynamic earth pressures for the harmonic excitation with $f = f_o$. As the level of applied acceleration increases the dynamic earth pressures decrease. In particular, the normalized dynamic earth force on the wall reduces to values ranging from 0.60 to 0.90 (corresponding to $A = 0.50\,g$ and $A = 0.12\,g$ levels, respectively) compared to the previously calculated value of three for the linear soil behavior case. It is evident that for excitations with dominant frequency nearly equal to the low-strain fundamental eigenfrequency of the retained soil, the material nonlinearity seems to act in a beneficial way. In Fig. 6 the height-wise distribution of the normalized induced dynamic earth pressures is plotted for the case of the low-frequency harmonic excitation

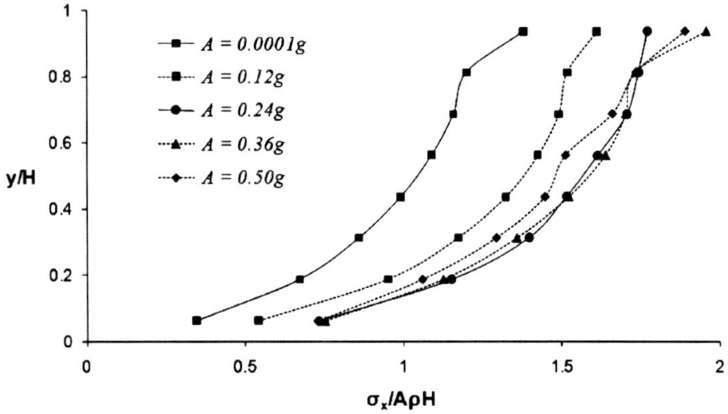

Figure 6. Height-wise distribution of the normalized induced dynamic earth pressures for the low-frequency harmonic excitation ($f = f_o/6$).

($f = f_o/6$). It is evident that the case of quasi-static excitation is of greater interest, especially for higher levels of soil nonlinearity.

5. Response to Ricker Pulse

As previously mentioned, apart from harmonic excitations a Ricker pulse with central frequency $f_R = 2\,\text{Hz}$ has also been used in the present study (Ricker, 1960). Despite the simplicity of its waveform, this wavelet covers smoothly a broad range of frequencies up to nearly $3f_R$ as shown in Fig. 7. The height-wise distribution of the normalized induced dynamic earth pressures in the case of the Ricker pulse excitation is plotted in Fig. 8, for three of the levels of peak base acceleration examined. The pattern revealed in Fig. 5 for the harmonic excitation is repeated in this case, due to the fact that the excitation pulse includes a broad range of frequencies close to the fundamental frequency of the retained soil layer ($f_o \approx 3\,\text{Hz}$). As a result, despite the lower levels of earth pressures in this case, the system response is quite similar to that caused by the harmonic excitation at resonance.

As the imposed Ricker pulse covers smoothly the range of frequencies between 1 and 5 Hz, it provides an efficient way to comprehend the effect of material nonlinearity on the wall distress in the frequency domain as well. Figure 9 shows the variation of the *Pressure Amplification Factor* (*PAF*) as a function of frequency. The aforementioned parameter is defined as:

$$PAF = \frac{FFT\left[\overline{\Delta P}_{AE}(t)\right]}{FFT[A(t)]} \tag{5}$$

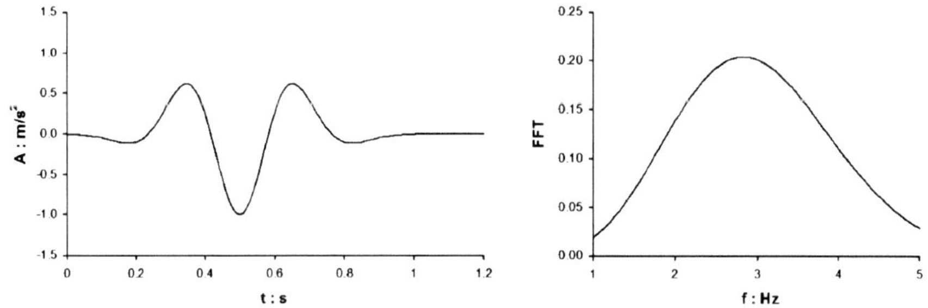

Figure 7. Acceleration time-history (A is scaled to peak acceleration of 1 m/s^2) and Fourier spectrum of the Ricker pulse excitation (with central frequency f_R = 2 Hz).

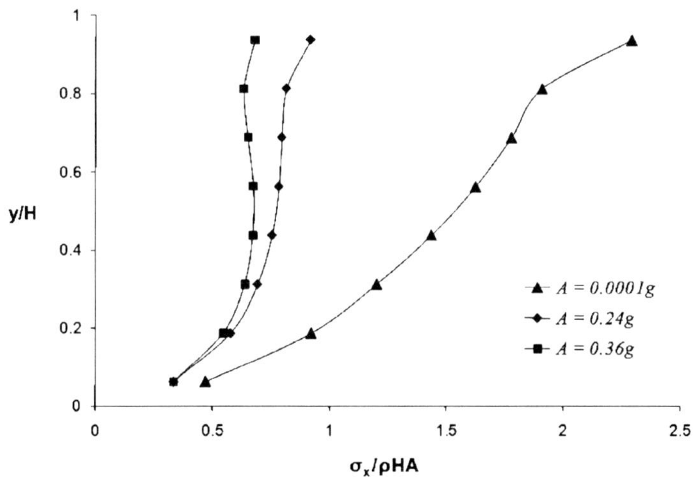

Figure 8. Height-wise distribution of the normalized induced dynamic earth pressures in the case of the Ricker pulse excitation.

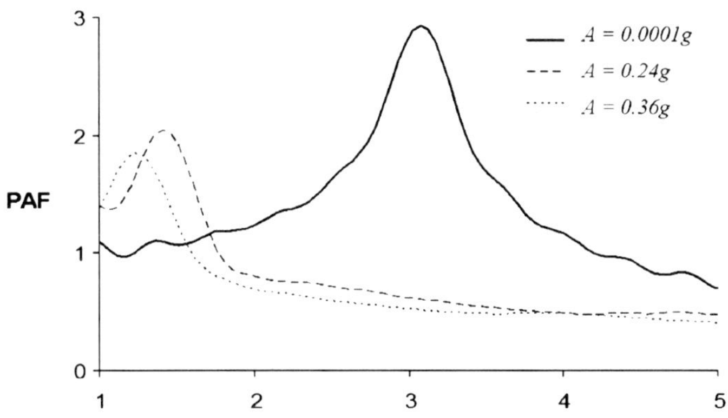

Figure 9. The Pressure Amplification Factors (PAF) calculated for the examined Ricker pulse.

where $FFT\left[\overline{\Delta P}_{AE}(t)\right]$ is the Fourier spectrum of the normalized induced dynamic earth force time history $\overline{\Delta P}_{AE}(t)$ and $FFT[A(t)]$ is the Fourier spectrum of the acceleration time history of the Ricker pulse excitation shown in Fig. 7.

It is evident that in the case of linear soil behavior, PAF reaches its maximum value for frequencies close to the fundamental frequency of the retained soil layer. This result matches the value calculated previously in the case of linear harmonic response at resonance. Additionally, for low-frequency excitations, the value of PAF converges to that proposed by Scott (1973) and Wood (1975) as calculated previously. For the case of increased levels of peak base acceleration (i.e., A = 0.24 or 0.36 g), the development of material nonlinearity not only affects the maximum value of PAF, but also shifts the range of its maximum values towards lower frequencies. This phenomenon can be either beneficial or detrimental, depending on the predominant frequency of the input motion.

Finally, Fig. 10 presents the maximum normalized dynamic earth force as a function of peak base acceleration A for the three excitations examined. Note that in the same plot, the values proposed by Wood and by Seed and Whitman are also included as references. It can be observed that in the case of linear response the wall distress is dominated by the frequency content of the excitation. More specifically, the earth force varies between the values of one and three, being thus always higher than the standard bounding values adopted from the seismic norms (noted as Wood and M-O). Nevertheless, as the degree of nonlinearity increases the distress decreases substantially, ranging between the aforementioned Wood and M-O bounds in the cases of harmonic resonant and Ricker pulses.

In contrast, the distress in the case of low-frequency (quasi-static) harmonic excitation and nonlinear response is always higher than the upper

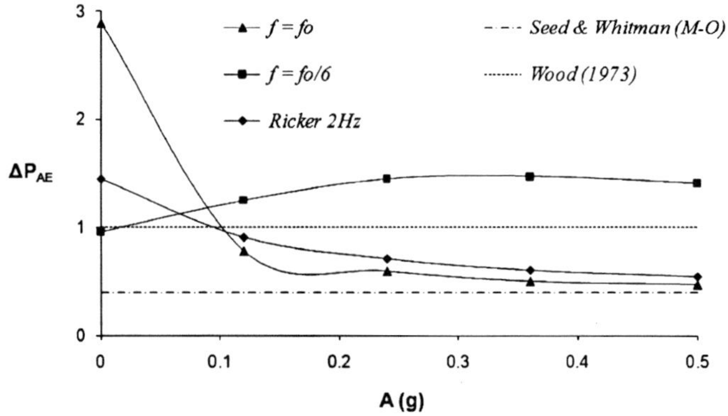

Figure 10. Maximum normalized dynamic earth force as a function of peak base acceleration A, for the three excitations examined. Graph also includes the proposals of Wood (1975) and of Seed and Whitman (1970).

bound (Wood's solution) for all levels of peak base acceleration, and is approximately 50% greater than the value of Wood's solution. An important conclusion resulting from Fig. 10 is that for high values of the imposed base acceleration the resulting force approximates in general the proposal of Seed and Whitman, even though the limiting-equilibrium conditions (imposed by the static theory of Coulomb, or its pseudo-static extension of M-O) are not valid in the specific retaining system. In other words, the force acting on the back of a yielding retaining wall (resulting from the weight of a rigid wedge of soil above a planar failure surface, according to M-O theory) coincides with the force that acts on the back of a rigid fixed-base wall (resulting from the earth pressures of a yielding soil material).

6. Conclusions

In the present study it was examined how and to what extent the potential soil nonlinearity, that a retained soil layer exhibits under moderate or severe seismic excitations, can possibly affect: (a) the dynamic distress of a rigid fixed-base retaining wall and (b) the seismic response of the retained soil layer itself. It was found that soil nonlinearity reduces in general the soil amplification of the retained soil and the dynamic earth pressures, leading thus to a lower wall distress. However, as soil nonlinearity alters the eigenfrequencies of the wall–soil system, there exists (under certain circumstances) the possibility that increased nonlinearity may lead to an amplified response. This phenomenon is more probable to occur when the frequency range of the excitation is narrow and concentrated around a fundamental frequency that is lower than the linear eigenfrequency of the soil layer. Thus, potential soil nonlinearity can be either beneficial or detrimental for the wall distress, depending on the circumstances. Conclusively, seismic design of retaining walls should consider more elaborately these interrelated phenomena as well as the impact of soil amplification for low to moderate intensity levels.

Acknowledgements

The comments and suggestions of Dr P. N. Psarropoulos are gratefully acknowledged.

References

EAK (2000) Greek Seismic Code, Greek Ministry of Public Works, Athens, Greece.

EC8 (2003) Eurocode 8: Design of structures for earthquake resistance, Part 1, European standard CEN-ENV-1998-1, European Committee for Standardization, Brussels.

Hudson, M., Idriss, I. M., and Beikae, M. (1994) *User's Manual for QUAD4M*, Center for Geotechnical Modeling, Department of Civil and Environmental Engineering, University of California, Davis, USA.

Iai, S. (1998) Seismic analysis and performance of retaining structures. In P. Dakoulas, M. Yegian, and R. D. Holtz (eds.), *Proc. of Geotechnical Earthquake Engineering and Soil Dynamics III*, Geotechnical Special Publ. No. 75, ASCE, Reston, VA, pp. 1020–1044.

Idriss, I. M. (1990) Response of soft soil sites during earthquakes. In J. M. Duncan (ed.), *Proc. of H. Bolton Seed Memorial Symposium*, Vol. 2, pp. 273–289.

Kramer, S. L. (1996) *Geotechnical Earthquake Engineering*, Prentice-Hall, New Jersey.

Mononobe, N. and Matsuo, H. (1929) On the determination of earth pressures during earthquakes. In *Proc. of the World Engineering Congress*, Vol. 9, Paper 388, Tokyo, Japan.

Okabe, S. (1926) General theory of earth pressures, *J. Japan Soc. Civil Eng.* **12**(1), 123–134.

Psarropoulos, P. N., Klonaris, G., and Gazetas, G. (2005) Seismic earth pressures on rigid and flexible retaining walls, *Soil Dynamics Earthquake Eng.* **25**(7–10), 795–809.

Ricker, N. (1960) The form and laws of propagation of seismic wavelets, *Geophysics* **18**, 10–40.

Scott, R. F. (1973) Earthquake-induced pressures on retaining walls, In *Proc. of the 5th World Conf. on Earthquake Engineering*, Vol. 2, pp. 1611–1620.

Seed, H. B. and Idriss, I. M. (1970) Soil moduli and damping factors for dynamic response analyses, Report EERC 70-10, Earthquake Engineering Research Center, University of California, Berkeley, CA.

Seed, H. B. and Whitman, R. V. (1970) Design of earth retaining structures for dynamic loads. In *Proc. of the Special Conf. on Lateral Stresses in the Ground and Design of Earth Retaining Structures*, ASCE, pp. 103–147.

Veletsos, A. S. and Younan, A. H. (1997) Dynamic response of cantilever retaining walls, *ASCE J. Geotechnical Geoenvironmental Eng.* **123**(2), 161–172.

Wood, J. H. (1975) Earthquake-induced pressures on a rigid wall structure, *Bullet. New Zealand National Earthquake Eng.* **8**, 175–186.

Wu, G. and Finn, W. D. L. (1999) Seismic lateral pressures for design of rigid walls, *Canadian Geotechnical J.* **36**(3), 509–522.

THE EFFECT OF SOIL-STRUCTURE INTERACTION AND SITE EFFECTS ON DYNAMIC RESPONSE AND STABILITY OF EARTH STRUCTURES

Varvara Zania and Yiannis Tsompanakis (jt@science.tuc.gr)*
Division of Mechanics, Department of Applied Sciences, Technical University of Crete, Chania GR-73100, Greece

Prodromos N. Psarropoulos (prod@central.rtua.gr)
Department of Infrastructure Engineering, Hellenic Air-force Academy, Themistocleous St. 43, 16674 Athens, Greece

Abstract. Seismic behaviour of earth structures, like soil embankments, earth dams and waste landfills, may be influenced by complicated phenomena related both to soil-structure interaction and site effects. The current study numerically investigates the main aspects of each of the aforementioned issues by taking into consideration even soil material nonlinearity. A detailed parametric investigation demonstrates the impact of the most important factors, like excitation characteristics and geometric/mechanical properties of an earth structure, on the evaluation of its dynamic response and stability. The results indicate that the impact of the examined issues on the seismic design of earth structures cannot be easily predefined or quantified, since it is dependent upon several interrelated factors.

Keywords: soil-structure interaction, site effects, landfills, material nonlinearity, finite element analyses, permanent deformation analyses

1. Introduction—Methodology

The role of dynamic soil-structure interaction (SSI) effects as well as the role of site effects on the seismic response of structures (buildings, bridges, etc) has been a subject of intense research activity during the last decades. Moreover, it has been demonstrated that the beneficial role of SSI is merely an oversimplification, which is only conditionally valid (Mylonakis and Gazetas, 2000). On the other hand, site effects degenerate to soil conditions (i.e., heterogeneity of materials) and to irregular geometry (i.e., regional topography), which results into a complex behavior lacking of the required generality for a seismic norm (Chavez-Garcia, 2007).

*Corresponding author.

T. Schanz and R. Iankov (eds.), *Coupled Site and Soil-Structure Interaction Effects with Application to Seismic Risk Mitigation,* NATO Science for Peace and Security Series C: Environmental Security,

The aforementioned issues are also related with the dynamic behavior of geostructures such as embankments, earth dams and waste landfills. These infrastructures have some special characteristics that distinguish them from other engineering structures. In general, they have large size and they are composed by soil or waste materials, thus, they are characterized by high flexibility which results to relatively large eigenperiod values. Moreover, their seismic behavior is strongly influenced by soil or waste material properties and potential nonlinearity. For instance, a wide range of shear-wave velocity values (Matasovic and Kavazanjian, 1998; Houston et al., 1995) has been proposed for waste materials.

Seismic design of earth structures should take into account the SSI and site effects considering also the impact of their large eigenperiod. The current study aims to provide an insight into the complex nature of the seismic behavior of such structures. For this purpose, the dynamic response is simulated by performing 2D parametric numerical time domain analyses by using the finite element software QUAD4M (Hudson et al., 1994) and frequency analyses using (ABAQUS, 2004). The finite element mesh size was tailored to the wavelengths of interest and the lateral boundaries were placed at a sufficiently far distance to avoid the impact of reflected waves on the computed response. Soil/waste material nonlinearity is modeled throughout an iterative equivalent linear procedure, while for the frequency analyses the waste material properties were assigned as provided by the results of the last iteration of the dynamic analyses. A more detailed nonlinear stress–strain relationship is rather prohibited for waste materials, due to numerous uncertainties related to the establishment of proper constitutive models.

Though there are certain shortcomings of the equivalent linear method, the parameters used to approximate material nonlinear behavior, are generally well established. This approach is accurate enough provided that the magnitude of maximum shear strain is of the order of 1% (Kramer, 1996), which is consistent with the results of the present study. The equivalent linear method is strongly affected by the curves that define soil/waste material nonlinearity via shear modulus reduction and damping increase relatively to the shear strain amplitude. Since a significant scatter on proposed curves for shear-modulus reduction and damping increase for waste materials is reported in the related studies in the literature (see Fig. 1), the effect of the selection of the proper material curves on the dynamic response is also evaluated in this work by comparing two characteristic sets.

The impact of the earthquake excitation characteristics is also estimated by considering three excitations with different mean period, T_m (as defined by Rathje et al., 1998), namely: (i) Ricker pulse, $T_m = 0.34$ s, (ii) Aegion, Greece record (1995), $T_m = 0.47$ s, (iii) Shin-Kobe, Japan record (1995), $T_m = 0.66$ s. The role of site effects on seismic slope stability is also highlighted by

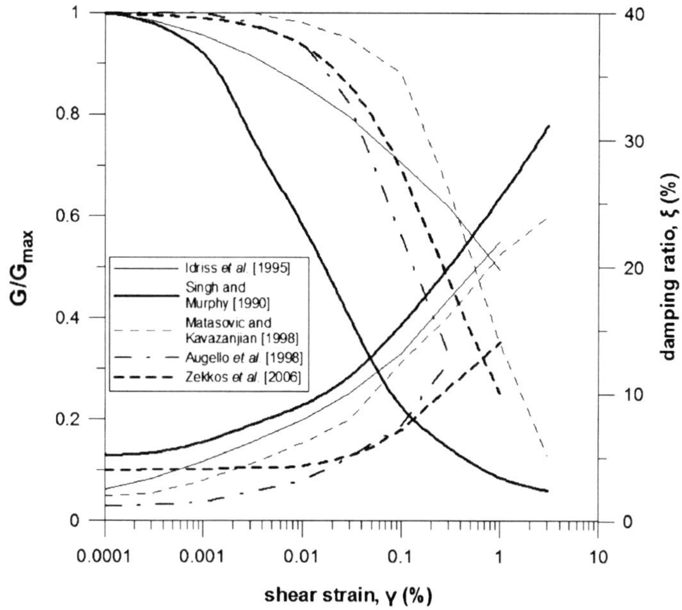

Figure 1. Shear modulus reduction and damping increase curves proposed in the literature for waste material.

applying two of the most common methods: the pseudostatic method and the permanent deformation method. The flexibility of the geostructure is considered in both methods, since initially the equivalent acceleration is calculated according to Makdisi and Seed (1978). The factor of safety is calculated by employing the simplified Bishop (1954) method by assigning to the seismic coefficient the maximum value of the horizontal equivalent acceleration (MHEA). In addition, permanent displacements are computed according to the sliding block method (Newmark, 1965) and by using the equivalent acceleration time history. The influence of each of the aforementioned parameters (mean period of the excitation and maximum acceleration, geometry and material properties) is proven to be significant not only for the dynamic response but also for the slope stability of the geostructures examined.

2. Soil-Structure Interaction Effects on Earth Structures

The aforementioned general remarks regarding the soil-structure interaction are further analyzed by calculating the dynamic response of each of the four models shown in Fig. 2 that represent a typical landfill configuration.

The influence of the soft soil layer at the foundation of the landfill is expected to be twofold, concerning not only the peak acceleration, but the frequency characteristics and the amplification levels of the dynamic response

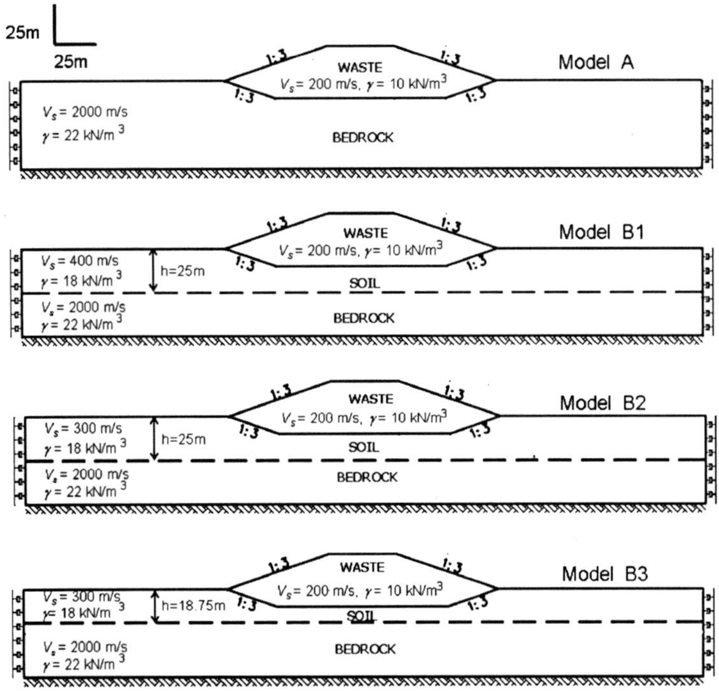

Figure 2. The four configurations whose dynamic response is numerically analyzed in order to investigate the SSI effects on earth structures.

as well. The effects, with respect to frequency characteristics, are presented in the sequence by plotting selected transfer functions of the models, an approach widely used (Borcherdt, 1970). The transfer functions are provided by the ratio of the Fourier amplitude at the middle of the top of the landfill to the corresponding at the bedrock.

Initially, the results of the seismic response of the four configurations are compared in terms of transfer functions when applying a Ricker pulse (Fig. 3). At first the linear case is examined, considering peak ground acceleration (PGA) equal to 0.01 g. The first eigenfrequency of Model A is around 2 Hz, while the second one is roughly observed around 4 Hz. As a result of the interaction between the landfill and the foundation soil layer, the first eigenfrequency of the soil–landfill system (Models B1, B2, and B3) is slightly reduced. In contrast, at the second eigenfrequency the amplification level increases substantially for Model B1, possibly due to the fact that it is close to the eigenfrequency of the soil layer at free field which is 4 Hz. Moreover, the maximum amplification values of Model B2 at the eigenfrequency of the soil layer at free field (3 Hz) are comparable to the corresponding at the first eigenfrequency of the soil–landfill system, implying a similar response of the system at both frequencies.

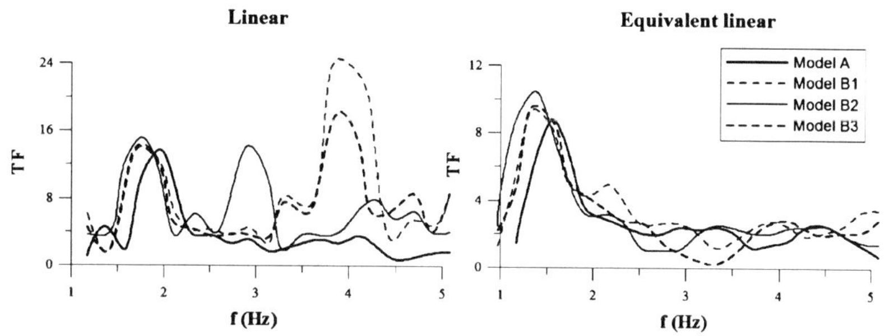

Figure 3. Transfer functions (TF) at the middle of the top of the examined models calculated after the evaluation of their linear and equivalent linear response for Ricker pulse.

Considering that waste material nonlinearity is characterized by the set of curves proposed by Zekkos et al. (2006) which have been used in equivalent linear analyses, the right plot of Fig. 3 presents the results for PGA equal to 0.36 g. The material nonlinearity is expected to affect the response characteristics by decreasing not only the eigenfrequencies of the structure, but the amplification levels as well. More specifically, it is evident that the first eigenfrequency of the landfill is reduced for all the four examined models. However, the first eigenfrequency of Models B1 and B2 receives values around 1.3 Hz, while the corresponding value for Model A is slightly higher. Additionally, the second eigenfrequency of model B1 seems to be close to 2.2 Hz. Note that in the equivalent linear case the amplification of the second eigenfrequency (attributed mainly to the soil–landfill interaction) is suppressed compared to the linear case, which may be the outcome of the higher damping ratio.

The sensitivity of the seismic response of the landfill to the characteristics of the soil layer has so far been examined for the small-strain shear-wave velocity (Model B2). However, the height of the soil layer contributes also to the dynamic SSI. The relevant effect is assumed to be related to the ratio of the shear-wave velocity (V_S) to height of the soil layer (H), or more precisely to the eigenfrequency of the soil layer at free field. The validity of this assumption is examined by comparing Model B1 and Model B3 cases in which the eigenfrequency of the soil layer at free field ($V_S/4H$) is identical. Figure 3 shows that for both linear and equivalent linear response the first eigenfrequency of the system does not seem to be affected by the height of the soil layer. The differences can be observed on the amplification levels at the second eigenfrequency of the system in the linear case and at high frequencies in the nonlinear case. Thus, the higher amplification values in Model B1 than in Model B3 may be possibly attributed to the higher values of V_S in the first one, since similarly to Model B3, lower amplification levels are observed for Model B2 also for the second eigenfrequency.

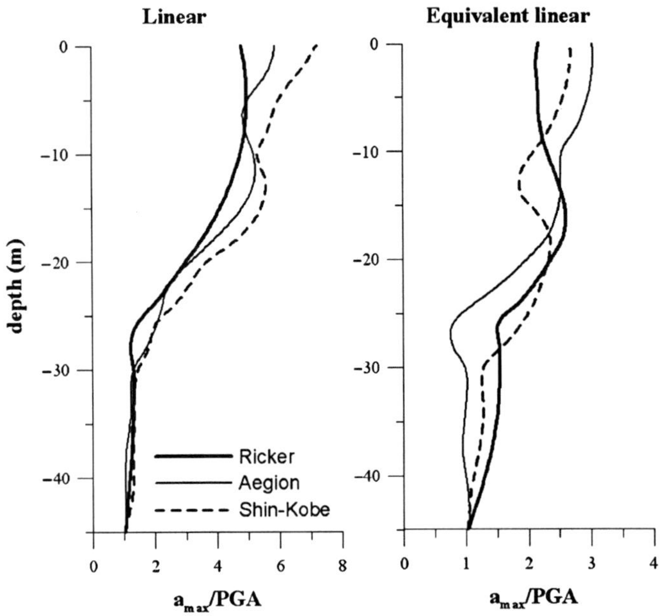

Figure 4. Variation of the amplification along the height of model B1 for linear and equivalent linear response to the three examined excitations. The selected cross-section is located at the axis of symmetry of the geostructure.

The response of the examined landfills is affected not only by the frequency characteristics of the structure, but of the seismic excitation as well. This is evident in Fig. 4, where the amplification along height (Model B1) for the three excitations considered in the current study is shown. Regarding the linear response, the maximum acceleration levels are higher as the mean period of the excitation increases (from Ricker pulse to Shin-Kobe excitation), possibly due to the decrease of the first eigenfrequency of the system resulting from the dynamic interaction with the soil layer. Moreover, it is evident that in the equivalent linear case the effect of the mean period of the excitation is more complex, mainly because it is also related with the decrease in the eigenfrequency of the system. However the observed amplification levels are substantially reduced.

3. Site Effects on Dynamic Response and Stability

As was already mentioned, site effects are related to the geometric characteristics of the earth structure and to the mechanical properties of the material. The mechanism of the response of three typical configurations of geostructures is evident through the snapshots shown in Fig. 5, obtained after the calculation of the linear response by performing finite element analyses.

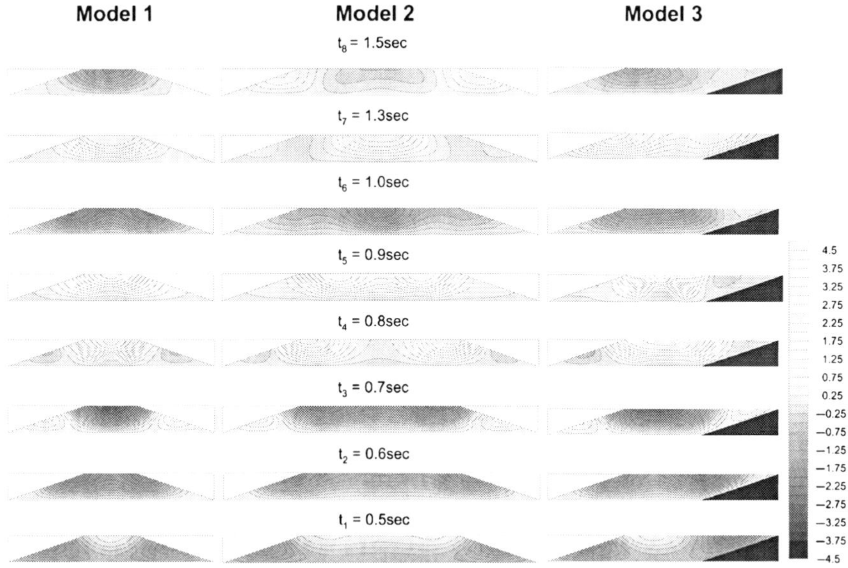

Figure 5. Snapshots of the normalized acceleration (to PGA) for the three examined configurations for linear response to Ricker pulse.

Initially, for the symmetrical models (Model 1 and Model 2) the reflected and diffracted waves generated at the slope of the landfill are evident by: (a) the concentration of the contours at time t_1 and (b) the resulting increase in the amplification in the upper corner of the two models at t_2. Consequently, a detrimental incidence of waves at the middle of the deck occurs (t_3 and t_4), that strongly amplifies the base motion, possibly due to the symmetry of the structure. Note that, despite the fact that the mechanisms of response in the two models are very similar, in Model 2 a slight delay in initiation of the aforementioned phenomena is observed due to the different geometry. Conversely, in Model 3, vertically and inclined incident waves are propagating from the two boundaries (base and side). The two areas of increased contour concentration at times t_1 and t_2 are characteristic of the generation of diffracted and reflected waves from the slope and the upper surface of the model. This phenomenon results into a detrimental incidence at a distance near to crest, observed as a high amplification level (t_3 and t_5).

Site effects can be quantified by means of the topographic aggravation factor (TAF) which is defined as the ratio of the 2D transfer function to the corresponding 1D. The TAF at three positions at the top of the three models is shown in Fig. 6 for the employed equivalent linear response. Waste material nonlinearity is considered to be characterized by two sets of curves of shear-modulus reduction and damping increase, the ones proposed by Singh and Murphy (1990) and those by Zekkos et al. (2006) (see Fig. 1).

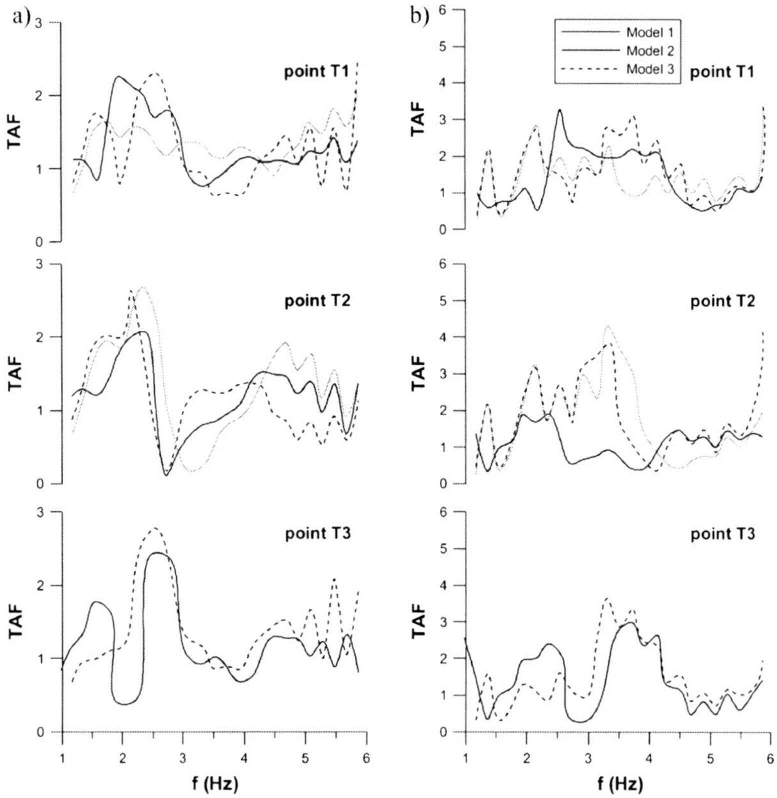

Figure 6. TAF variation at three characteristic positions (T1 is the crest of each model, T2 is located 20 m from crest and T3 60 m from crest). Results are shown for equivalent linear response to Ricker pulse considering the shear modulus reduction and damping increase curves proposed by: (a) Singh and Murphy (1990) and (b) Zekkos et al. (2006).

The selection of these two sets of curves aims at evaluating the response of a landfill for both strongly (Fig. 6a, Singh and Murphy, 1990) and moderately (Fig. 6b, Zekkos et al., 2006) nonlinear material behavior. Subsequently, the impact of the selection of the relationship of shear modulus degradation on the dynamic response of the landfill, simulated via the employed equivalent linear approach, is qualitatively estimated. It is evident that the amplitude of the TAF is strongly frequency dependent, and furthermore the degree of shear modulus reduction affects the frequency content of the TAF variation as well. Moreover, for strongly nonlinear curves the maximum values of TAF are observed at lower frequencies compared to moderate nonlinear curves. In addition, it is shown that the stronger the nonlinear behavior is, the lower the amplitude of the maximum TAF values is. Conclusively, the selection of the relationship of shear modulus reduction and damping increase, relatively to shear strain, alters not only the frequency

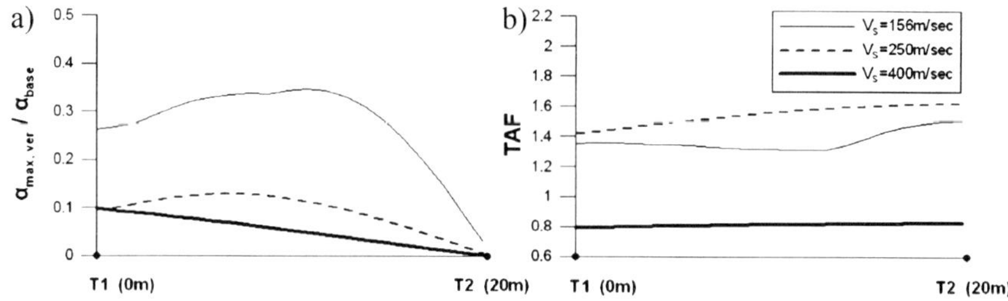

Figure 7. Effect of shear wave velocity on site effects. (a) Maximum normalized parasitic vertical acceleration and (b) maximum TAF variations along the top of Model 1 are presented for equivalent linear response (PGA = 0.36 g) to Ricker pulse.

characteristics of the response, but also the developed amplification levels. Since this impact is also related to the characteristics of the excitation, it cannot be easily quantified.

Nevertheless, a noteworthy consequence of site effects related to topographic irregularities consists in the generation of vertical parasitic acceleration. Observing Fig. 7a, it is evident that the variation of the parasitic vertical acceleration follows a different pattern depending on the value of small-strain V_S. Though the acceleration was applied only horizontally (with a PGA equal to 0.36 g), the equivalent linear response resulted to a parasitic vertical acceleration of the order of 30% of the PGA. The increase in V_S is related to a decrease in the normalized parasitic vertical acceleration, whose maximum value is attained at the crest for the highest V_S value. As shown in Fig. 7b, TAF is decreased for high values of V_S, while it is increased for lower values of V_S.

It has been shown that site effects are strongly related to the geometry of the earth structure, to the material properties, and to the degree of nonlinearity in conjunction to the characteristics of the excitation (mean period and PGA). Thus, a further investigation was conducted to relate these characteristics with the equivalent linear eigenperiod of the earth structure. This was achieved by performing frequency analyses, after applying the updated material properties. The variation of the normalized maximum horizontal acceleration (Fig. 8a) shows that only for Model 3, which is characterized by a more complicated response, the amplification is somewhat higher. Similarly, Model 3 exhibits also the highest levels of parasitic vertical acceleration (Fig. 8b). Moreover, amplification levels for ratio of structure's eigenperiod to mean period of excitation (T_{str}/T_m) lower than the value of two attain a significant scatter obtaining values between two and six. In addition, the results for Model 1 and Model 2 show that the parasitic vertical acceleration may receive values of the order of 60% of the PGA, with insignificant variation relatively to T_{str}/T_m.

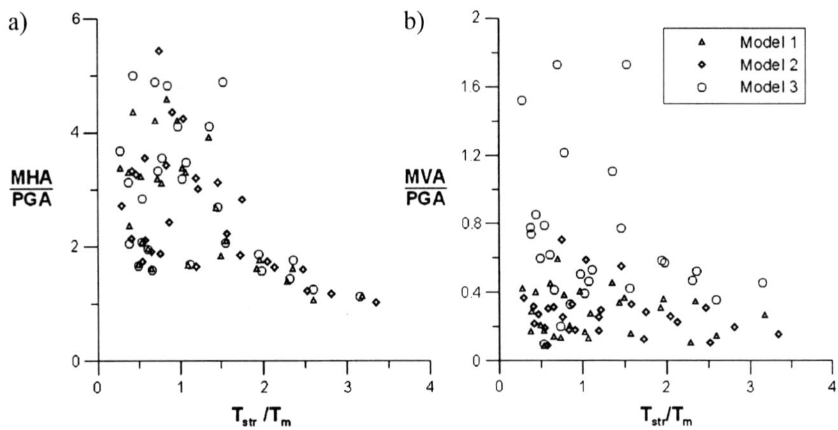

Figure 8. Variation of (a) normalized maximum horizontal acceleration (MHA/PGA) and (b) normalized maximum parasitic vertical acceleration (MVA/PGA) at the top of the earth structure, with respect to the ratio (T_{str}/T_m).

It is expected that the trends observed for the dynamic response, are affecting also the slope stability of the investigated earth structures. The safety factor (SF) corresponding to the critical failure surface was calculated following the Bishop simplified method (1954) for each case. The seismic coefficient was computed as the maximum value of the ratio of the earthquake induced force along the slip circle to the mass of the failure surface. Results of linear analyses cannot be considered to provoke any instability, but are included for completeness. The higher values of SF (of the order of two) correspond to low PGA values, while it was observed that instability (SF < 1) is related to ratios of T_{str}/T_m lower than two and PGA equal to 0.36 g. The general trend of the results was that the SF value reduced as the PGA level increased and as the ratio T_{str}/T_m decreased. In contrast, no significant trend with respect to the characteristics of the examined models can be observed.

For the failure cases (SF < 1) the permanent displacements were calculated either with or without the vertical component of the equivalent acceleration (MVEA). The seismic displacements were obtained numerically after a double integration of the relative acceleration time history. This is defined as the difference between the horizontal seismic coefficient and the yield acceleration (value of acceleration corresponding to SF equal to unity). By inspecting Fig. 9b it is evident that neglecting MVEA may result to an underestimation of the displacement magnitude and even to an incorrect estimation of the stability. Moreover, excessive seismic displacements may develop when the ratio T_{str}/T_m ranges between 0.5 and 1.

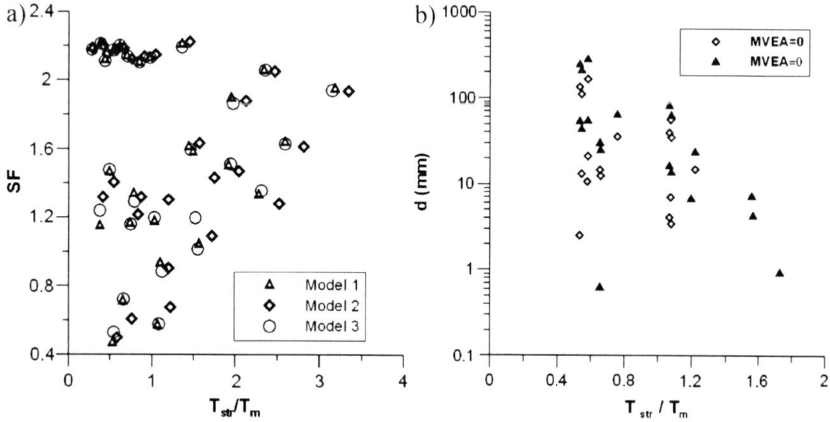

Figure 9. Variation of (a) pseudostatic safety factor (SF) and (b) permanent displacements relatively to the ratio of structure's eigenperiod to mean period of excitation (T_{str}/T_m).

4. Conclusions

The current study has demonstrated that SSI may substantially affect the response of an earth structure. This impact is strongly related to the eigenperiod of the underlying soil layer at free field and the characteristics of the excitation. Moreover, material nonlinearity, which was taken into account utilizing the efficient equivalent linear approach, along with the peak ground acceleration and the mean period of the excitation are critical for the significance of site effects, not only on the response but also on the stability of large-scale earth structures.

Acknowledgements

This paper is part of the 03ED454 research project, implemented within the framework of the *Reinforcement Programme of Human Research Manpower* (PENED) and co-financed by National and Community funds (75% from E.U. —European Social Fund and 25% from the Greek Ministry of Development —General Secretariat of Research and Technology).

References

ABAQUS (2004) *Analysis User Manual Version 6.4*, Abaqus Inc., USA.

Augello, A. J., Bray, J. D., Abrahamson, N. A., and Seed, R. B. (1998) Dynamic properties of solid waste based on back-analysis of OII landfill, *ASCE J. Geotechnical Geoenvironmental Eng.* **124**(3), 211–222.

Bishop, A. W. (1954) The use of the slip circle in the stability analysis of slopes, *Geotechnique* **5**(1), 7–17.

Borcherdt, R. D. (1970) Effects of local geology on ground motion near San Francisco Bay, *Bullet. Seismol. Soc. Am.* **60**, 29–61.

Chavez-Garcia, F. J. (2007) Site effects: from observation and modeling to accounting for them in building codes. In K. D. Pitilakis (ed.), *Earthquake Geotechnical Engineering, 4th Int. Conf. on Earthquake Geotechnical Eng.—Invited Lectures*, Vol. 6 of *Geotechnical, Geological, and Earthquake Engineering*, pp. 53–72, Springer, Netherlands.

Houston, W. N., Houston, S. L., Liu, J. W., Elsayed, A., and Sanders, C. O. (1995) In-situ testing methods for dynamic properties of MSW landfills. In M. K. Yegian and W. D. L. Finn (eds.), *Earthquake Design and Performance of Solid Waste Landfills*, No. 54 of *Geotechnical Special Publication*, New York, pp. 73–82, ASCE.

Hudson, M., Idriss, I. M., and Beikae, M. (1994) *User's Manual for QUAD4M*, Center for Geotechnical Modeling, Department of Civil and Environmental Engineering, University of California, Davis, USA.

Idriss, I. M., Fiegel, G., Hudson, M. B., Mundy, P .K., and Herzig, R. (1995) Seismic response of the Operating Industries Landfill. In M. K. Yegian and W. D. L. Finn (eds.), *Earthquake Design and Performance of Solid Waste Landfills*, No. 54 of *Geotechnical Special Publication*, New York, pp. 83–118, ASCE.

Kramer, S. L. (1996) *Geotechnical Earthquake Engineering*, New Jersey, Prentice-Hall.

Makdisi, F. I. and Seed, H. B. (1978) Simplified procedure for estimating dam and embankment earthquake induced deformations, *ASCE J. Geotechnical Eng. Division* **104**, 849–867.

Matasovic, N. and Kavazanjian, E. Jr. (1998) Cyclic characterization of OII landfill solid waste, *ASCE J. Geotechnical Geoenvironmental Eng.* **124**(3), 197–210.

Mylonakis, G. and Gazetas, G. (2000) Seismic soil-structure interaction: beneficial or detrimental? *J. Earthquake Eng.* **4**(3), 277–301.

Newmark, N. M. (1965) Effect of earthquakes on dams and embankments, *Géotechnique* **15**(2), 139–160.

Rathje, E. M., Abrahamson, N. A., and Bray, J. D. (1998) Simplified frequency content estimates of earthquake ground motions, *ASCE J. Geotechnical Geoenvironmental Eng.* **124**(2), 150–159.

Singh, S. and Murphy, B. (1990) Evaluation of the stability of sanitary landfills. In A. Landva and G. D. Knowles (eds.), *Geotechnics of Waste Fills: Theory and Practice*, Philadelphia, pp. 240–258, American Society for Testing and Materials.

Zekkos, D. P., Bray, J. D., and Riemer, M. (2006) Laboratory evaluation of dynamic properties of municipal solid waste. In: *Proc. of the 5th Hellenic Conf. on Geotechnical and Geoenvironmental Eng.*, Vol. 1, Xanthi, Greece, pp. 513–520.

CYCLIC AND DYNAMIC MECHANICAL BEHAVIOUR OF SHALLOW FOUNDATIONS ON GRANULAR DEPOSITS

Claudio di Prisco*, Andrea Galli
and Mauro Vecchiotti (claudio.diprisco@polimi.it)
Department of Structural Engineering, Politecnico di Milano
Piazza Leonardo da Vinci 32, 20133 Milano, Italy

Abstract. This paper concerns the problem of dynamic soil-structure interaction and in particular rigid shallow footings placed on homogeneous sand strata are taken into consideration. The role played by the development of irreversible strains is discussed on a theoretical point of view and in the light of displacement based design approaches. For achieving the aforementioned objective, small scale experimental test results as well as numerical data, theoretical issues as well as simplified modelling methods are briefly taken into account. The topic is analysed by employing the so-called Macro-element theory. The limitations of this approach as well as its potentialities are considered and critically tackled.

Keywords: shallow foundations, cyclic and dynamic load, numerical modeling, Macroelement approach

1. Introduction

Even if many authors have recently put in evidence the actual role of the foundation on the overall seismic capacity of the system (e.g., Pecker, 2006), in particular when strong earthquakes take place, the seismic/dynamic response of superstructures, in standard civil engineering design approaches, is usually numerically analyzed by disregarding the mechanical interaction between foundations and underlying soil strata. This is partially justified by the lack of well established and calibrated methods for studying the post-yielding behaviour of soil-foundation systems. Standard design approaches start from the idea of solving uncoupled problems by artificially separating geotechnical from structural issues. Dynamic effects concerning the underlying soil strata are either neglected or solved separately by employing sophisticated numerical FEM codes. Once the structural problem is solved, foundation settlements are generally computed by performing finite element numerical analyses under load controlled conditions.

*Corresponding author.

T. Schanz and R. Iankov (eds.), *Coupled Site and Soil-Structure Interaction Effects with Application to Seismic Risk Mitigation,* NATO Science for Peace and Security Series C: Environmental Security,

The dynamic soil-structure interaction problem can be approached in a fully coupled manner even by performing either small/large scale and centrifuge experimental tests (Paolucci et al., 2007; PWRI, 2005; Zeng and Steedman, 1998; Gajan et al., 2005) or dynamic finite element numerical analyses of large spatial domains, including the superstructure, the foundation and the surrounding soil. These numerical analyses allow to take both the local site dynamic effects induced by the presence of the superstructure and the soil-structure cyclic/dynamic interaction into account. Nevertheless, since these numerical analyses are very time consuming, they cannot be used to perform extensive parametric numerical analyses and be employed like heuristic tools. A further alternative strategy, which will be discussed hereafter, consists in employing the Macro-element concept (Nova and Montrasio, 1991; Paolucci, 1997; Cremer et al., 2001; Cremer et al., 2002; Le Pape and Sieffert, 2001). This allows to interpret in a partially coupled manner the soil-structure interaction and to evaluate the consequences of the cyclic loads transmitted by the superstructure to the foundation. This approach is based on the idea of subdividing the entire domain in three sub-structures: (i) the far field, (ii) the near field and (iii) the superstructure. The "far field" is the zone of the soil stratum which is not influenced by the presence of the superstructure; the displacement field can be there assumed to be known. The "near-field" is the soil stratum zone where the effects induced by soil-structure interaction become dominant and irreversibilities are mainly expected to develop. Here in the following the "near field" mechanical behaviour will be analysed by assuming the footing to be rigid, whilst both superstructure and "far field" will not be taken into consideration.

2. Generalized Stress–Strain Variables: Shallow Foundations under Static Loading

The Macro-element theory was initially conceived for rigid strip footings placed on homogeneous dry sand strata under monotonously increasing both inclined and eccentric loads, and more recently it was extended to simulate cyclic tests (di Prisco et al., 1998; Cremer et al., 2001; di Prisco et al., 2003a; di Prisco et al., 2003b; di Prisco et al., 2003c). Since the footing is assumed to be rigid, the static mechanical interaction can be described by employing (under plane strain conditions) only three generalised stresses (the vertical load component V, the horizontal load component H, and the overturning moment M) and three generalised strains (the vertical displacement v, the horizontal displacement u and the foundation rotation θ). For convenience, the generalised stress variables can be summarised in a three dimensional vector Q, whereas the associated generalised strain variables can be collected in a corresponding q vector. The relative incremental constitutive relationship can be thus written as follows:

$$\dot{Q}^t = D(Q^t, \dot{Q}^t, \psi)\dot{q}, \quad (1)$$

where D is the incremental 3×3 stiffness matrix, ψ stands for a hardening parameter vector, whereas dots stand for time derivation and subscript index t for transposition.

2.1. ELASTOPLASTIC APPROACHES

If an elasto-perfectly plastic approach was chosen, for obtaining the symbolic relation of Eq. (1), it would be sufficient to define an appropriate failure locus F, a plastic potential G and an elastic stiffness matrix.

Hereafter the analysis will be exclusively focussed on the plastic mechanisms and the definition of both the elastic stiffness and radiation damping factors (Sieffert and Cevaer, 1991) is not discussed. As was experimentally shown by many authors (Butterfield and Ticof, 1979; Georgiadis and Butterfield, 1988; Nova and Montrasio, 1991; Nova and Montrasio, 1997; Butterfield and Gottardi, 1994), at each point belonging to F a well defined failure mechanism corresponds. Shallow footings placed on loose sand strata constitute an exception, since, as it is well known, their experimental mechanical response is characterised by a continuous hardening due to the progressive sinking of the foundation level.

From experimental results and theoretical analyses these additional remarks can be also derived: (i) the embedment prevalently causes a leftward translation of F along the V axis (Calvetti et al., 2004; di Prisco et al., 2004), i.e., even negative values of V can be sustained in this case by the foundation; (ii) the definition of the interaction domain can be extended to the case of rectangular shallow footings (Grange et al., 2007; Grange, 2008), (iii) by simply changing the set of constitutive parameters either a heterogeneous stratigraphy or the presence of geo-reinforcements within the soil can be taken into account (di Prisco et al., 2003a).

A more realistic simulation of the mechanical behaviour of rigid shallow footings under monotonously increasing loading can be obtained by employing an elasto-plastic strain hardening constitutive model. Both the loading function f and the plastic potential g can be in this case conveniently defined by assuming their shape to be coincident with that of F and G, respectively. It follows that the failure locus is a special yield locus for which the hardening parameter $V_c = V_M$ (Nova and Montrasio, 1991), where V_M is the footing bearing capacity. In order to define more precisely the loading function shape, in nineties some authors (Butterfield and Gottardi, 2003) performed on rigid shallow footings the so-called "swipe-tests", during which vertical displacements are inhibited, whilst either horizontal displacement or rocking angle is increased.

2.2. SHALLOW FOUNDATIONS UNDER CYCLIC LOADING

Both the previously defined approaches, capable of reproducing quite satisfactorily the mechanical response of this type of structures under monotonously increasing loading, fail totally in simulating the experimental evidence when cyclic tests are taken into account. As far as these are concerned, two different types are available in literature: (i) the symmetric load controlled cyclic tests (in this case the cycles are symmetric with respect to the V axis) (Pedretti, 1998; Shirato et al., 2007) and (ii) the asymmetric load controlled tests. In this latter case the cyclic perturbation induces a loading oscillation about a generalised state of stress not belonging to the V axis (di Prisco et al., 2003c). The first class of tests mainly causes an accumulation with time of vertical displacements that progressively stabilise reaching a sort of shake down condition. The second class of tests, on the contrary, induces a marked accumulation both of vertical and horizontal/rotational displacements. Analogously to what is usually done for the representative volume of soil, a way of overcoming both the limitations of the previously defined class of models, consists in introducing either generalised plastic constitutive relationships or multi-mechanism plastic approaches. The constitutive model of Nova and Montrasio (1991) cited here above was for instance modified (di Prisco et al., 1998), by introducing within the yield locus, which becomes a bounding surface, a subloading surface. A convenient mapping rule connects any point within the yield locus belonging to the subloading surface with an appropriate point on the bounding surface. In this way, it is possible to account for the occurrence of permanent generalised strains even when the stress point is within what is usually considered a purely elastic region and the accumulation of plastic distortions during cyclic or transient loading can therefore be described (di Prisco et al., 2003a; di Prisco et al., 2003b). The model becomes inevitably more complex and the determination of further parameters becomes necessary. An example of application is given below.

Figure 1 concerns the behaviour of a plinth, 1 m wide, founded on a dense sand stratum: during the cyclic phase of the tests, the vertical load is kept constant whilst both horizontal force H and overturning moment M are varied by keeping constant their ratio (experimental data after Pedretti, 1998). Loading cycles were applied at low frequency, so that dynamic effects could be neglected. It is apparent that with the proposed method it is possible to capture in a satisfactory way the essential features of the experimental response.

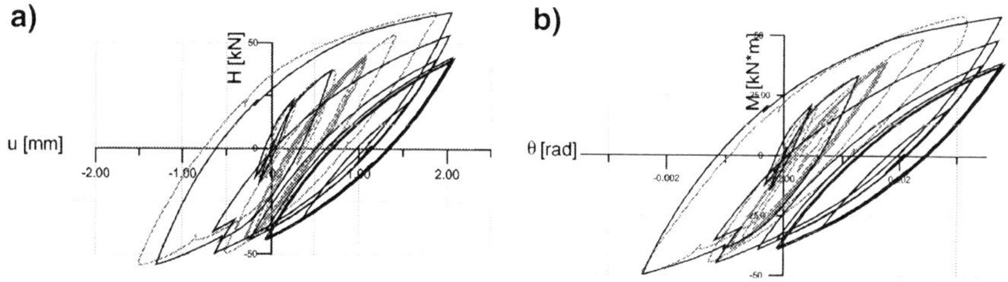

Figure 1. Comparison of measured (dotted lines) and calculated (full lines) displacements of a real scale foundation under cyclic horizontal loading and overturning moment with constant vertical load (after Pedretti, 1998): (a) horizontal displacement; (b) rotation.

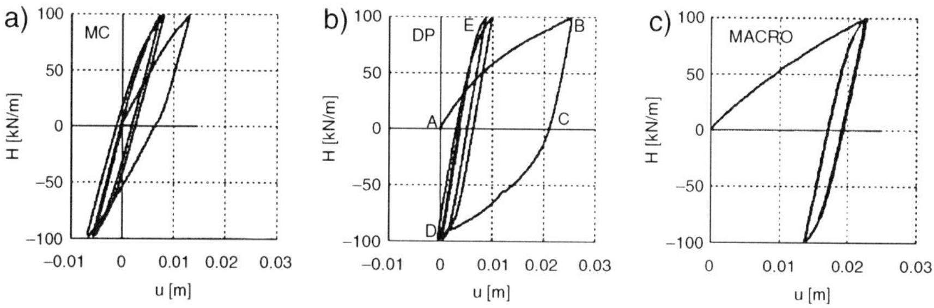

Figure 2. Comparison among numerical data obtained by employing the FEM numerical analyses with (a) the MC, (b) the DP constitutive models and (c) the Macro-element approach.

2.3. FURTHER MODELLING DEVELOPMENTS

The previously mentioned class of models, which are characterised by an isotropic yield function, unfortunately fails in reproducing three very important aspects of the mechanical behaviour of rigid shallow foundations under repeated loading that have not yet been cited: (i) the large settlements induced by the first unloading, (ii) the loss of stiffness during the first phase of the unloading when the footing is previously largely tilted, (iii) the overestimation of ratcheting when asymmetric loading paths are imposed.

In order to clarify the first issue, we can start from the discussion of Fig. 2, where the numerical results (curves MACRO) obtained by employing the previously cited cyclic constitutive model (di Prisco et al., 2002) are compared with those obtained by performing small strain and displacement finite element numerical analyses (FEAT, 2004). In particular curves MC are obtained by assigning to the soil an elasto-perfectly plastic Mohr-Coulomb constitutive model, whilst curves DP have been obtained by assigning to the soil an anisotropic strain-hardening constitutive relationship.

These results are relative to a rigid shallow strip foundation placed on a loose sand stratum initially subjected to a vertical load V which is kept constant when the horizontal load is cyclically varied. It is evident that the numerical data obtained by employing the Macro-element approach reproduce quite well the FEM numerical simulations during all the successive cycles but not during the first one. This is essentially due to the use of an isotropic hardening for the bounding/yield surface. In fact, when the load is rightward directed, irreversible strains mainly develop close to the right corner of the foundation, whilst at the opposite corner when the load is inclined leftward. This implies that even the two associated failure mechanisms can be assumed to be totally independent and the same can be inferred when overturning moments are applied. As already assumed by Cremer et al. (2001), to simulate such a behaviour, the hardening of the plastic surface will have to be assumed anisotropic.

The second aspect previously cited is more important and it can be clarified for instance by discussing the experimental test results, obtained on footing large scale models at Public Works Research Institute (PWRI, 2005; Shirato et al., 2007), plotted in Fig. 3. These data concern a rigid steel structure placed on a caisson filled with sand cyclically loaded under displacement controlled conditions so that even the reduction of generalised loads due to strain localisation can be observed.

During the cyclic phase, V is kept constant, whilst both H and M vary. As is evident by comparing Fig. 3a and b, during the cyclic phase, the footing mechanical response of the system is severely affected by the soil relative density. In particular, when the soil density is sufficiently high, for cycles characterised by large values of the rocking angle, during the unloading a typical "s" shaped trend (Fig. 3a) is observed. On the contrary, this trend disappears when loose sands are tested (Fig. 3b). The mechanical response during the unloading in case of dense sands is due to the uplift of the foundation:

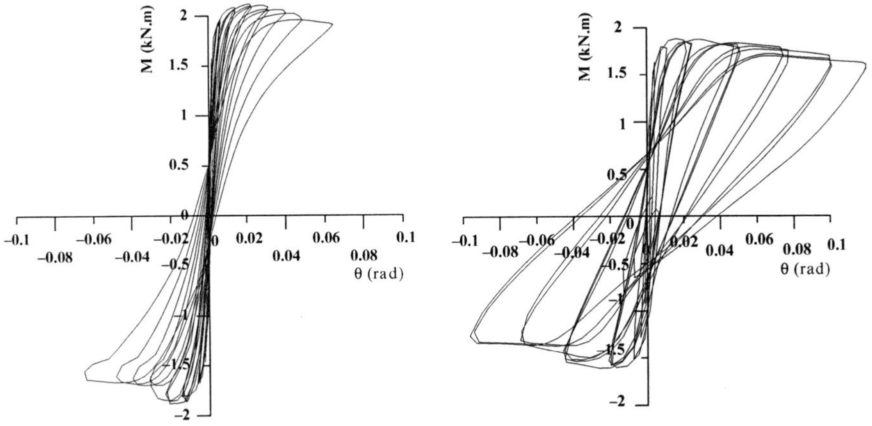

Figure 3. Experimental data concerning horizontal displacement controlled cyclic tests on model foundation: (a) dense sand; (b) loose sand (after PWRI, 2005).

the reduction in the contact surface between the footing and the soil generates a sort of damage of the system that could be described coherently for instance by introducing an elasto-plastic coupling. Some efforts in this direction, although not totally satisfying, are in Cremer et al. (2001), Shirato et al. (2007) and Grange (2008).

The third aspect previously cited, crucial for footings subjected to inclined and/or eccentric loads even in static conditions, regards asymmetric loading paths. When this type of loading is numerically simulated by using both the aforementioned generalised plastic and anisotropic strain hardening elasto-plastic constitutive models, accumulated irreversible generalised strains are dramatically overestimated and the numerical evaluation of ratcheting becomes totally unrealistic. Just to highlight this aspect one of the authors has recently performed (di Prisco et al., 2003c) an extensive experimental test campaign by employing a small scale rigid strip footing placed on a loose sand stratum.

2.4. A SIMPLIFIED STRUCTURAL APPROACH

The mechanical response of rigid strip foundations can be further schematised by employing the well-known concepts of secant stiffness K and damping factor η. An example of the dependency of K (rotational stiffness) on rocking angle is illustrated in Fig. 4. The points reported in this figure are obtained by processing the aforementioned experimental data relative to dense sands from Pedretti (1998) and PWRI (2005). Stiffness K is the tangent of the inclination angle of the experimental loop plotted in the M-θ plane. In Fig. 4a the values of ratio K/K_0 are plotted, where K_0 is the initial rotational stiffness of the system. In the same plots, the data obtained by performing numerical analyses throughout the previously cited bounding surface constitutive model are also reported. It is worth noting that even for relatively small values of the rocking angle, for instance 1 mrad, the reduction in foundation stiffness ranges between about 50% and 60%.

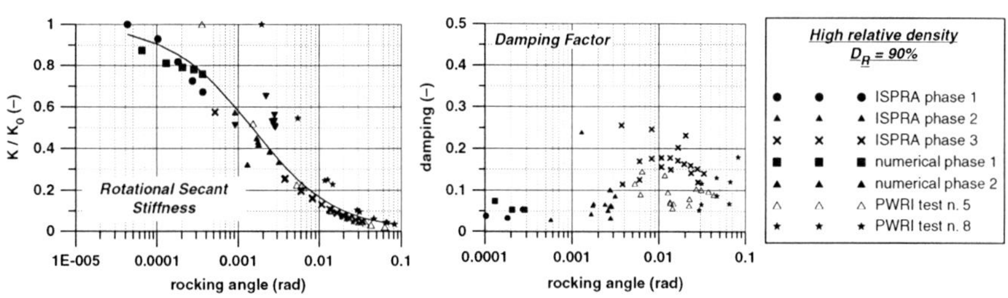

Figure 4. Normalized stiffness and damping factor for $Dr = 90\%$ (after Paolucci et al., 2007).

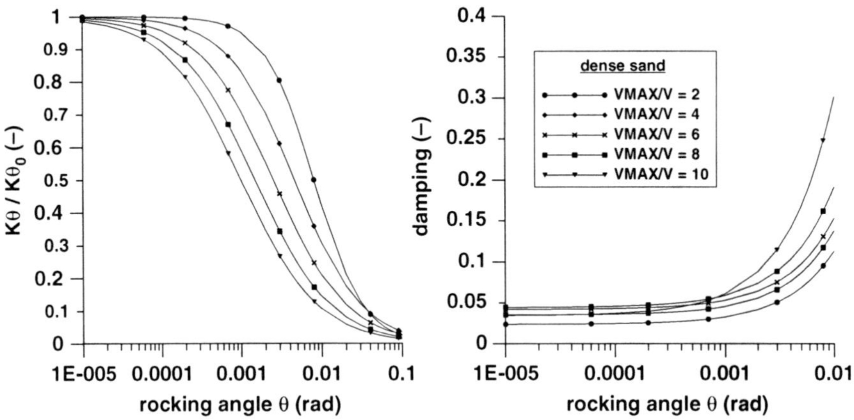

Figure 5. Influence of the loading path on the values of the secant rotational stiffness K_θ and of the damping factor for dense sand.

The equivalent damping ratio in the rocking mode η has been also plotted in Fig. 4b. η is computed as the ratio between the dissipated energy D (area of the hysteresis loop) and the stored elastic energy ΔW. η values range from 5% to 10% for rocking values up to 1 mrad, whilst η significantly increases for larger rocking angles, up to 20% for dense sands and 30% for medium dense sands. For highlighting the influence of the loading path on the dispersion of points of Fig. 4 some further numerical simulations were performed by employing the previously cited bounding surface constitutive model. In Fig. 5, numerical data obtained by performing cyclic tests on a dense sand stratum, during which the vertical load V is kept constant, $H = 0$ and M is varied, are illustrated. The trend of both $K_\theta/K_{\theta 0}$ and η is plotted for different values of V/V_M. These results justify the large dispersion of both the experimental and numerical data collected in Fig. 4, where no distinction among the generalised stress paths imposed has been done. Due to the already commented limitations of the model employed, in Fig. 5 only numerical damping factors corresponding to small values of θ ($\theta < 0.01$ rad) are plotted. In fact, as the model is not capable neither of accounting for the foundation uplift nor of reproducing the "s" shaped response of Fig. 3a, for larger values of θ numerical results would be meaningless.

3. Macro-Element-Dynamic Approach

The use of the Macro-element theory for reproducing the mechanical behaviour of rigid shallow foundations is meaningful when the soil-structure interaction is considered under quasi-static conditions and excess pore water pressure is nil in the entire soil stratum. Under either seismic or dynamic conditions its employment becomes rather ambiguous: since the soil stratum is deformable, inertial forces are distributed in the spatial domain and their role cannot be apparently taken by the Macro-element constitutive relationship into account.

When the "far field" can be assumed to be still, we could try to simulate numerically the dynamic system response by employing the Macro-element constitutive relationship, already validated and calibrated in quasi-static conditions. This approach is appropriate if, and only if, inertial forces within the soil stratum are negligible. On the contrary, when applied loading becomes very fast and/or the loading frequency sufficiently high, the previous hypothesis may become mechanically meaningless. The influence of inertial terms on the dynamic soil-structure interaction phenomenon can be estimated for instance, by considering impacts of rock boulders on sand strata. When either small or large scale impact experimental tests are performed, in fact, impact loading, which is measured by accelerometers placed within the boulder, can largely overcome the bearing capacity quasi-statically evaluated for the equivalent circular footing (di Prisco and Vecchiotti, 2006). This discrepancy is essentially due to the arising of inertial forces within the soil stratum. For this reason, the Macro-element constitutive relationship has been recently modified and the flow rule has been renewed by following the standard visco-plastic Perzyna approach (Perzyna, 1963).

A challenging task for the next future research on this subject consists therefore in introducing either the time factor or the frequency variable into the Macro-element constitutive relationship even when cyclic/dynamic tests are performed. In this perspective, the authors have very recently performed cyclic numerical analyses under quasi static conditions and in dynamic conditions on the same footing by changing the loading frequency. The domain was characterised by absorbing boundaries and the constitutive relationship for the soil element implemented in the employed finite element code (FEAT, 2004) was elasto-perfectly plastic with a Mohr-Coulomb failure criterion and a non-associated flow rule. The data obtained up to now suggest a reduction in the damping factor at increasing values of the loading frequency but even a not yet clear increase in the computed displacements. An extensive numerical parametric analysis is therefore foreseen at least as far as simple generalised stress paths are concerned.

Finally, when the superstructure is loaded by seismic actions (Paolucci, 1997), the problem becomes dramatically more complex (Paolucci et al., 2007; Grange, 2008). In this case, the presence of the superstructure can locally modify the response of the system, even if the mass of the superstructure is disregarded. This phenomenon is usually cited as local dynamic site effect. This has been disregarded in all the works cited here above, but, as it well known, if the stratum is saturated, this can sometimes lead even to soil liquefaction.

4. Concluding Remarks

The problem of soil-structure cyclic/dynamic interaction has been tackled by accounting for the development of irreversible strains within the soil stratum. Shallow rigid footings have been considered; both experimental data and finite element numerical analyses have been very briefly discussed. The topic has been heuristically analysed throughout the Macro-element theoretical approach. A schematic critical presentation of constitutive relationships available in literature has been provided. The problem of ratcheting taking place when asymmetric stress paths are imposed, as well as the phenomenon of uplift of shallow foundations placed on very dense sand strata have been discussed. Whether the problem of the dynamic response of the system is taken into consideration by employing either an uncoupled approach or a coupled one, the Macro-element theory seems to be very promising in clarifying on a methodological point of view the main features of the problem. The paper has been particularly aimed at stressing the use of this theory as a useful tool both for defining the main factors governing the response of the system and for planning either experimental and numerical test series. A simplified interpretation of the soil-structure interaction problem for reproducing the dynamic response of the superstructure within the framework of displacement based design approaches has been suggested. The dependency of rotational stiffness as well as of damping factor on the rocking angle of the footing has been described by plotting the data experimentally available in literature. The dependency of the shape of the curves obtained by interpolating the aforementioned data on the generalised stress path imposed has been put in evidence.

The problem of dynamic coupling deriving from the arising of inertial forces within the soil stratum has been cited. This becomes enormously complex in particular when coupling between the two dynamic effects cited above (the first one deriving from the dynamic actions coming from the superstructure and the second associated to the local dynamic amplification of the seismic signal) is considered.

Acknowledgements

This research was conducted within the framework of a Five-year joint research agreement between Public Works Research Institute, Tsukuba, Japan, and Politecnico di Milano, Italy, 2003–2007, on the seismic design methods for bridge foundations. The research was partly supported by the Executive cooperation program between the governments of Italy and Japan 2002–2006, Project No. 13B2 and by the DPC-RELUIS National Research Project No. 4 *Sviluppo di approcci agli spostamenti per il progetto e la valutazione della vulnerabilità*, Framework Programme 2005–2008.

References

Butterfield, R. and Gottardi, G. (1994) A complete three dimensional failure envelope for shallow footings on sand, *Géotechnique* **44**(1), 181–184.

Butterfield, R. and Ticof, J. (1979) Discussion: design parameters for granular soils. In *Proc. 7th European Conf. Soil Mech. Fndn Eng.*, Vol. 4, Brighton, pp. 259–262.

Butterfield, R. and Gottardi, G. (2003) Determination of yield curves for shallow foundation by "swipe" testing. In Magnan, J.P. and Droniuc, N. (eds.), *Proc. Int. Symp. on Shallow Foundations FONDSUP*, Vol. 1, Paris, 5–7 Novemeber 2003, pp. 111–118.

Calvetti, F., di Prisco, C., and Nova, R. (2004) Experimental and numerical analysis of pipeline-landslide interaction, *J. Geotech. Geoenv. Eng. ASCE* **12**, 1292–1299.

Cremer, C., Pecker, A., and Davenne, L. (2001) Cyclic Macro-element for soil-structure interaction: material and geometrical non-linearities, *Int. J. Num. Anal. Meth. Geomechanics* **25**, 1257–1284.

Cremer, C., Pecker, A., and Davenne, L. (2002) Modelling of nonlinear dynamic behaviour of a shallow strip foundation with Macro-element, *J. Earthquake Eng.* **6**, 175–212.

di Prisco, C. and Vecchiotti, M. (2006) A rheological model for the description of boulder impacts on granular strata, *Géotechnique* **56**(7), 469–482.

di Prisco, C., Fornari, B., Nova, R., and Pedretti, S. (1998) A constitutive model for cyclically loaded shallow foundations. In *Proc. Euromech. Coll. Inelastic Analysis Structures under Variable Loads*, pp. 107–111.

di Prisco, C., Nova, R., and Sibilia, A. (2002) Analysis of soil-structure interaction of towers under cyclic loading. In G. N. Pande and S. Pietruszczak (eds.), *Proc. NUMOG 8*, Rome, Balkema, pp. 637–642.

di Prisco, C., Montanelli, F., Caloni, G., and Savoldi, A. (2003a) Shallow foundations on geo-reinforced sand layers: experimental results and theoretical observations. In Magnan, J.P. and Droniuc, N. (eds.), *Proc. Int. Symp. on Shallow Foundations FONDSUP*, Vol. 1, Paris, 5–7 Novemeber 2003, pp. 185–192.

di Prisco, C., Nova, R., Perotti, F., and Sibilia, A. (2003b) Analysis of soil-foundation interaction of tower structures under cyclic loading. In M. Maugeri and R. Nova (eds.), *Geotechnical Analysis of Seismic Vulnerability of Historical Monuments*, Bologna, Pàtron.

di Prisco, C., Nova, R., and Sibilia, A. (2003c) Shallow footings under cyclic loading: experimental behaviour and constitutive modeling. In M. Maugeri and R. Nova (eds.), *Geotechnical Analysis of Seismic Vulnerability of Historical Monuments*, Bologna, Pàtron.

di Prisco, C., Nova, R., and Corengia, A. (2004) A model for landslide-pipe interaction analysis, *Soils Foundations* **44**, 1–12.

FEAT (2004) *Tochnog Professional User's Manual*, Finite Element Application Technology.

Gajan, S., Kutter, B., Phalen, J., Hutchinson, T. C., and Martin, G. R. (2005) Centrifuge modeling of load-deformation behaviour of rocking shallow foundations, *Soil Dynamics and Earthquake Eng.* **25**, 773–783.

Georgiadis, M. and Butterfield, R. (1988) Displacements of footings on sand under eccentric and inclined loads, *Can. Geotech. J.* **25**, 192–212.

Grange S. (2008) *Modélisation simplifiée 3D de l'interaction sol-structure: application au genie parasismique*, PhD Thesis, Institut Polytechnique de Grenoble, Grenoble, France.

Grange, S., Kotronis, P., and Mazars, J. (2007) 3D Macro element for soil structure interaction. In *4th Int. Conf. Earthquake Geotechnical Eng.*, Thessaloniki, Greece, June 25–28.

Le Pape, Y. and Sieffert, J. P. (2001) Application of thermodynamics to the global modelling of shallow foundations on frictional material, *Int. J. Num. Anal. Meth. Geomechanics* **25**, 1377–1408.

Nova, R. and Montrasio, L. (1991) Settlement of shallow foundations on sand, *Géotechnique* **41**(2), 243–256.

Nova, R. and Montrasio, L. (1997) Settlements of shallow foundations on sand: geometrical effects, *Géotechnique* **47**(1), 46–60.

Paolucci, R. (1997) Simplified evaluation of earthquake induced permanent displacements of shallow foundations, *J. Earthquake Eng.* **1**, 563–579.

Paolucci, R. and Pecker, A. (1997) Seismic bearing capacity of shallow strip foundations on dry soils, *Soils Foundations* **37**, 95–105.

Paolucci, R., di Prisco, C., Vecchiotti, M., Shirato, M., and Yilmaz, M. (2007) Seismic behyaviour of shallow foundations: large scale experiments vs. numerical modelling and implications for performance based design. In *1st US Italy Sesimic Bridge Workshop*, Eucentre, Pavia, Italy.

Pecker, A. (2006) Enhanced seismic design of shallow foundations: example of the Rion Antirion bridge, *4th Athenian Lecture on Geotechnical Engineering*, Athens, Greece.

Pedretti, S. (1998) *Non-linear soil-foundation interaction: analysis and modelling methods*, PhD Thesis, Politecnico di Milano, Milano, Italy.

Perzyna, P. (1963) The constitutive equations for rate sensitive plastic materials, *Quart. Appl. Math.* **20**, 321–332.

Public Work Research Institute (2005) Experimental study on the residual displacements of shallow foundations, Technical Note, Tsukuba, Japan.

Shirato, M., Nakatani, S., Fukui, J., and Paolucci, R. (2007) Large-scale model tests on shallow foundations subjected to earthquake loads. In *Proc. 2nd Japan–Greece Workshop on Seismic Design, Observation and Retrofit of Foundations*, Tokyo, Japan, 3–4 April.

Sieffert, J. G. and Cevaer, F. (1991) *Handbook of Impedance Functions*, Quest Editions, Nantes, France, Presses Academiques.

Zeng, X. and Steedman, R. S. (1998) Bearing capacity failure of shallow foundations in earthquakes, *Géotechnique*, **48**, 235–256.

SOIL-STRUCTURE INTERACTION IN NONLINEAR SOIL

Vlado Gicev (vgicev@gmail.com)
Department of Computer Science,
Goce Delcev University, Toso Arsov St. 14, 2000 Stip,
F.Y.R. of Macedonia

Abstract. A two-dimensional (2-D) model of a building supported by a semi-circular flexible foundation embedded in nonlinear soil is analyzed. The building, the foundation, and the soil have different physical properties. The model is excited by a half-sine SH wave pulse, which travels toward the foundation. The results show that the spatial distribution of permanent, nonlinear strain in the soil depends upon the incident angle, the amplitude, and the duration of the pulse. If the wave has a large amplitude and a short duration, a nonlinear zone in the soil appears immediately after the reflection from the half-space and is located close to the free surface. This results from interference of the reflected pulse from the free surface and the incoming part of the pulse that still has not reached the free surface. When the wave reaches the foundation, it is divided on two parts—the first part is reflected, and the second part enters the foundation. Further, there is separation of this second part at the foundation-building contact. One part is reflected back, and one part enters the building. After each contact of the part of the wave that enters the building with the foundation-building contact, one part of the wave energy is released back into the soil. This process continues until all of the energy in the building is released back into the soil. The work needed for the development of nonlinear strains spends part of the input wave energy, and thus a smaller amount of energy is available for exciting the building.

Keywords: soil-structure interaction; non-linear wave propagation; energy distribution

1. Introduction

Field reconnaissance of the effects of many earthquakes has provided numerous examples of different types of soil failure and permanent deformations caused by strong shaking. Examples include settlement of cohesionless soils, liquefaction of saturated sands, flow slides due to liquefaction of cohesionless soils, bulkhead failures due to backfill liquefaction, slides caused by liquefaction of thin sand layers, failures of fills on weak foundations, and lateral movement of bridge abutments. Many structures settle, tilt, or overturn on liquefied soil. Some of the best-known examples of this occurred during the 1964 Alaska and 1964 Niigata earthquakes (Seed, 1970). The sequence of the

T. Schanz and R. Iankov (eds.), *Coupled Site and Soil-Structure Interaction Effects with Application to Seismic Risk Mitigation,* NATO Science for Peace and Security Series C: Environmental Security,

soil-structure interaction (SSI) phenomena, which led to the overturning of apartment buildings in Kawagishi-cho during the Niigata earthquake, is complicated, and its complete modelling and analysis are still a major challenge for any nonlinear numerical simulation. It probably started with development of nonlinear strain zones in the soil close to the foundation, which in turn expanded the trapped nonlinear energy to initiate liquefaction. We are assuming here that the large energy of earthquake waves trapped in the zones of strain localization initiated liquefaction (Trifunac, 1995), which then spread all around the foundation, causing the buildings to tilt and overturn. Analysis of this sequence is well beyond the scope of this paper, however. We will describe only the early stages, which involve the creation of the nonlinear zones of soil response.

Trifunac (1972) presented the analytical solution for interaction of the wall sitting on an embedded semi-circular rigid foundation. Wong and Trifunac (1975) studied the wall–soil–wall interaction with the presence of two or more shear walls, and Abdel-Ghaffar and Trifunac (1977) studied the soil–bridge interaction with a semi-cylindrical rigid foundation and an input plane-SH wave. Other studies have been conducted to analyze the influence of the shape of a rigid foundation on the interaction. Wong and Trifunac (1974) solved the interaction of the shear wall erected on an elliptical rigid foundation for shallow and deep embedment, and Westermo and Wong (1977) studied different boundary models for the soil-structure interaction of an embedded, semi-circular, rigid foundation. They concluded that without a transmitting boundary all of the models develop resonant behaviour and that the introduced damping in the soil cannot adequately model the radiation damping. Luco and Wong (1977) studied a rectangular foundation welded to an elastic half-space and excited by a horizontally propagating Rayleigh wave. Lee (1979) solved a 3-D interaction problem consisting of a single mass supported by an embedded, hemispherical, rigid foundation for incident plane P, SV, and SH waves in spherical coordinates. The recent publications deal with a flexible foundation. Todorovska et al. (2001) solved an interaction of a dike on a flexible, embedded foundation, and Hayir et al. (2001) described the same dike but in the absence of a foundation. Aviles et al. (2002) analyzed the in-plane motion of a 4-degrees-of-freedom model and Gicev (2005) studied the soil-flexible foundation-structure interaction for incident-plane SH waves with a numerical model using finite differences.

The soil-structure interaction phenomenon includes several features, among them wave scattering, radiation damping, damping in the structure, and the presence of different frequencies (system frequency, apparent frequency, rocking frequency, horizontal frequency, and fixed-base frequency). In this paper, in the presence of the interaction, the development of the nonlinear zones in the soil is studied for incident pulses representing the near-field destructive strong ground motion. The problems that must be addressed in the numerical study of the nonlinear soil-structure interaction include heterogeneities and discontinuities in the medium, the modelling of the free surface, artificial boundaries, and keeping track of the nonlinear constitutive law at each point in the soil. According to Moczo (1989) and Zahradnik et al. (1993), the computational FD schemes that are used in applications of wave propagation can be divided into homogenous and heterogeneous. Alterman and Karal (1968) used the homogeneous formulation to solve elastic wave propagation in layered media, and Boore (1972) proposed the heterogeneous scheme. Tsynkov (1998) reviewed the existing global and local artificial boundaries. The global boundaries are perfect absorbers, but they cannot be readily applied in "marching-in-time" procedures because of their non-locality, both in time and space. The main advantage of the local (imperfect) artificial boundaries is that they are local in space and time and are not frequency dependent.

2. Model

During the wave passage, the soil, the foundation, and the superstructure undergo nonlinear deformations and permanent strains. Because the aim of this paper is to study the nonlinear zones in the soil only—for simplicity—only the soil is modeled as nonlinear, while the foundation and the building are assumed to remain linear. The model is shown in Fig. 1. The incoming wave is a half-sine pulse of a plane SH wave. A dimensionless frequency

$$\eta = \frac{2a}{\lambda} = \frac{a}{\beta_s \cdot t_{d0}} \tag{1a}$$

is introduced as a measure of the pulse duration, where a is the radius of the foundation, λ is the wavelength of the incident wave, β_s is the shear-wave velocity in the soil, and t_{d0} is duration of the pulse.

To set up the grid spacing, the pulse is analysed in space domain (s), and the displacement in the points occupied by the pulse is

$$w(s) = A \sin \frac{\pi \cdot s}{\beta_s \cdot t_{d0}}, \tag{1}$$

where A is the amplitude of the pulse and s is the distance of the considered point to the wave front in initial time in the direction of propagation. Using the fast Fourier transform algorithm, the half-sine pulse Eq. (1) is transformed in wave number domain (k) as follows:

$$w(k) = \mathfrak{F}(w(s)). \tag{2}$$

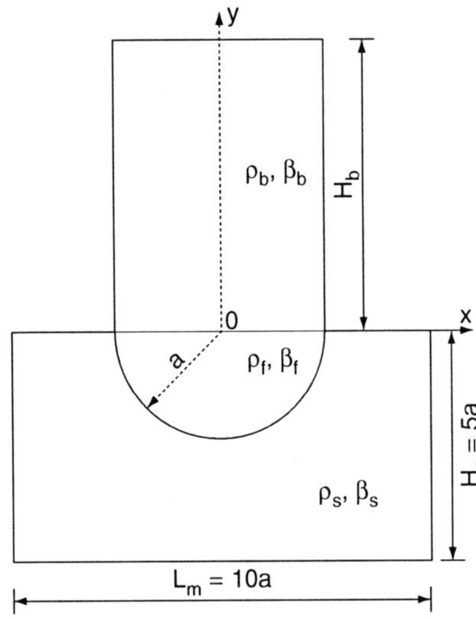

Figure 1. Soil-flexible foundation-structure system.

The maximum response occurs for $k = 0$ (rigid-body motion). As k increases, the response decreases and goes asymptotically to zero as k approaches infinity. We selected the largest wave number, $k = k_{max}$, for which the k-response is at least 0.03 of the maximum response (dashed lines in Fig. 2a). Then, for this value of k_{max}, the corresponding frequencies and the corresponding wavelengths are computed:

$$\lambda_{min} = \frac{2\pi}{k_{max}} = \frac{2\pi\beta}{\omega_{max}}. \tag{3}$$

It can be seen from Fig. 2a that $\omega_{max} \approx 245$ rad/s for $\eta = 0.5$, while $\omega_{max} \approx 980$ rad/s for $\eta = 2$.

A measure of the numerical accuracy of the grid is related to the ratio between the numerical and physical velocity of propagation, $r = c/\beta$, which ideally should be 1. The parameters that influence this accuracy are:

- the density of the grid $m = \lambda/\Delta x$ (m is the number of points per wavelength λ and Δx is the spacing between the grid points);
- the Courant number, $\chi = \beta_s \Delta t/\Delta x$;
- the angle of the wave incidence, θ.

It has been shown (Alford et al., 1974; Dablain, 1986; Fah, 1992) that the error increases when m decreases, χ decreases, and θ is close to 0 or $\pi/2$. For second-order approximation, the above authors recommend $m = 12$.

To compare hysteretic energies and the nonlinear zones in the soil, the soil box should have the same dimensions for any dimensionless frequency of the

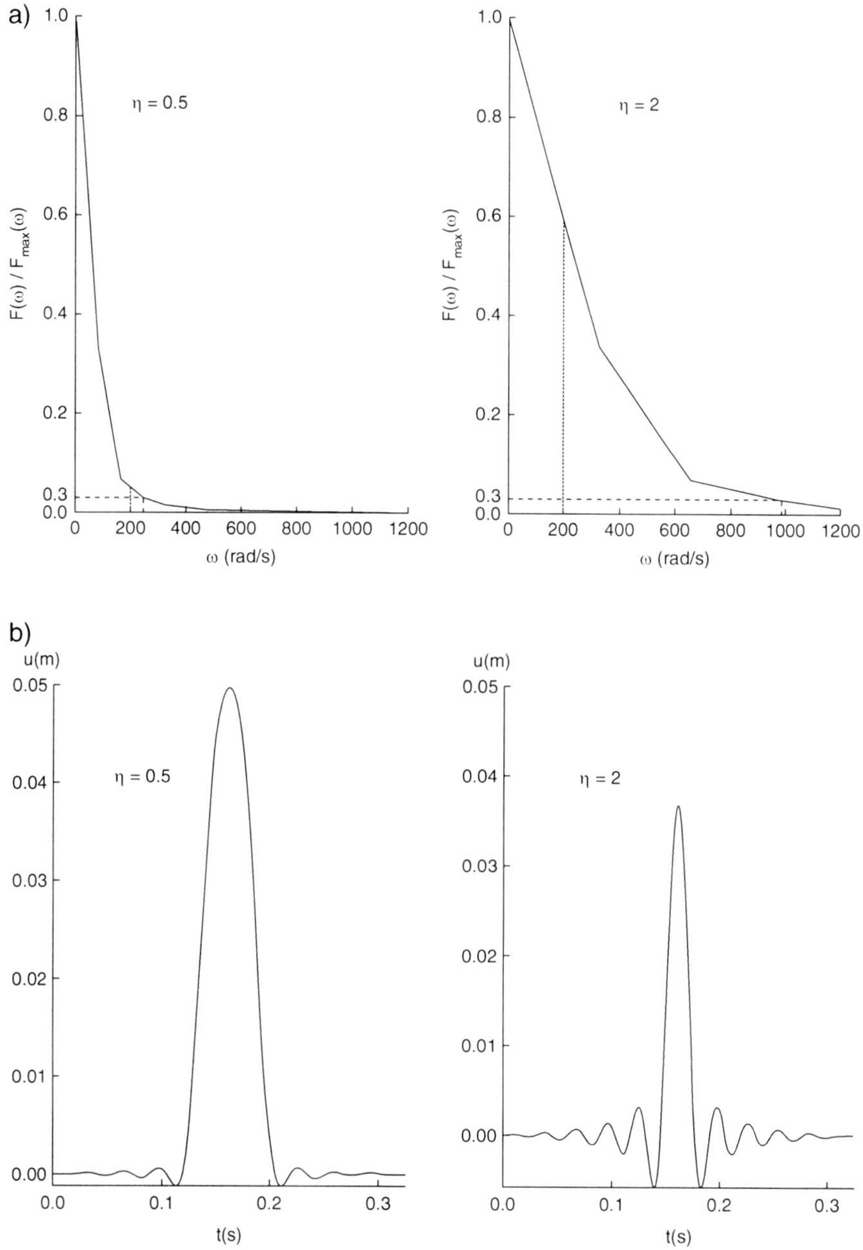

Figure 2. (a) Normalized one-side frequency response: η = 0.5 (left), η = 2 (right); (b) filtered pulse: η = 0.5 (left), η = 2 (right).

pulse, η. For that reason, we chose a rectangular soil box with dimensions $L_m = 10 \cdot a$ and $H_s = L_m/2 = 5 \cdot a$ (Fig. 1). Also, for merely practically reasons, the maximum number of space intervals in the grid in the horizontal (x) direction is set at 250 and in the vertical (y) direction at 400 (125 in the soil box and 275 in the building). The minimum spatial interval for this setup is $\Delta x_{\min} = L_m/250 = 95.5/250 = 0.382$ m. For a finer grid, the computational time increases rapidly. Having this limitation in mind, from Eq. (3) and for $\eta = 2$ ($\omega_{\max} = 980$ rad/s), the shortest wavelength is $\lambda_{\min} = 1.603$ m, and the finest grid density for this wavelength is $m = \lambda_{\min}/\Delta x_{\min} = 1.603/0.382 \approx 4$ points/$\lambda_{\min} < m_{\min}$.

Our numerical scheme is $O\left(\Delta t^2, \Delta x^2\right)$, so from the above recommendations we should have at least $m = 12$ points/$\lambda_{\min}$ to resolve for the shortest wavelength, $\lambda_{\min}$. This implies that the pulse should be low-pass filtered. A cut-off frequency $\omega_c = 200$ rad/s was chosen, and the pulse was low-pass filtered (Fig. 2b). This implies that $\lambda_{\min} = 7.854$ m and then the grid density is

$$m = \frac{\lambda_{\min}}{\Delta x_{\min}} = \frac{7.854}{0.382} \approx 20 \text{ points}/\lambda_{\min} > m_{\min}. \tag{4}$$

It can be seen in Fig. 2a (dotted lines) that for $\eta = 0.5$ only a negligible amount of the total power is filtered out, while for $\eta = 2$ a considerable amount is filtered out. Also, it can be seen in Fig. 2b that for $\eta = 2$ the amplitude of the filtered pulse is smaller than the amplitude of the non-filtered pulse, which is $A = 0.05$ m, while for $\eta = 0.5$ the amplitude is almost equal with the amplitude of the non-filtered pulse. From numerical tests, it has been shown that the viscous absorbing boundary rotated toward the centre of the foundation reflects only a negligible amount of energy back into the model (Gicev, 2005).

For 2-D problems, the numerical scheme is stable if the time increment (Mitchell, 1969) is:

$$\Delta t \leq \min\left(\frac{1}{\beta\sqrt{\frac{1}{\Delta x^2} + \frac{1}{\Delta y^2}}}\right). \tag{5}$$

Further, we assume that the shear stress in the x direction depends only upon the shear strain in the same direction and is independent of the shear strain in the y direction (and vice versa for shear stress in the y direction). The motivation for this assumption comes from our simplified representation of layered soil, which is created by deposition (floods and wind) into more or less horizontal layers. The soil is assumed to be ideally elastoplastic, and the constitutive $\sigma - \varepsilon$ diagram is shown in Fig. 3. Further, it is assumed that the contacts remain bonded during the analysis and the contact cells C, D, E, F, G, and H in Fig. 4 remain linear, as does the zone next to the artificial boundary (the bottom four rows and the left-most and right-most four columns).

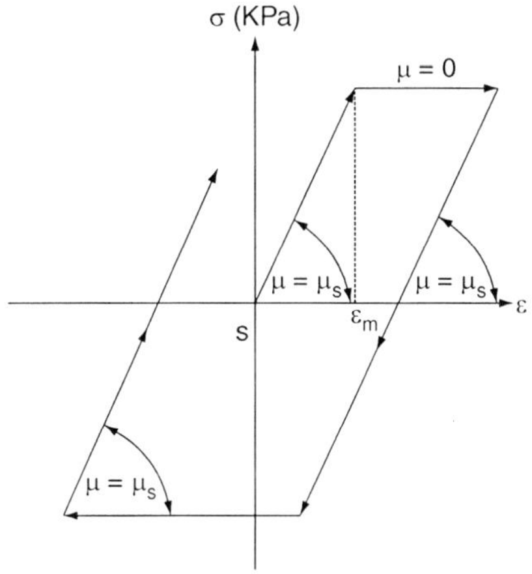

Figure 3. The constitutive law, $\sigma - \varepsilon$, for the soil.

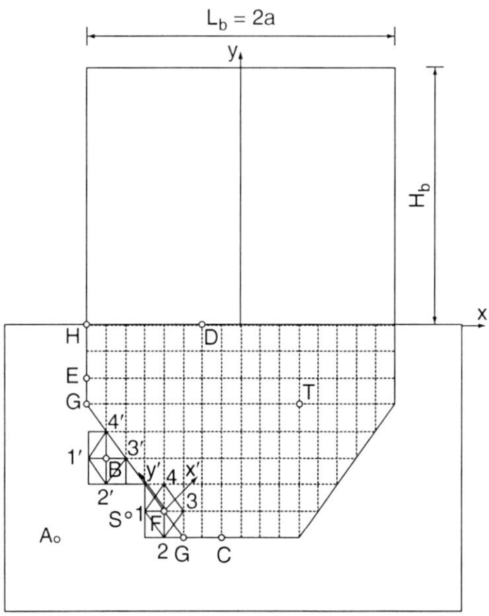

Figure 4. Numerical model with nonlinear soil. The points A, S, B, 1, 2, 1′, and 2′ can undergo permanent strains.

For our problem, the system of three partial differential equations (for u, v, and w) describing the dynamic equilibrium of an elastic body is reduced to the third equation only (because $u = v = \frac{\partial}{\partial z} = 0$). Neglecting the body forces in the z direction ($F_z = 0$), this equation is:

$$\rho \frac{\partial^2 w}{\partial t^2} = \left(\frac{\partial \tau_{xz}}{\partial x} + \frac{\partial \tau_{yz}}{\partial y} \right). \tag{6}$$

Introducing the new variables $v = \partial w/\partial t$, $\varepsilon_{xz} = \partial w/\partial x$, and $\varepsilon_{yz} = \partial w/\partial y$, and dividing (5) by ρ, the order of (6) is reduced to the system of three first-order partial differential equations (PDE)

$$\underline{U}_{,t} = \underline{F}_{,x} + \underline{G}_{,y}, \tag{7}$$

where

$$\underline{U} = \begin{Bmatrix} v \\ \varepsilon_{xz} \\ \varepsilon_{yz} \end{Bmatrix}, \quad \underline{F} = \underline{F}(\underline{U}) = \begin{Bmatrix} \frac{1}{\rho}\tau_{xz} \\ v \\ 0 \end{Bmatrix}, \quad \underline{G} = \underline{G}(\underline{U}) = \begin{Bmatrix} \frac{1}{\rho}\tau_{yz} \\ 0 \\ v \end{Bmatrix}. \tag{8}$$

The first equation in (7) represents the dynamic equilibrium of forces in the z direction with neglected body force F_z, while the second and third equations give the relations between the strains and the velocity. The abbreviations $\varepsilon_x = \varepsilon_{xz}$, $\sigma_x = \tau_{xz}$, $\varepsilon_y = \varepsilon_{yz}$, and $\sigma_y = \tau_{yz}$ are used later in the text. The Lax–Wendroff computational scheme (Lax and Wendroff, 1964) is used for solving Eq. (7) (Gicev, 2005).

3. Energy and Permanent Strain Distribution

As a test example, the properties of the Holiday Inn hotel in Van Nuys, California in the east–west direction are considered (Blume and Assoc., 1973). A question arises about how to choose the yielding strain ε_m (Fig. 3) to study permanent strain distribution. The displacement, the velocity, and the linear strain in the soil ($\beta_s = 250$ m/s) during the passage of a plane wave in the form of a half-sine pulse are:

$$w = A \sin\left[\frac{\pi}{t_{d0}}\left(t - \frac{s}{\beta_s}\right)\right], \tag{9}$$

$$v = \dot{w} = \frac{\pi}{t_{d0}} A \cos\frac{\pi t}{t_{d0}}, \tag{10}$$

$$|\varepsilon| = \frac{v_{\max}}{\beta_s} = \frac{\pi A}{\beta_s t_{d0}}. \tag{11}$$

If, for a given input plane wave, we choose the yielding strain ε_m given by (11) multiplied by some constant between 1 and 2, the strains in both directions will remain linear before the wave reaches the free surface or the foundation. This case can be called “intermediate nonlinearity”. If we want to analyze only the nonlinearity due to scattering and radiating from the foundation, we should avoid the occurrence of the nonlinear strains caused by reflection from the half-space boundary. Then we may choose $\varepsilon_m = \max(2\pi A \sin\gamma/\beta_s t_{d0}; 2\pi A \cos\gamma/\beta_s t_{d0})$. We call this case "small nonlinearity".

If the soil is allowed to undergo permanent strains only due to wave passage of incident waves in the full space, then we may choose the maximum strain $\varepsilon_m < \max(\pi A \sin\gamma/\beta_s t_{d0}; \pi A \cos\gamma/\beta_s t_{d0})$. This condition guarantees that in either the x or y direction the soil will undergo permanent strains during the passage of the plane wave. Generally, the yielding strain can be written as

$$\varepsilon_m = C\frac{v_{\max}}{\beta_s} = C\frac{\pi A}{\beta_s t_{d0}}, \tag{12}$$

where C is a constant that controls the yielding stress (strain) in the soil. We then consider the following cases of nonlinearity, depending upon C:

1. $C \geq 2$: **Small nonlinearity**. Permanent strain does not occur until the wave hits the foundation with any angle of incidence.
2. $1 \leq C < 2$: **Intermediate nonlinearity**. Permanent strain does not occur until the wave is reflected from the free surface or is scattered from the foundation, for any angle of incidence. Permanent strain will or will not occur after the reflection of the incident wave from the free surface, depending upon the angle of incidence.
3. $C < 1$: **Large nonlinearity**. Permanent strain occurs after reflection from the free surface. Permanent strain may or may not occur before the wave reflects from the foundation surface, depending upon the angle of incidence.

4. Energy Distribution in the System

The energy flow through a given area can be defined, in terms of a plane-wave approximation (Aki and Richards, 1980), as:

$$E_{in}^{a} = \rho_s \cdot \beta_s \cdot A_{sn} \int_0^{t_{d0}} v^2 \cdot dt, \tag{13}$$

where ρ_s and β_s are density and shear-wave velocity in the soil and v is a particle velocity, which, for the excitation considered in this paper, is given by Eq. (10). A_{sn} is the normal area through which the wave is passing. For our geometrical settings of the soil (Fig. 1), the area normal to the wave passage is:

$$A_{sn} = 2 \cdot H_s \cdot \sin\gamma + L_m \cdot \cos\gamma = L_m \cdot (\sin\gamma + \cos\gamma). \tag{14}$$

Inserting Eqs. (10) and (14) into (13) and integrating, the analytical solution for the input wave energy into the model is

$$E_{\text{in}}^{a} = \rho_s \cdot \beta_s \cdot L_m \cdot (\sin\gamma + \cos\gamma) \cdot \left(\frac{\pi \cdot A}{t_{d0}}\right)^2 \cdot \frac{t_{d0}}{2}. \tag{15}$$

As can be seen from Eq. (15), for the defined size of the soil island, L_m, and the defined angle of incidence, γ, the input energy is reciprocal with the duration of the pulse and is a linear function of the dimensionless frequency η (Eq. (1a)). Because the short pulses are low-pass filtered up to $\omega_c = 200$ rad/s (Fig. 2b), the analytical and the numerical solutions (13) for input wave energy do not coincide (Fig. 5). Since our system is conservative, the input energy is balanced by:

- Cumulative energy going out from the model, E_{out}, computed using Eq. (13); cumulative hysteretic energy (energy spent for creation and development of permanent strains in the soil), computed from:

$$E_{\text{hys}} = \sum_{t=0}^{T_{\text{end}}} \Delta t \cdot \sum_{i=1}^{N} \left(\sigma_{xi}(\Delta\varepsilon_{xpi} + 0.5 \cdot \Delta\varepsilon_{xei}) + \sigma_{yi}(\Delta\varepsilon_{ypi} + 0.5 \cdot \Delta\varepsilon_{yei})\right), \tag{16}$$

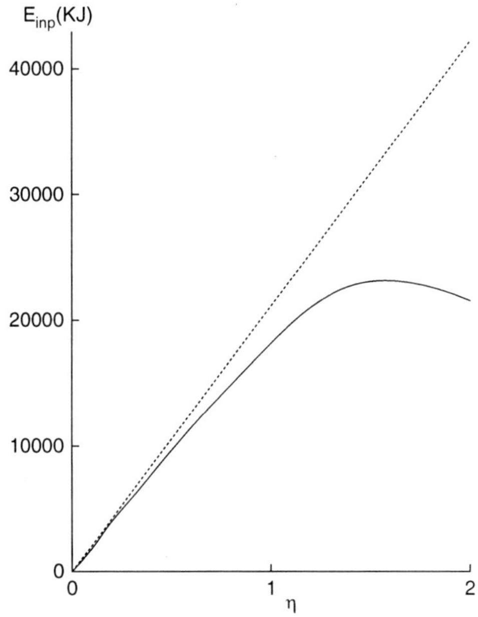

Figure 5. Input energy in the model: from analytic half-sine pulse (dashed line); from low-pass filtered half-sine pulse (solid line).

where N is the total number of soil points; σ_{xi}, σ_{yi} are the stresses at the point i in the x and y directions, respectively; $\Delta\varepsilon_{xpi} = \varepsilon_{xpi}^{t+\Delta t} - \varepsilon_{xpi}^{t}$ is the increment of the permanent strain in the x direction at point i; and $\Delta\varepsilon_{ypi} = \varepsilon_{ypi}^{t+\Delta t} - \varepsilon_{ypi}^{t}$ is the increment of the permanent strain in the y direction at point i

- Instantaneous energy in the building, consisting of kinetic and potential energy, which can be computed from:

$$E_b = E_k + E_p = 0.5 \cdot \Delta x \cdot \Delta y_b \cdot \sum_{i=1}^{N} \left(\rho \cdot v_i^2 + \mu \cdot (\varepsilon_x^2 + \varepsilon_y^2)\right). \tag{17}$$

In Fig. 6, this balance is shown for a pulse with $\eta = 1.5$, for incident angle $\gamma = 30°$, and a yielding strain defined by $C = 1.5$ (Eq. (12)).

To study the effect of scattering from the foundation only, the building is considered to be high enough so that the reflected wave from the top of the building cannot reach the building-foundation contact during the analysis. The analysis is terminated when the wave completely exits the soil island. In this study, the hysteretic energy in the soil and the energy in the building are the subjects of interest. In Fig. 7, these two types of energy are presented as

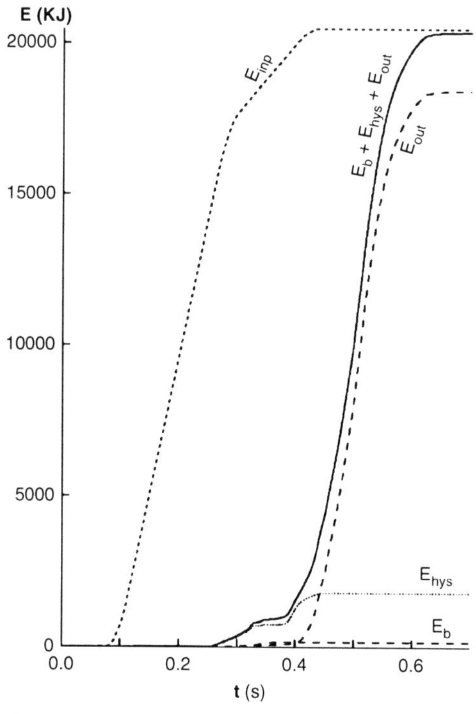

Figure 6. Energy balance in the model for $\gamma = 30°$ and $\eta = 1.5$.

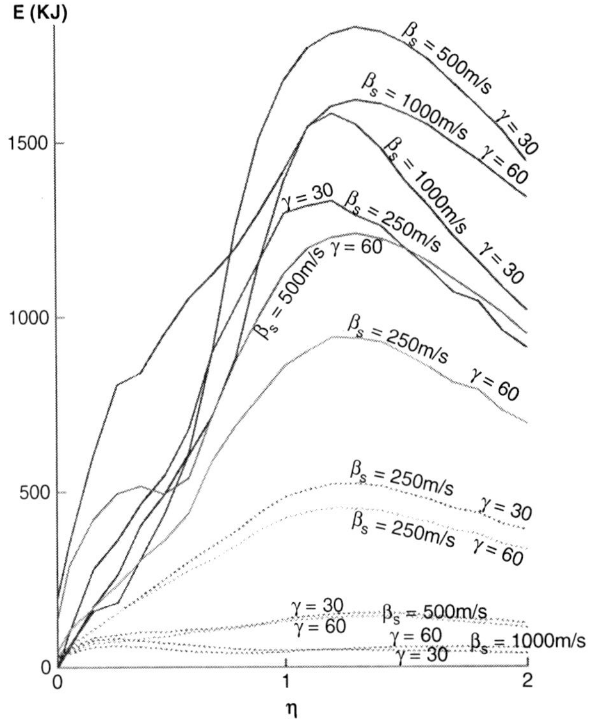

Figure 7. Hysteretic energy (solid lines) and energy entering the building (dashed lines) vs. dimensionless frequency for intermediate nonlinearity $C = 1.5$.

functions of the dimensionless frequency η. Considering the energy entering the building (dashed lines), the results confirm the expectations that as the foundation becomes stiffer, a larger part of the input energy is scattered and less energy enters the building. In contrast, the results for hysteretic energy in the soil are not so straightforward. For an angle of incidence $\gamma = 60°$, the results are as would be expected—e.g., the hysteretic energy increases as the foundation stiffness increases. However, this is not the case for the angle of incidence $\gamma = 30°$. It can be seen that at some frequency intervals the hysteretic energy can be larger for softer foundations. For example, for $\eta \leq 0.7$ and an angle of incidence $\gamma = 30°$, as the foundation becomes softer the hysteretic energy becomes larger (the softest foundation we considered had $\beta_f = 250$ m/s). Similarly, a foundation with medium stiffness, $\beta_f = 500$ m/s, for $\eta > 0.8$ and an angle of incidence $\gamma = 30°$, causes the largest hysteretic energy in the soil, where the maximum occurs at $\eta = 1.5$. This unexpected behaviour of the soil can be explained by the destructive interference that may occur with stiffer foundations, which decreases the released hysteretic energy in the soil.

5. Distribution of the Permanent Strain in the Soil

Considering Fig. 8, and starting from dynamic equilibrium of the differential pentahedron shown in the figure, we can find the principal stress at a point and its direction as:

$$\tau_{zp} = \tau_{zx} \cos\gamma + \tau_{zy} \sin\gamma \quad \text{and} \quad \gamma = \tan^{-1} \frac{\tau_{zy}}{\tau_{zx}}.$$

The principal permanent strain in the soil is illustrated in Fig. 9a, b, c, for the case of small nonlinearity ($C = \sqrt{3}$) for two angles of incidence, $\theta = 30°$ and $60°$, and for three foundation stiffnesses, $\beta_f = 2500$, 500, and 1000 m/s. This value of C guarantees that for angles of incidence $30° \leq \gamma \leq 60°$ there is no occurrence of permanent strain until the wave hits the foundation.

The principal permanent strain is illustrated in Fig. 10a, b, c, for the case of intermediate nonlinearity ($C = 1.5$) for the same angles of incidence, $\theta = 30°$ and $\theta = 60°$ and for three foundation stiffnesses, $\beta_f = 250$, 500, and 1000 m/s. In this case, permanent strain occurs before the wave hits the foundation but after it reflects from the free surface.

For long pulses $\eta = 0.1$, it can be seen from Fig. 9a that for an angle of incidence $\gamma = 30°$ there is a small, permanent strain for the stiffest foundation ($\beta_f = 1000$ m/s) only, while for softer foundations the soil remains linear after the pulse has left the model. For intermediate nonlinearity, shown in Fig. 10a, for an angle of incidence $\gamma = 30°$ it can be seen that after the creation of nonlinear zones the effect of the interaction is negligible compared with the effects of interference of the incoming wave and the reflected wave from the

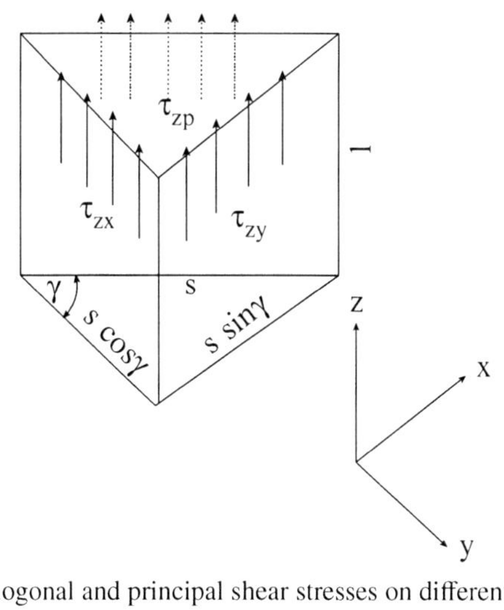

Figure 8. Orthogonal and principal shear stresses on differential pentahedron.

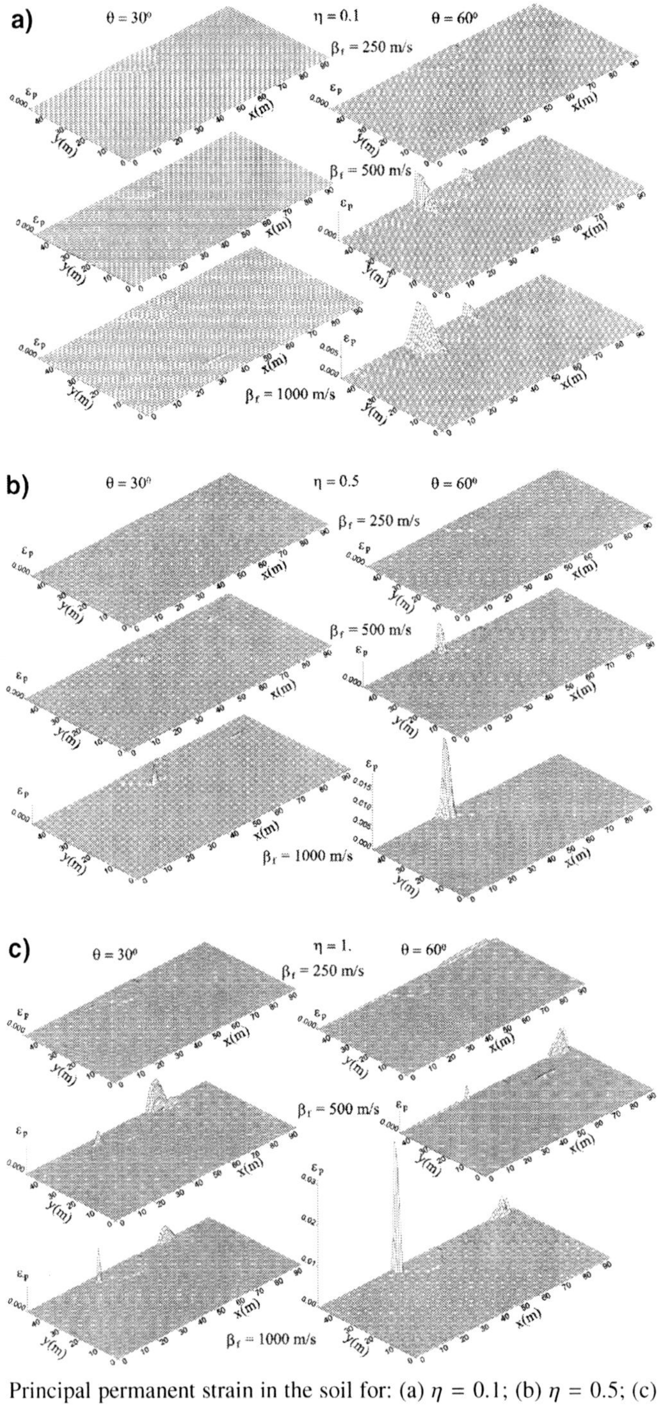

Figure 9. Principal permanent strain in the soil for: (a) $\eta = 0.1$; (b) $\eta = 0.5$; (c) $\eta = 1$, two angles of incidence, and three foundation stiffness. Small nonlinearity in the soil $C = 1.73$.

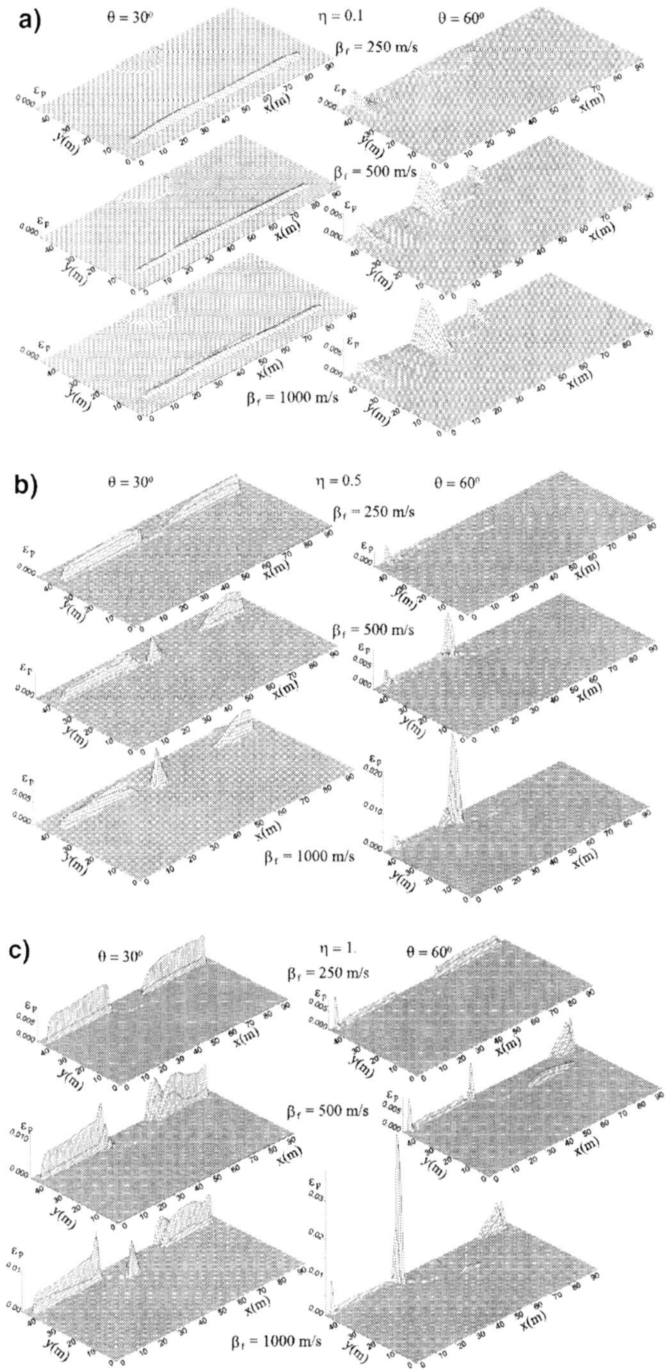

Figure 10. Principal permanent strain in the soil for: (a) $\eta = 0.1$; (b) $\eta = 0.5$; (c) $\eta = 1$, two angles of incidence, and three foundation stiffness. Small nonlinearity in the soil $C = 1.5$.

free surface. This is not the case for $\gamma = 60°$. It can be concluded from Figs. 9a and 10a that for stiffer foundations the effect of interaction is more dominant than the effect of the interference. For the softest considered foundation, the effect of the interaction on creation of nonlinear strains is small.

The observations are similar for a five-times-shorter pulse $\eta = 0.5$. It can be seen from Figs. 9b and 10b that for the softest foundation the effect of the interaction is negligible and that as the foundation becomes stiffer the nonlinear zones are created and developed in the soil next to the front of the foundation.

As the pulse becomes shorter, $\eta = 1$, it can be seen that nonlinear zones are also formed behind the foundation. This can be explained by the interference of waves reflected from the free surface and diffracted around the foundation. Again, the permanent strain in front of the foundation increases as the stiffness of the foundation increases.

6. Conclusions

Numerical methods are powerful tools for studying nonlinear soil-structure interaction problems. Because of grid dispersion, the selection of the grid spacing must be done carefully. Short waves cannot be reconstructed even with very fine grids, and the incident wave (pulse) should be low-pass filtered to utilize numerical methods effectively.

In the presence of a foundation and small angles of incidence (close to vertical incidence), the permanent strains in the y direction are dominant, while for large angles of incidence (close to horizontal incidence) the permanent strains in the x direction are dominant.

For long waves and small angles of incidence (Figs. 9a and 10a for $\theta = 30°$, the effect of the interaction on the nonlinear response in the soil is small. For soft foundations, $\beta_f = 250$ m/s, and small incident angles (the top left plots in Figs. 9a, b, c and 10a, b, c), the effect of the interaction on the nonlinear response of the soil is also small. As the foundation becomes stiffer, zones of large permanent strains develop around the foundation. For stiff foundations, short waves ($\eta = 1$ and $\eta = 2$, and large incidence angles, a zone of permanent strains develops behind the foundation, which appears to be due to the concentration of rays associated with diffraction of the waves from the foundation. The zones of large permanent strains illustrated in Figs. 9a, b, c and 10a, b, c are responsible for the damage and failures in the shallow infrastructure (water and gas pipes, underground cables, etc.) that accompany large earthquakes and cause interruptions of gas and water supplies (Trifunac and Todorovska, 1997, 1998a, 1998b).

As the large and permanent strains develop along the foundation-soil interface, the effective foundation compliances are reduced, which decreases the equivalent rocking stiffness of the foundation-structure system. With simultaneous action of in-plane wave motions, which are always present in 3-D settings during earthquake excitation and which will excite the in-plane rocking of the model we studied in this paper, it is easy to see how the nonlinear zones in the soil (as illustrated in Figs. 9 and 10) will take the structure one step closer to overturning or even eventual collapse, as in the examples mentioned in the introduction.

References

Abdel-Ghaffar, A. M. and Trifunac, M. D. (1977) Antiplane dynamic soil–bridge interaction for incident plane SH-waves. In *Proc. 6th World Conf. Earthquake Eng.*, Vol. 2, New Delhi, India.

Aki, K. and Richards, P. (1980) *Quantitative Seismology, Theory and Methods*, San Francisco, W. H. Freeman & Co.

Alford, R. M., Kelly, K. R., and Boore, D. M. (1974) Accuracy of finite-difference modeling of the acoustic wave equation, *Geophysics* **39**, 834–842.

Alterman, Z. and Karal, F. C. (1968) Propagation of elastic waves in layered media by finite difference methods, *Bull. Seism. Soc. of Am.* **58**(1), 367–398.

Aviles, J., Suarez, M., and Sanchez-Sesma, F. J. (2002). Effects of wave passage on the relevant dynamic properties of structures with flexible foundation, *Earthq. Eng. Struct. Dynam.* **31**, 139–159.

Blume J. A. and Assoc. (1973) Holiday Inn. In L. M. Murphy (ed.), *San Fernando, California, Earthquake of February 9, 1971*, U.S. Dept. of Commerce, National Oceanic and Atmospheric Administration, Washington, D.C., pp. 359–394.

Boore, D. M. (1972) Finite difference methods for seismic wave propagation in heterogeneous materials. In B. A. Bolt (ed.), *Seismology, Surface Waves and Earth Oscillations*, Vol. 11 of *Methods in Comp. Physics*, New York, pp. 1–37, Academic Press.

Dablain, M. A. (1986) The application of high-order differencing to the scalar wave equation, *Geophysics* **51**(1), 54–66.

Fah, D. J. (1992) A hybrid technique for the estimation of strong ground motion in sedimentary basins, Ph.D. Dissertation, Swiss Federal Institute of Technology, Zurich, Switzerland.

Gicev, V. (2005) Investigation of soil-flexible foundation–structure interaction for incident plane SH waves, Ph.D. Dissertation, Dept. of Civil Engineering, Univ. Southern California, Los Angeles, California.

Hayir, A., Todorovska, M. I., and Trifunac, M. D. (2001) Antiplane response of a dike with flexible soil-structure interface to incident SH waves, *Soil Dynam. Earthq. Eng.* **21**, 603–613.

Lax, P. D. and Wendroff, B. (1964) Difference schemes for hyperbolic equations with high order of accuracy, *Comm. Pure Appl. Math.* **XVII**, 381–398.

Lee, V. W. (1979) Investigation of three-dimensional soil-structure interaction, Report No. CE 79-11, Dept. of Civil Engineering, Univ. of Southern California, Los Angeles, CA.

Luco, J. E. and Wong, H. L. (1977) Dynamic response of rectangular foundations for Rayleigh wave excitation. In *Proc. 6th World Conf. Earthq. Eng.*, Vol. 2, New Delhi, India.

Mitchell, A. R. (1969) *Computational Methods in Partial Differential Equations*, New York, Willey.

Moczo, P. (1989) Finite-difference technique for SH-waves in 2-D media using irregular grids—application to the seismic response problem, *Geophys. J. Int.* **99**, 321–329.

Seed, H. B. (1970) Soil problems and soil behavior. In R. L. Wiegel (ed.), *Earthquake Engineering*, Englewood Cliffs, N.J., Prentice Hall, pp. 227–251.

Todorovska, M. I., Hayir, A., and Trifunac, M. D. (2001) Antiplane response of a dike on flexible embedded foundation to incident SH-waves, *Soil Dynam. Earthq. Eng.* **21**, 593–601.

Trifunac, M. D. (1972) Interaction of a shear wall with the soil for incident plane SH waves, *Bull. Seism. Soc. Am.* **62**(1), 63–83.

Trifunac, M. D. (1995) Empirical criteria for liquefaction in sands via standard penetration tests and seismic wave energy, *Soil Dynam. Earthq. Eng.* **14**(6), 419–426.

Trifunac, M. D. and Todorovska, M. I. (1997) Northridge, California, earthquake of 17 January 1994: Density of pipe breaks and surface strains, *Soil Dynam. Earthq. Eng.* **16**(3), 193–207.

Trifunac, M. D. and Todorovska, M. I. (1998a) Nonlinear soil response as a natural passive isolation mechanism—the 1994 Northridge, California earthquake, *Soil Dynam. Earthq. Eng.* **17**(1), 41–51.

Trifunac, M. D. and Todorovska, M. I. (1998b) The Northridge, California, earthquake of 1994: Fire ignition by strong shaking, *Soil Dynam. Earthq. Eng.* **17**(3), 165–175.

Tsynkov, S. V. (1998) Numerical solution of problems on unbounded domains: A review, *Appl. Numerical Math.* **27**, 465–532.

Westermo, B. D. and Wong, H. L. (1977) On the fundamental differences of three basic soil-structure interaction models, *Proc. 6th World Conf. of Earth. Eng.*, Vol. 2, New Delhi, India.

Wong, H. L. and Trifunac, M. D. (1974) Interaction of a shear wall with the soil for incident plane SH waves: Elliptical rigid foundation, *Bull. Seism. Soc. Am.* **64**(6), 1825–1884.

Wong, H. L. and Trifunac, M. D. (1975) Two-dimensional antiplane, building-soil–building interaction for two or more buildings and for incident plane SH waves, *Bull. Seism. Soc. Am.* **65**(6), 1863–1885.

Zahradnik, J., Moczo, P., and Hron, F. (1993) Testing four elastic finite-difference schemes for behaviour at discontinuities, *Bull. Seism. Soc. Am.* **83**, 107–129.

SEPARATION OF THE EFFECTS OF SOIL-STRUCTURE INTERACTION IN FREQUENCY ESTIMATION OF BUILDINGS FROM EARTHQUAKE RECORDS

Maria I. Todorovska (mtodorov@usc.edu)*
Department of Civil Engineering, University Southern California, Los Angeles, California 9089-2531, U.S.A.

Abstract. Fourier type of analyses give the frequencies of vibration of the *soil-structure system*, which depend on the properties of the soil, structure and foundation. This paper shows how the *fixed-base* frequency of a building, which depends only on its properties, and *rigid-body rocking* frequency can be estimated data from only two sensors recording horizontal motion — at ground level and at the roof. The method is based on Fourier analysis, wave travel time analysis and a relationship between fixed-base, rigid-body and system frequencies. Results are shown and discussed from an application to the NS response of Millikan Library in Pasadena, California, during four earthquakes between 1970 and 2003.

Keywords: soil-structure interaction, structural identification, fixed-base frequency of vibration, impulse response analysis, rigid body frequency of vibration, Millikan library, structural health monitoring

1. Introduction

Long term seismic monitoring of structures has demonstrated that their resonant frequencies of vibration (as determined by Fourier analysis, which are those of the soil-structure system, and depend on the properties of the structure, soil and foundation) can vary significantly from one earthquake to another and with time. Udwadia and Trifunac (1974) showed that these frequencies drop during strong shaking, but recover partially or totally. Recoverable changes, not related to damage, appear to reach and exceed 20% (Trifunac et al., 2001a,b, 2008; Todorovska et al., 2006). Unfortunately, for the majority of significant recordings, it has been difficult to tell to what degree the observed changes have been due to change of the properties of the soil and foundation as opposed to the structure, because of inadequate instrumentation to separate the effects of the soil-structure interaction. For the same reason, it has been also difficult to estimate, directly from earthquake

*http://www.usc.edu/dept/civil_eng/Earthquake_eng/

T. Schanz and R. Iankov (eds.), *Coupled Site and Soil-Structure Interaction Effects with Application to Seismic Risk Mitigation,* NATO Science for Peace and Security Series C: Environmental Security,

observations, the fixed-base frequencies of the structure. Distinguishing between fixed-base and system frequency is important because the observed frequencies are used to calibrate the stiffness of structural models (leading to underestimation of the structural stiffness), and to infer the change in the health of a structure.

This paper presents a method that can determining both the building fixed-base frequency f_1 and rigid-body rocking frequency f_R using earthquake response data from only two horizontal sensors, one at the roof and the other one at the base. Being able to do this using data from such minimum instrumentation is significant because it will extend the usability of structural response data from past earthquakes. It also summarizes an application to the NS earthquake response of Millikan library (Fig. 1) during four earthquakes (1970–2003), which revealed to what degree the observed "wander" of this building NS system frequency from one earthquake to another and with time

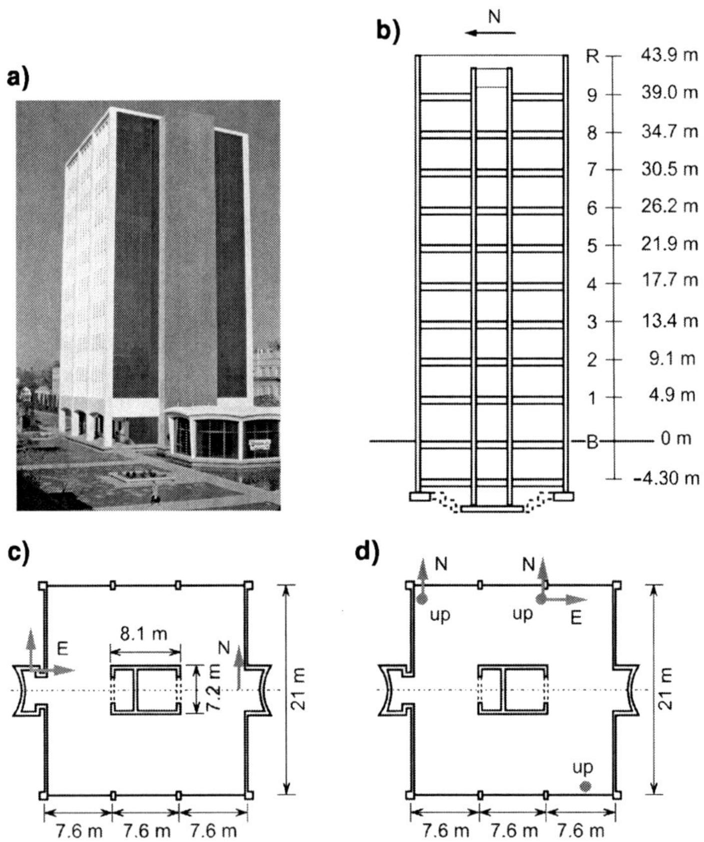

Figure 1. Millikan library: (a) photo (taken by M. Trifunac); (b) vertical cross-section; (c) typical floor layout (redrawn from Snieder and Şafak, 2006); (d) sensor locations at basement.

has been due to changes in the building vs. changes in the soil (Clinton et al., 2006). More detail about these studies can be found in Todorovska (2009a,b).

2. Methodology

The method based on Fourier and wave travel time analyses of the recorded building response, and on a relationship between *fixed-base frequency* f_1, *apparent frequency* $f_{1,\text{app}}$ and *rigid-body rocking frequency* f_R (Todorovska, 2009a). The steps are the following.

1. *Fourier analysis*: $f_{1,\text{app}}$ is determined from the transfer-function (TF) between the roof and base accelerations (Udwadia and Trifunac, 1974).
2. *Wave travel-time analysis*: wave travel time τ from the base to the top of the building is measured using impulse response (IR) functions. Such functions, e.g., obtained by inverse Fourier transform of the TF in step 1 (with some regularization), emulate propagation vertically of an input impulse at the base. Then τ can be measured from the arrival of that pulse at the roof (Snieder and Şafak, 2006).
3. *Fixed-base frequency*: f_1 is computed from wave travel time as $f_1 = 1/(4\tau)$ based on the assumption that the building as a whole deforms in shear (Todorovska and Trifunac, 2008a,b). This differs from the Snieder and Şafak (2006).
4. *Rigid-body rocking frequency*: f_R is computed from $f_{1,\text{app}}$ and f_1 using the relationship (for structures on rigid foundations).

$$\frac{1}{f_{1,\text{app}}^2} \approx \frac{1}{f_R^2} + \frac{1}{f_1^2} \tag{1}$$

which is a special case of (Luco et al., 1987)

$$\frac{1}{f_{1,\text{sys}}^2} \approx \frac{1}{f_H^2} + \frac{1}{f_R^2} + \frac{1}{f_1^2}, \tag{2}$$

where f_H is the *horizontal* rigid-body frequency, and $f_{1,\text{sys}}$ is the soil-structure system frequency (special case when $f_H \to \infty$ and $f_{1,\text{sys}} \to f_{1,\text{app}}$).

Todorovska (2009a) demonstrated the method on numerically simulated data using a 2D soil-structure interaction model, with a shear beam representing the structure and with coupled horizontal and rocking motion (as result of soil-structure interaction), and showed that, despite the contribution from rigid body rocking to the roof total response, the measured wave travel time reflected only the properties of the structure.

3. Results and Analysis

Millikan library (Fig. 1) is a 9-story reinforced concrete building in Pasadena, California, instrumented over a period of 40 years, and tested extensively, in particular for soil-structure interaction studies (Udwadia and Trifunac, 1974; Foutch et al., 1975; Luco et al., 1886, 1987, 1988; Wong et al., 1988). Structural deformation patterns obtained from detailed forced vibration tests showed that, for its NS response, rigid foundation model may be appropriate, and that as much as 30% of its roof response can be accounted for by rigid body rocking, suggesting significant soil-structure interaction effects (Foutch et al., 1975; Luco et al., 1987). A soil-structure interaction system identification, using forced vibration test data, was also carried out (Luco et al., 1986, 1987, 1988; Wong et al., 1988). The San Fernando, 1971, earthquake produced a significant drop of $f_{1,\mathrm{app}}$ (Udwadia and Trifunac, 1974), which did not recover completely to its pre-earthquake value. The cause of the permanent change has been attributed to degradation of the structural stiffness based on ambient and forced vibration tests before and after this earthquake (Luco et al., 1987). Since the 1971 earthquake, smaller drops of its system frequencies have been observed over time in data from many forced vibration tests, documented most recently by Clinton et al. (2006). They also documented the variability of the resonant (i.e. system) frequencies of the building, for ambient noise excitation, due to environmental effects (strong winds, heavy rainfall, and temperature), and changes in mass. An explanation of such changes after heavy rainfall in terms of soil-structure interaction and poroelasticity was proposed by Todorovska and Al Rjoub (2006, 2008). Finally, wave propagation through the building has also been studied for small earthquake excitation (Loma Linda 2002, earthquake), and wave travel times through the structure was measured by Snieder and Şafak (2006).

The building is 21 × 23 m in plan, and vertically extends 43.9 m above grade and 48.2 m above basement level (Fig. 1). Resistance to lateral forces in the NS direction is provided by RC shear walls on the east and west sides of the building. The RC central core houses the elevators and provides resistance to lateral forces in the EW direction. The local soil can be characterized as alluvium, with average shear wave velocity in the top 30 m of about 300 m/s, and depth to "bedrock" of about 275 m. The alluvium consists of medium to dense sands mixed with gravels, and the water table appears to be at about 11 m depth (Luco et al., 1986).

The first earthquake recorded in the building was the Lytle Creek of 1970, which produced small amplitude response. It was followed by the San Fernando of 1971, which caused a significant drop of its resonant frequencies (Udwadia and Trifunac, 1974). Many other earthquakes were recorded in the building over the past 40 years, most notably the Whittier-Narrows earthquake of 1987 (M = 5.9, R = 19 km), the Northridge earthquake of 1994 (M_L = 6.4, R = 34 km) and their aftershocks.

At the time of this study, only data from five of the earthquakes were available for this study both at basement and roof level. These earthquakes are listed in Table I. The last two columns show roof peak acceleration, and rocking angle $\theta(t)$ (sum of rigid body rocking and rocking due to deformation of the structure). The peak drift during the San Fernando earthquake was 0.052% and during Whittier-Narrows earthquake it was 0.187%, which approaches but is less than the drift considered to cause damage of moment resistant frames (0.2%; Ghobarah, 2004). Figure 2 shows plot of $\theta(t)$ for these earthquakes.

TABLE I. Earthquakes analyzed in this study

No.	Name	Data Source	Date	M	R km	PGA cm/s^2	θ_{max} 10^{-3} rad
1	Lytle Creek	Caltech	12 Sep 1970	5.3	57	54	0.065
2	San Fernando	Caltech	9 Feb. 1971	6.6	31	301	0.516
3	Whittier-Narrows	CDMG	1 Oct. 1987	6.1	19	543	1.866
4	Yorba Linda	USGS	3 Sep. 2002	4.8	40	.8	0.012
5	San Simeon	USGS	22 Dec. 2003	6.4	323	14.2	0.035

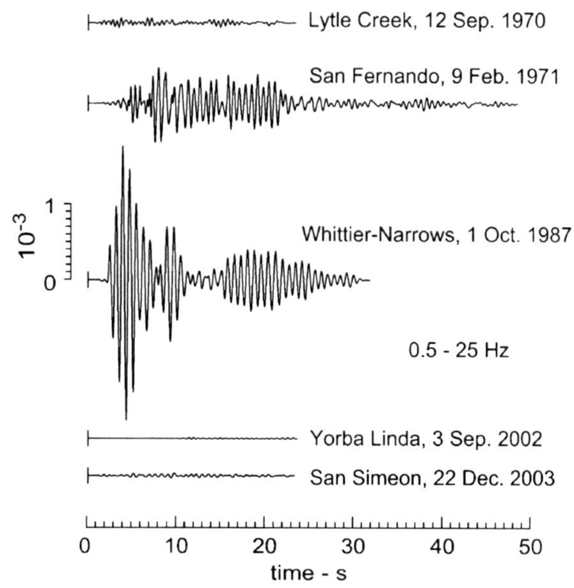

Figure 2. Roof NS rocking angle in miliradians.

For the San Fernando earthquake, the analysis was carried out for six segments (SF1: 0–3.5 s; SF2: 3–6.5 s; SF3: 6.5–10 s; SF4: 10–15 s; SF5: 15–22 s; and SF6: 22–50 s), and for the Whittier-Narrows earthquake for three segments (WN1: 2.2–7.2 s; WN2: 7.2–15 s and WN3: 15–30 s).

The results are presented graphically in Fig. 3a, b, c, d. Parts a, b and c show f_1, f_R and $f_{1,\mathrm{app}}$ versus $\theta_{\max}$, while part d shows a correlation plot of the percentage changes in f_R and f_1 (relative to their values during the Lytle Creek earthquake). Each data point is represented by the corresponding symbol for the earthquake/segment. For this set of earthquakes, 0.07 mrad $< \theta_{\max} <$ 1.87 mrad, 2.12 Hz $< f_1 <$ 3.05 Hz (30% variation), 1.44 Hz $< f_R <$ 2.47 Hz (42% variation), and 1.19 Hz $< f_{1,\mathrm{app}} <$ 1.92 Hz (38% variation).

Figure 3a shows that f_1 dropped by 24% during the first 10 s of shaking by the San Fernando earthquake (segments SF2 and SF3), and increased slightly during the subsequent smaller amplitude response. It further dropped

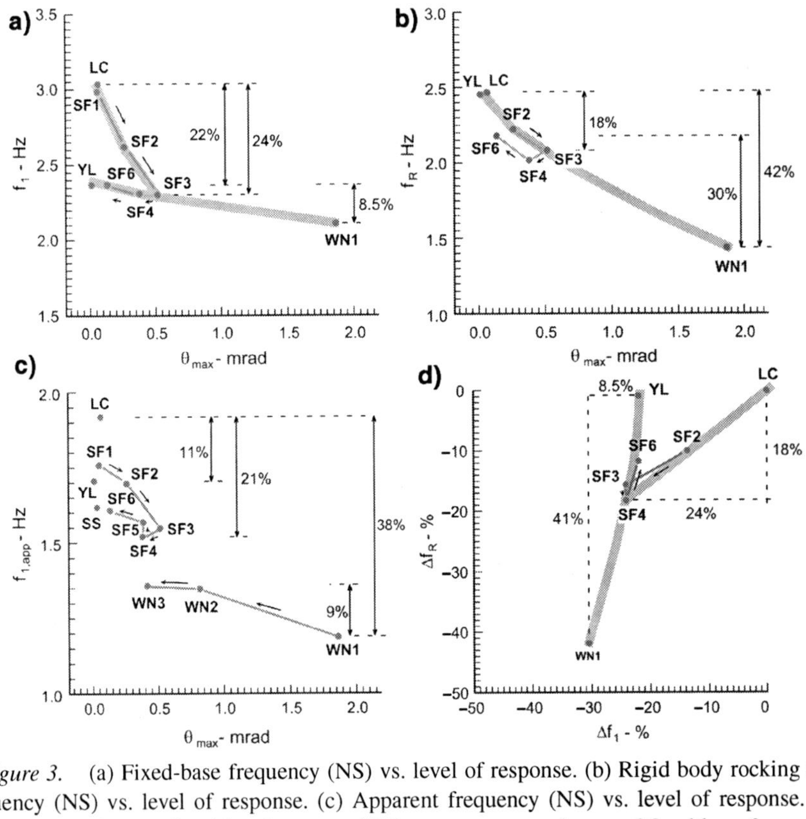

Figure 3. (a) Fixed-base frequency (NS) vs. level of response. (b) Rigid body rocking frequency (NS) vs. level of response. (c) Apparent frequency (NS) vs. level of response. (d) Percentage change of rocking frequency (NS) vs. percentage change of fixed-base frequency (NS) for four earthquakes.

during the first 7 s of the Whittier-Narrows earthquake, but much less (by 8.5%) considering the large increase in amplitudes of response (θ_{max} reaching 1.87 mrad), and recovered with decreasing response during the small amplitude Yorba Linda earthquake following practically the same path. The two trends of variation of f_1 vs. θ_{max} are shown by thick fuzzy lines, drawn by hand (the first through points LC, SF1, SF2 and SF3, and the second through points SF4, SF6, WN1 and YL), which indicate permanent change in f_1 due to structural degradation caused by the San Fernando earthquake, and amplitude dependent and mostly recoverable (within the accuracy of the estimates) change of f_1 for the cracked structure during the subsequent earthquakes. Therefore, no *significant* additional degradation of stiffness occurred during the Whittier-Narrows earthquake.

Figure 3b shows that f_R decreased by 18% with increasing θ_{max} during the first 10 s of shaking by the San Fernando earthquake (segments SF2 and SF3), continued to decrease during segment SF4 although θ_{max} decreased, and recovered partially during segments SF6, as θ_{max} continued to decrease. During the Whittier-Narrows earthquake, it dropped markedly, by 30%, and recovered during the Yorba Linda earthquake approximately to its value during the Lytle Creek earthquake. This pattern suggests that, in the long term, f_R recovered completely, but the recovery was not instantaneous (as for f_1, see Fig. 3a) following the strong shaking during the initial 10 s of the response to the San Fernando earthquake. The total change of f_R was 42%.

Figure 3c shows that $f_{1,app}$ dropped during the San Fernando earthquake and recovered partially towards the end of shaking. Then it dropped further during the initial 7 s of the response to the Whittier-Narrows earthquake, recovered partially towards the end of the shaking, and further recovered during the smaller Yorba Linda and San Simeon earthquakes. The total change of $f_{1,app}$ was 38%.

The data from the Lytle Creek and Yorba Linda earthquakes, which caused similar amplitude responses, give an opportunity to examine permanent changes in f_1, f_R and $f_{1,app}$ over the period 1970 to 2002. The data shows no change of f_R, change in f_1 of −22%, and change in $f_{1,app}$ of −11%. The much smaller change in $f_{1,app}$, which represents the combined effect of change in f_1 and f_R, compared to the change in f_1 alone, is due to the nature of their combination rule (see Eq. (1)). These changes in f_1 and $f_{1,app}$ correspond to about −39% change in the overall structural stiffness and abut −21% change in the equivalent stiffness of the structure and the "rocking soil spring". Hence, observing change in $f_{1,app}$ instead of f_1 will *underestimate* the changes in structural stiffness.

An interesting question is to what degree the observed changes in $f_{1,\mathrm{app}}$ of this building have been due to changes in f_1 as opposed to f_R, during the San Fernando earthquake and after. Figure 3d shows that, between the Lytle Creek earthquake and San Fernando earthquake (segment SF4), f_1 dropped by 24% and f_R dropped by 18%, while $f_{1,\mathrm{app}}$ dropped by 21% (see Fig. 3c). Between the Whittier-Narrows earthquake (segment WN1) and the Yorba Linda earthquake, f_1 changed (recovered) by 8.6%, f_R by 41%, while $f_{1,\mathrm{app}}$ by 27%. This suggests that f_1 and f_R changed comparably during the San Fernando earthquake, when degradation of the structural stiffness occurred (with the change in f_1 being slightly larger), while after the San Fernando earthquake, the observed changes in $f_{1,\mathrm{app}}$ have been to a much larger degree (by a factor of almost 5) due to changes in f_R.

4. Conclusions

The trends in the variations of the NS fixed-base (f_1) and rigid body rocking (f_R) frequencies of Millikan library during four earthquakes (1970–2002) suggest the following. All variations are relative to the values during the Lytle Creek earthquake of 1970. (1) Both f_1 and f_R are amplitude dependent, (2) significant permanent reduction of frequency occurred over the years, ~22% for f_1 and 11% for $f_{1,\mathrm{app}}$, mostly caused by the San Fernando earthquake of 1971, while (3) the changes of f_R have been amplitude dependent and recoverable. (4) During the San Fernando earthquake, both f_1 and f_R dropped, respectively by ~24% and ~18%, resulting in 21% drop of $f_{1,\mathrm{app}}$. (5) After this earthquake, the changes in the observed resonant frequencies (which are those of the system) have been to a much larger degree (4–5 times) due to changes of f_R than of f_1. (6) The small permanent changes in f_1 that appear to have occurred after the San Fernando earthquake cannot be deciphered with certainty because of the small number of earthquakes recorded since 1971, and because strong motion records from the period 1988 to 2002 have not been released.

The analysis and conclusions of this study are preliminary and based on data from only four earthquakes. Other earthquakes have also been recorded, e.g., Sierra Madre, 1988, M = 5.8, R = 18 km; Northridge, 1994, M = 6.7, R = 34 km and its aftershocks; Beverly Hills, 2001, M = 4.2, R = 26 km, listed in Clinton et al. (2006). The Landers (M = 7.5) and Big Bear (M = 6.5) earthquakes of 1992 should have also been recorded. Unfortunately, no data recorded by t he CR-1 array after 1987 has been digitized and released. Once these and other future earthquake records become available, it will be possible to refine and to verify the trends discussed in this paper.

Acknowledgements

The digitized records of the Lytle Creek and San Fernando earthquakes were made available for this study by M. D. Trifunac, of the Whittier-Narrows earthquake—by the California Geological Survey, and of the Loma Linda and San Simeon earthquakes by the U.S. Geological Survey.

References

Clinton, J. F., Bradford, S. K., Heaton, T. H., and Favela, J. (2006) The observed wander of the natural frequencies in a structure, *Bull. Seism. Soc. Am.* **96**(1), 237–257.

Foutch, D. A., Luco, J. E., Trifunac, M. D., and Udwadia, F. E. (1975) Full scale three-dimensional tests of structural deformations during forced excitation of a nine-story reinforced concrete building. In *Proc. U.S. National Conf. on Earthq. Eng.*, Ann Arbor, Michigan, pp. 206–215.

Ghobarah, A. (2004) On drift limits associated with different damage levels. In P. Fajfar and H. Krawinkler (eds.), *Proc. Int. Workshop on Performance-Based Seismic Design Concepts and Implementation*, Bled, Slovenia, 28 June–1 July 2004, PEER Report 2004/05, Pacific Earthquake Engineering Research Center, University of California, Berkeley, California.

Luco, J. E., Trifunac, M. D., and Wong, H. L. (1987) On the apparent change in the dynamic behavior of a nine-story reinforced concrete building, *Bull. Seism. Soc. Am.* **77**(6), 1961–1983.

Luco, J. E., Trifunac, M. D., and Wong, H. L. (1988) Isolation of soil-structure interaction effects by full-scale forced vibration tests, *Earthq. Eng. Struct. Dyn.* **16**, 1–21.

Luco, J. E., Wong, H. L., and Trifunac, M. D. (1986) Soil-structure interaction effects on forced vibration tests, Technical Report 86-05, University of Southern California, Department of Civil Engineering, Los Angeles, California.

Snieder, R. and Şafak, E. (2006) Extracting the building response using interferometry: theory and applications to the Millikan Library in Pasadena, California, *Bull. Seism. Soc. Am.* **96**(2), 586–598.

Todorovska, M. I. (2009a) Seismic interferometry of a soil-structure interaction model with coupled horizontal and rocking response, *Bull. Seism. Soc. Am.* **99**(2A), 611–625, doi:10.1785/0120080191.

Todorovska, M. I. (2009b) Soil-structure system identification of Millikan Library North-South response during four earthquakes (1970–2002): what caused the observed wandering of the system frequencies? *Bull. Seism. Soc. Am.* **99**(2A), 626–635, doi:10.1785/0120080333.

Todorovska, M. I. and Al Rjoub, Y. (2006) Effects of rainfall on soil-structure system frequency: examples based on poroelasticity and a comparison with full-scale measurements, *Soil Dyn. Earthq. Eng.* **26**(6–7), 708–717.

Todorovska, M. I. and Al Rjoub, Y. (2008) Environmental effects on measured structural frequencies—model prediction of short term shift during heavy rainfall and comparison with full-scale observations, *Struct. Control Health Monitor.* in press, doi:10.1002/stc.260.

Todorovska, M. I. and Trifunac, M. D. (2008a) Earthquake damage detection in the Imperial County Services Building III: Analysis of wave travel times via impulse response functions, *Soil Dyn. Earthq. Eng.* **28**(5), 387–404.

Todorovska, M. I. and Trifunac, M. D. (2008b) Impulse response analysis of the Van Nuys 7-storey hotel during 11 earthquakes and earthquake damage detection, *Struct. Control Health Monit.* **15**(1), 90–116.

Todorovska, M. I., Trifunac, M. D., and Hao, T. Y. (2006) Variations of apparent building frequencies-lessons from full-scale earthquake observations. In *Proc. First European Conf. on Earthquake Engineering and Seismology* (a joint event of the 13th ECEE and 30th General Assembly of the ESC), Geneva, Switzerland, 3–8 Sept. 2006, Paper No. 1547, pp. 9.

Trifunac, M. D., Ivanović, S. S., and Todorovska, M. (2001a) Apparent periods of a building I: Fourier analysis, *J. Struct. Eng., ASCE* **127**(5), 517–526.

Trifunac, M. D., Ivanović, S. S., and Todorovska, M. I. (2001b) Apparent periods of a building II: time-frequency analysis, *J. Struct. Eng., ASCE* **127**(5), 527–537.

Trifunac, M. D., Todorovska, M. I., Manić, M. I., and Bulajić, B. Đ. (2008) Variability of the fixed-base and soil-structure system frequencies of a building—the case of Borik-2 building, *Struct. Control Health Monitor.*, `doi:10.1002/stc.277`, in press.

Udwadia, F. E. and Trifunac, M. D. (1974) Time and amplitude dependent response of structures, *Earthq. Eng. Struct. Dyn.* **2**, 359–378.

Wong, H. L., Trifunac, M. D., and Luco, J. E. (1988) A comparison of soil-structure interaction calculations with results of full-scale forced vibration tests, *Soil Dyn. Earthq. Eng.* **7**(1), 22–31.

ANALYSIS OF SEISMIC INTERACTIONS SOIL-FOUNDATION–BRIDGE STRUCTURES FOR DIFFERENT FOUNDATIONS

Boris Folić
"Gardi", Temerinska 112, 21000 Novi Sad, Serbia

Radomir Folić (folic@uns.ns.ac.yu)*
Faculty of Technical Sciences, Trg D. Obradovica 6, 21000 Novi Sad, Serbia

Abstract. The interaction of soil, foundations and structure (SFSI) depends on numerous parameters, although the effect of vibrations due to earthquake is dominant: foundation manner, and properties of soil, foundation and structure. Important are the characteristics of vibration caused by earthquake (frequency contents, pick acceleration, etc.) depending on soil quality. The paper describes the kinematics and inertia interactions and their influence on the behaviour of the structure. A different design model for SFSI is described and analysed. Consequently, the decomposition of the soil–pile–structure response is presented and steps in modelling and analysis are given. The emphasis is on the simplified model and foundation on piles.

Keywords: soil, bridge structure, foundations, pile, interaction, model, linear, nonlinear, elastodynamic, *p-y*, Winkler springs, dashpot, FEM, BEM

1. Introduction

The damage caused to the foundation of bridges in earthquakes has emphasized the importance of understanding soil structure interaction (SSI). The SSI study is initiated by detrimental actions of several earthquakes, particularly the one in Kobe, 1995. It changed the previous opinion on positive effects of the interaction (Tseng and Penzien, 2003). The process in which the response of the soil influences the motion of the structure and the response of the structure influences the motion of the soil, is referred to as soil-structure interaction. The analysis of seismic soil-structure interactions (SSI) induced by seismic actions deals with the question: how does the interplay between soil, foundation and structure affect the respond of the structure?

*Corresponding author.

T. Schanz and R. Iankov (eds.), *Coupled Site and Soil-Structure Interaction Effects with Application to Seismic Risk Mitigation,* NATO Science for Peace and Security Series C: Environmental Security, © Springer Science+Business Media B.V. 2009

The interaction of soil, foundations and structure (SFSI) depends on numerous parameters: foundation manner, properties of soil, foundation and structure. Important are the characteristics of ground motion (frequency contents, pick acceleration, etc.) surrounding soil, and the structure itself. The dynamic response of a structure founded on rock or very stiff soils is considered as the one in fixed/base structures. To evaluate the seismic response of a structure at a given site, the dynamic properties if the combined soil-structure system must be understood. The seismic excitation at bedrock is modified during transmission through the overlying soil to the foundation, causing attenuation or amplification which is also influenced by the presence of the structure (Dowrick, 2005). Ground motions not influenced by the presence of structures are referred to as *free-field* motions.

Dynamic SSI can be divided into the kinematic part and the inertial part. The former is caused by foundation motion which is different from free-field motions. In the latter, inertial forces developed in the structure cause displacements of the foundation relative to the free field.

In this paper several relevant theoretical and numerical models for the analysis of soil-foundation–bridge structures are reviewed. The accent of the review is on the simplified models applicable in practice and pile foundations that are most frequently used in bridges. Possible resonance effects between the soft soil and a flexible superstructure may cause a significant amplification of the responses.

2. Soil Structure Interaction Phenomena and Methods of Analysis

Two physical phenomena describe the interaction between the structure, foundation and soil: *kinematic interaction* results from the presence of the stiff foundation structure within or on the soil causing the difference in motion between the foundation and the surrounding soil; and *inertial interaction* causes foundation displacements in relation to the surrounding soil due to the action of inertial forces developed in the structure as a result of its oscillations, which increases forces and moments in the foundations (Gazetas and Mylonakis, 1998). The effect of these phenomena is usually described by a complex transfer function (TF) relating the motion of soil and foundation, or by a complex function of *impedance* that quantifies the stiffness and damping so characteristic for the soil-foundation interaction (SFI). Both are functions of stiffness and damping of the surrounding soil. In the assumption of perfectly stiff soil, the stiffness of soil is infinite, the amplitude of TF is 1, and the phase 0, which means the motions of the foundation and the soil are identical. Impedance function has an infinitely big real part, while the imaginary part is zero. It is best to create one *model* that includes both the soil and the structure, because it takes into account the separate influence of the soil on seismic motions, as well as the joint soil-structure action.

According to EN 1998-5 (Fardis et al., 2005) SSI shall be introduced into design of the following: (a) structures with P-δ (2nd order) play a significant role; (b) structures with massive or deep-seated foundations (bridge piers, caissons, and silos); (c) slender tall structures (towers and chimneys); (d) structures supported on very soft soils, with average shear wave velocity $v_{s,\max} < 100\,\mathrm{m/s}$, ground type S1.

In most cases in the past, the evaluation of SFSI effects for bridges has been regarded as a part of bridge foundation design problem, i.e., evaluation of load-resisting capacities of foundations without evaluation of DFSI effects on seismic response of the complete bridge system. Two different methods have been used: (1) the "elastodynamic" method developed in the nuclear power industry for large foundations and (2) the so-called empirical *p-y* method developed and practiced in the oil industry for pile foundations. Both methods have restrictions in use to different types of bridge foundations. Often, a hybrid method is used in practice minimising their weak features. It is important to adequately include seismic inputs, properties of the soil-foundation systems, conducting force-deformation demand analysis by using substructuring approach, performing force-deformation capacity evaluations, and judging overall bridge performance (Tseng and Penzien, 2003).

Construction of the *p-y* curve depends mainly on soil material strength parameters (friction angle for sand and cohesion for clays), at a specific depth. For shallow soil depths where it is important, this curve depends on the local failure mechanisms, such as failure by a passive soil resistance wedge. The same model of *p-y* curve is used for the lateral response analysis of piles, the method can be extended to treating axial resistance of the soil to piles per unit length of pile, t, as a nonlinear function of the corresponding axial displacement, z, resulting in axial *t-z* curve, and by treating the axial resistance of the soil at the pile tip, Q, as a nonlinear function of the pile tip displacement, d, can obtain *Q-d* curve. Using a set of *p-y*, *t-z*, and *Q-d* curves developed for pile foundations, the response of the pile subjected to 3D loading applied at the pile head can be obtained by using the model of a 3D beam supported on discrete sets of nonlinear lateral *p-y*, *t-z*, and *Q-d* springs (Tseng and Penzien, 2003; Stewart et al., 1998).

The elasto-dynamic theory is based on wave propagation in a linear elastic, viscoelastic, or constant-hysteresis-damped elastic half-space soil medium. This method of SSI analysis is currently being practiced in the nuclear power industry. For foundations on slab having dimensions in the base smaller than $11 \times 15\,\mathrm{m}$ the interaction is insignificant and it is less obvious in shallow foundations (Stewart et al., 1998). Therefore, herein the accent is on foundations on piles. FE and finite boundary element methods are used in the analyses.

The effects of SSI in pile foundations shall be assessed for all structures, so that they can resist the following two types of action effects:

- Inertia forces from the super-structure combined with static loads. Inertia developed in structure causing displacement of the foundations relative to the free field. Frequency dependent foundation impedance functions are introduced to describe the flexibility of the foundation support, as well as the radiation damping associate with soil-foundation interaction. In the absence of large, rigid foundation slabs or of deep embedment, inertial interaction tends to be more important.
- Kinematic forces arise from the deformation of the surrounding soil due to the passage of seismic waves. Stiff, slab-like or deeply embedded foundation elements cause foundation motions to be different from free-field motions because of wave-scattering phenomena, wave inclination or embedment.

To completely understand the seismic behaviour of a soil–pile–structure system it is necessary to conduct the three following analyses (Gazetas and Mylonakis, 1998; Stewart et al., 1998):

1. **Soil Response Analysis** gives a realistic picture of the seismic environment during the design earthquake. It defines the seismic excitation and provides information for assessing possible loss of strength resulting from pore-water pressure in the saturated liquefied cohesionless layers.
2. **Kinematic Pile Response Analysis** provides the response of the piled foundation in the absence of inertia forces from the super-structure.
3. **Inertial Soil-Structure Interaction Analysis** serves to determine the dynamic response of the super-structure and the loads imposed on the foundation by the response. To determine and describe precisely the role of this interaction it is necessary to evaluate the *inelastic* behaviour of the super-structure.

3. Lamped Mass System

In the analysis of an interaction system, the soil, foundation and superstructure must be adequately modelled. Soil models for dynamic analysis are: equivalent static springs and viscous damping located at the base of the structure; shear beam analogy using continua or lumped masses and sprigs distributed vertically through the soil profile; elastic half-space (EHS);

finite elements (FE); and hybrid models EHS and FE. In practice are most frequently used lumped mass systems since they consider the ground and the structure having extension in space discretely. Because shear vibration is predominant in the soil during earthquake the ground can be considered as a one-dimensional shear vibration model. The ground is considered as an appropriate number of layers of discrete masses connected by springs. The mass of each point m_i and the corresponding shear spring constant k_i are given by:

$$m_i = \frac{1}{2}\left(\rho_{i-1} \cdot h_{i-1} + \rho_i \cdot h_i\right), \qquad k_i = \frac{G_i}{h_i}, \tag{1}$$

where h_i, ρ_i and G_i are the mass density, the thickness and the shear modulus of the i-th layer, respectively.

The methods based on the finite element method (FEM) assume a mass matrix as a mass lumped matrix. Structures, comprising discrete masses (Fig. 1), can also be modelled in the same manner as the free field. The masses are considered connected by a spring, and depending on the structure we can select the shear spring, the rotational spring or the linear combination of these two types.

If a foundation is considered rigid in comparison with ground, it is usually modelled as a single mass or a rigid body having two degrees of freedom. A pile having high flexibility is modelled as a beam element but sometimes it is also considered as a system with multiple degrees of freedom similar to the superstructure (Fig. 2). The interaction spring is introduced between

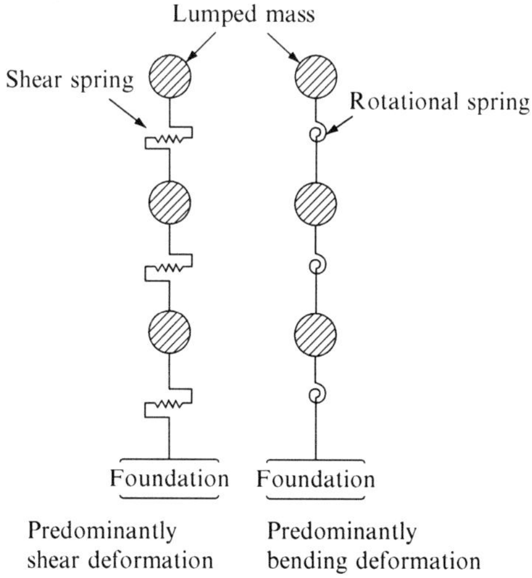

Figure 1. Discrete model of superstructures (after Dynamic Analysis and Earthquake Resistant Design, 2000).

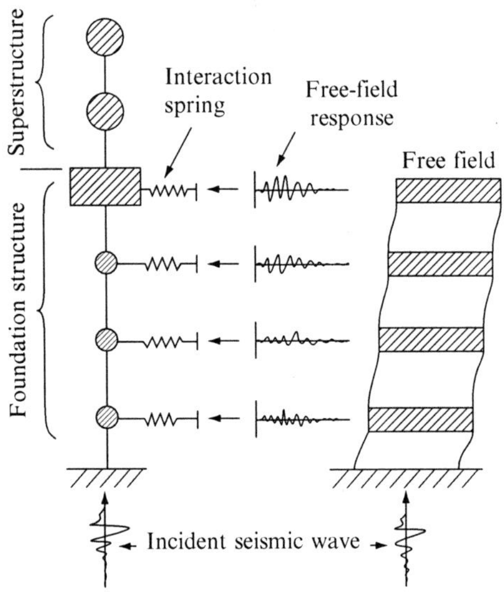

Figure 2. Modelling soil-structure interaction system (after Dynamic Analysis and Earthquake Resistant Design, 2000).

the foundation and the soil. This spring is a function of the frequency of vibration but generally some static value that is independent on the frequency of vibration is used for this purpose. Mindlin analysis based on the theory of elasticity or the coefficient of subgrade reaction from experiment can be used.

4. Analytical Models for Dynamic Interaction

Assuming that the response displacement of natural ground as a result of an earthquake is given by $\{u_A t\}$, Fig. 3, the displacement of massless foundation as a result of natural ground motion may be assumed as $\{u_C\}$. If we denote the additional displacement due to inertial force as $\{u_D\}$, the dynamic displacement of foundation with mass $\{u_B\}$ will be given as (Dynamic Analysis and Earthquake Resistant Design, 2000):

$$\{u_B\} = \{u_C\} + \{u_D\} \tag{2}$$

with the impedance matrix $[K]$ dynamic reactions from the ground and the mass and stiffness of foundation $[M]$ and $[K_F]$; then the following equation from equilibrium of force is obtained:

$$[M]\{\ddot{u}_B\} + [K_F]\{u_D\} + [K]\{u_D\} = 0. \tag{3}$$

The first part represents inertial force, the second restoration force based on foundation rigidity, and the third ground reaction. If we express equation (3)

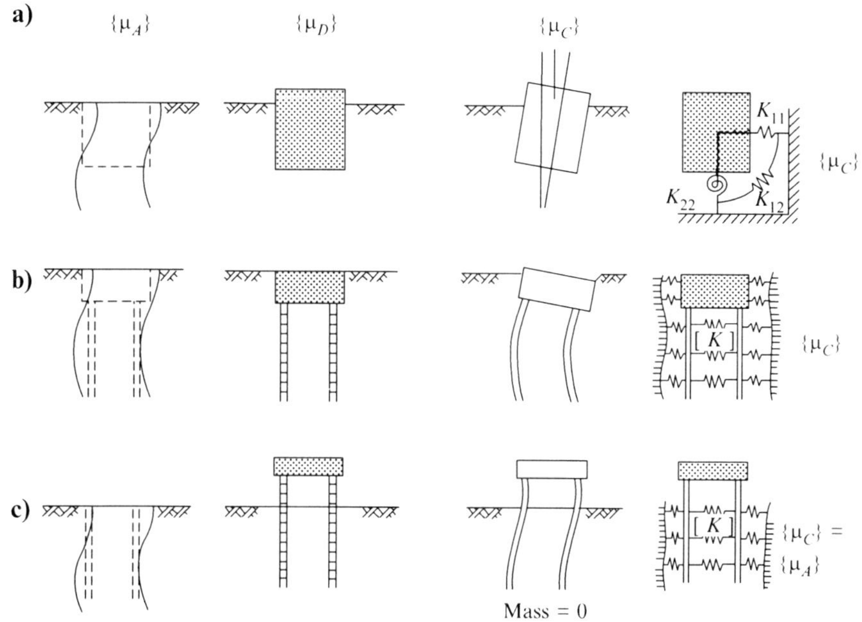

Figure 3. Analytical models for dynamic interaction in case of rigid foundation and pile foundation (after Dynamic Analysis and Earthquake Resistant Design, 2000).

in term $\{u_D\}$ using (2) we get:

$$[M]\{\ddot{u}_B\} + [K_F + K]\{u_D\} = -[M]\{\ddot{u}_C\}, \tag{4}$$

where $\{\ddot{u}\} = d^2u/dt^2$ represents the acceleration. For the rigid foundation, there is no restoration force due to foundation stiffness and in (4) $[K_F]\{u_D\} = 0$. The displacement of the foundation $\{u_B\}$ (Fig. 3) can be determined from the mass and stiffness of the foundation after assigning the characteristics and value of impedance $[K]$ and effective earthquake input motion $\{u_C\}$ that is effective in the response analysis of the foundation and it is used to distinguish this from the earthquake motion of the natural ground $\{u_A\}$ used in the conventional response analysis (Dynamic Analysis and Earthquake Resistant Design, 2000).

For the pile foundation (Fig. 3c), the difference in $\{u_C\}$ and $\{u_A\}$ is considered to be small and hence one can use value $\{u_A\}$ in place of $\{u_C\}$. The equation of motion becomes:

$$[M]\{\ddot{u}_D\} + [K_F + K]\{u_D\} = -[M]\{\ddot{u}_A\}. \tag{5}$$

This is similar to the equation used in the conventional earthquake response analysis.

Using an adequate discretised model that corresponds to the geometry and arrangement of soil layers, and observing seismic motions as displacements of stiff support, the response on the soil surface can be obtained starting from the equation of motion of the system, through direct integration (in frequency domain as it is the most efficient), i.e., a *time-history*. The obtained record (accelerograms) describes *free-field* motions in place where the structure is built. By generating accelerograms the contribution of more than one possible earthquake can be described by varying the frequency contains (Fardis et al., 2005). A set of artificial accelerograms is better than a set of real accelerograms for it enables a better usage of the techniques. Accelerograms can be scaled, i.e., their ordinates can be multiplied by a corresponding scale factor, which according to EN 1998/2 can be from 0.5 to 2, in order to adjust them to the problem in question. In the evaluation of SFSI effects on bridges the elastoplastic model can be used for large foundations, and the empirical *P-y* method that has been developed for pile foundations.

5. Idealized Nonlinear Model of a Pile Foundation

In a pile foundation, the foundation is so idealized that the rigid footing is supported by piles which are supported by the soils. Figure 4 shows the idealized nonlinear model of a pile foundation. The nonlinearities of soil and piles are considered in the analysis. The safety of the foundation shall be checked so that: the foundation shall not reach the yield point of foundation; if the primary nonlinearity is developed in the foundation, the response displacement shall be less then the displacement ductility limit; and the displacement developed in the foundation shall be less than the allowed limit (Earthquake Resistant Design Codes in Japan, 2000).

The structural model for an overall structural system for viaducts with the pushover analysis (Fig. 5) is similar. The model includes the nonlinear properties of both subgrade and structures. The springs expressing the subgrade reaction are attached to the nodal points, and the part connecting the pile to the spread footing and the piles to the embedded lateral beams are assumed to be rigid. The properties of ground resistance of the pile foundation are assumed to be represented by an elasto-plastic (bilinear) model (Fig. 6).

Nonlinear Winkler model for the investigation of the soil–pile group interaction is shown in Fig. 8. The theoretical model (Tahgihighi and Konagai, 2006) (Fig. 7) provides a rich information base for the study of soil–pile–structure interaction. The soil–pile group interaction using nonlinear Winkler foundation (or *p-y* model) analysis method was evaluated against two sets of physical experimental data (the field pushover test, as well as, the centrifuge model test). The elasto-dynamics theory yields a Winkler type expression for the nonlinear soil–pile group interaction analysis.

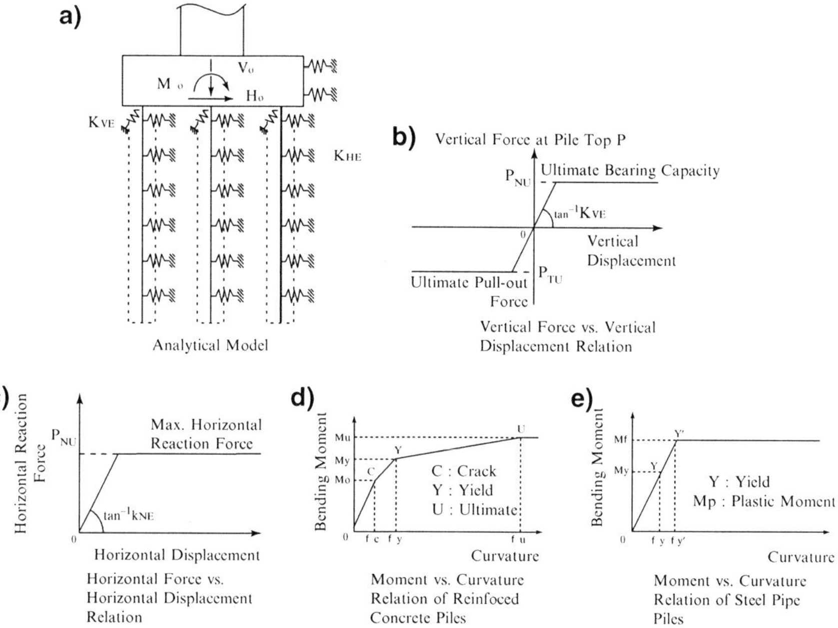

Figure 4. Idealized nonlinear model of a pile foundation (after Earthquake Resistant Design Codes in Japan, 2000).

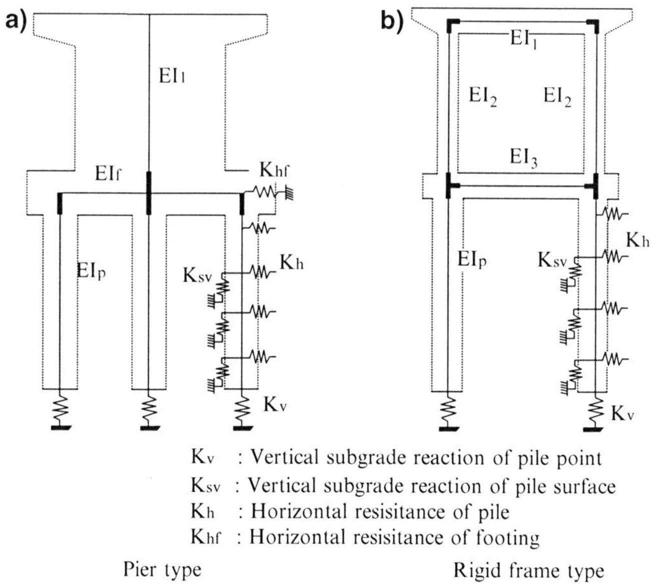

Figure 5. Structural model for viaducts (Earthquake Resistant Design Codes in Japan, 2000).

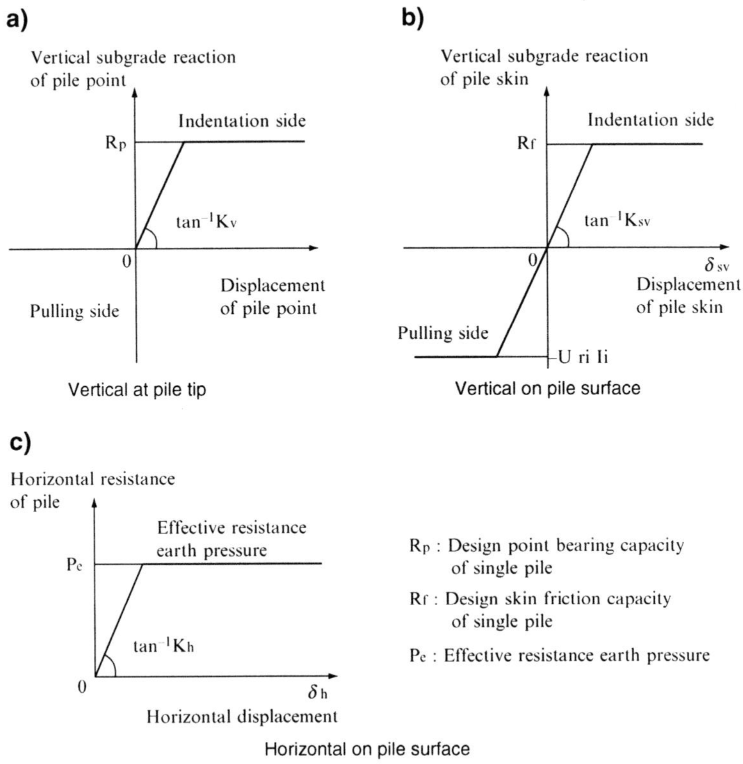

Figure 6. Model for ground resistance.

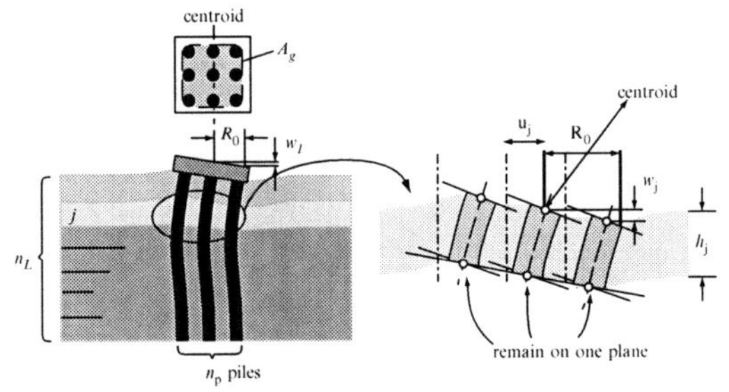

Figure 7. Assumptions for evaluation of equivalent upright beam: soil-grouped pile (left), sliced elements (right) (Tahgihighi and Konagai, 2006).

A group piles, supported by an elasto-plastic side soil, is viewed as an equivalent upright beam connected to the superstructure through a right pile cap. The idealization of pile group as a single beam analogy with the active pile length has enabled the derivation of this efficient perspective. The obtained results showed a fairly good correlation with rigorous 3D simulations and also with the measured data. The proposed *p-y* model may conventionally

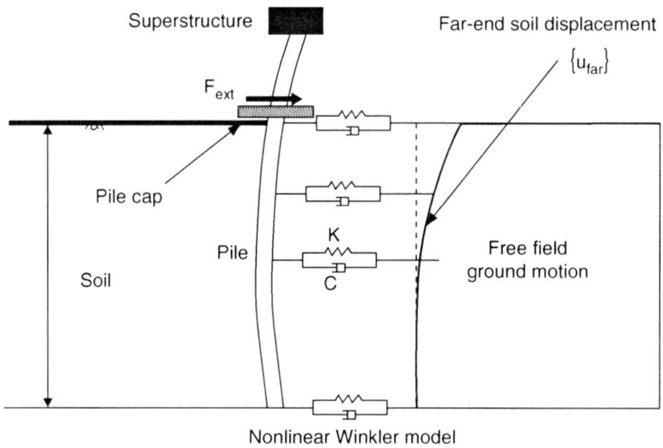

Figure 8. Scheme of dynamic nonlinear *p-y* element for soil–pile–structure interaction.

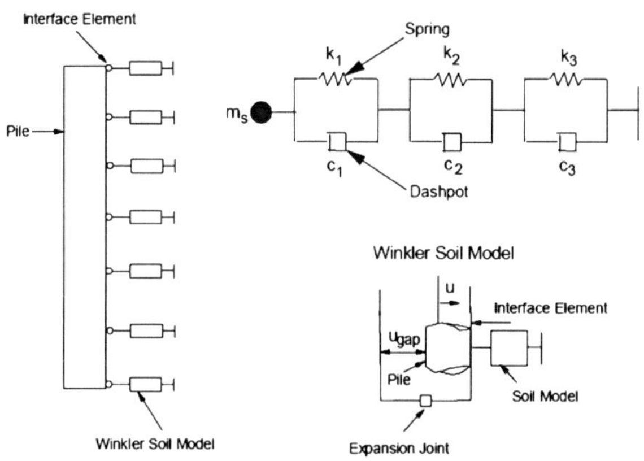

Figure 9. Model of soil–pile system for separation (Maheshwari and Watanabe, 2000).

be used for the design of pile foundation in engineering practice. This method makes possible the prediction of distributions of lateral loads, shear and bending moment in each individual pile (Tahgihighi and Konagai, 2006).

During strong ground motions the soil surrounding the pile behaves nonlinearly, and large inertia forces cause a separation between the soil and the pile (Fig. 9). The soil–pile system is divided into a number of layers. Load is assumed to act at the pile head, and Winkler's hypothesis is used to analyse each layer of the soil–pile system. The interface model works like a rigid frame with an expansion joint that can change its size as well as, position at different instants of time (Maheshwari and Watanabe, 2000). The investigation (Finn, 2004) of the bridge with the parametric analysis, whose computing model is in Fig. 10, is of interest. The methodology of analysis of interaction is detailed in (Nogami, 1987) and the analysis of a set of models for foundations on piles in (Maymond, 1998).

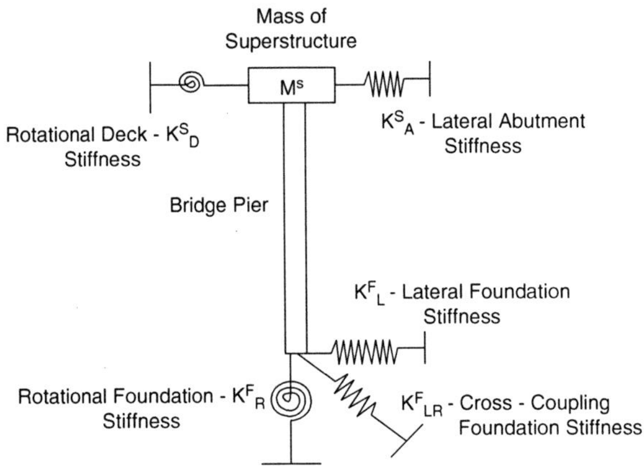

Figure 10. Computational bridge model for parametric study.

Recent investigations (Balendra, 2005) identified the conditions under which soil nonlinearity, soil–pile separation and pile diameter effects become important in the treatment if kinematic and inertial effects. Ground motion intensity does not change the value TF in nonlinear soil medium. An increase in input motion intensity decreases the soil stiffness and increases the hysteretic damping in the soil so TF is not affected. The influence of the pile diameter on TF is significant. Separation between soil and pile has no significant effects on the transfer of a pile embedded in an elastic medium. Maximum bending moment envelope along the pile increases with the pile both for rigid and flexible piles. Soil–pile separation increases the normal stress along the lower half of a flexible pile. For flexible piles in an elastic soil medium, the real part of the impedance function (IF) is nearly constant and does not change much with pile diameter. For the rigid piles, the real part of the IF reduces with increasing frequency. Soil–pile separation reduces the imaginary and real part of the IF, as well as increases the normal stress along the pile. Nonlinearity does not affect the normal stress along the pile.

The seismic analyses of the soil–pile interaction given in Balendra (2005) are useful. The basic principles of electrodynamics and 1D problems: wave propagation in layers due to earthquakes in the underlying layers, and basics of the dynamic problem in elastic continua are shown in Verruijt (2008).

6. Concluding Remarks

Because of the limited space in this paper, only the basic principles for the seismic design and some results are reviewed. The methodology for seismic design with interaction SFSI is rather complicated due to the consideration of both nonlinearities in the structures and the subgrade.

The advances in computer technology justify the use of rigorous SFSI analysis for important bridges. The FEM is a convenient tool to study the effects of various variables on the kinematic transfer function, as well as inertial impedance function. Numerical models for dynamic soil-structure interaction are often developed using FEM well-known software, SASI 2000, SHAKE, ABAQUS, FLIP, DYNAFLOW, PILE-3D-EFF (a program for non-linear dynamic effective stress analysis of pile foundations), and finite difference program FLAC 3D based on the finite difference method. Recently seismic loading of model pile foundations in centrifuge tests has provided data that allows a more realistic evaluation of various methods for seismic analysis of pile foundations.

References

Balendra, S. (2005) Numerical modeling of dynamic soil–pile–structure interaction, M. S. Thesis in civil engineering, Washington State University, Washington, http://hdl.handle.net/2376/414.

Dowrick, D. (2005) *Earthquake Risk Reduction*, Chichester, England, Wiley.

Earthquake Resistant Design Codes in Japan (2000) Ch. 2, *Japan Society of Civil Engineers*, January 2000, pp. 2.1–2.21.

Fardis, M. N., Carvalho, E., Alnashai, A., Faccioli, E., Pinto, P., and Plumier, A. (2005) *Designers' Guide to EN 1998-1 and 1998-5 Eurocode 8: Design of Structures for Earthquake Resistance*, London, Thomas Telford.

Finn, W. D. L. (2004) Characterizing Pile Foundations for Evaluation of Performance Based Seismic Design of Critical Lifeline Structures, *13WCEE*, Pap. 5002.

Gazetas, G. and Mylonakis, G. (1998) Seismic Soil-Structure Interaction: New Evidence and Emerging Issues, *Geo-Institute ASCE Conf.*, Seattle, 1–56.

Japanese Society of Civil Engineers (2000) *Dynamic Analysis and Earthquake Resistant Design: Methods of Dynamic Analysis*, Vol. 2, A. A. Balkema, Rotterdam, p. 304.

Maheshwari, B. K. and Watanabe, H. (2000) Nonlinear dynamic analysis of pile foundation: effects of separation of pile soil interface. In *Proc. 12th World Conf. Earthquake Eng.*, Auckland, New Zealand, Pap. 0494.

Maymond, P. J. (1998) Shaking table scale model test of nonlinear soil–pile–superstructure interaction in soft clay, UC, Berkley.

Nogami, T. (ed.) (1987) *Dynamic Response of Pile Foundations—Experimental, Analysis and Observation*, ASCE, Geotechnical SP GEDivision, No. 11.

Stewart, J., Seed, R., and Fenves, G. (1998) Empirical evaluation of inertial soil-structure interaction effects, Pacific EERC, University of California, Berkeley.

Tahgihighi, H. and Konagai, K. (2006) Numerical study of soil–pile group interaction in sand, *First ECEES*, Geneva, 3–8 September 2006, Paper 1333.

Tseng, W.-S. and Penzien, J. (2003) Soil-foundation–structure interaction, In W.-F. Chen and L. Duan (eds.), Chapter 42 of *Bridge Engineering Handbook*, Boca Raton, Florida, USA, CRC Press LLC.

Verruijt, A. (2008) Soil dynamics and offshore soil mechanics, Delft University of Technology, (September 2008, http://geo.verruijt.net).

Part III

The Role of Site Effects and of Soil Structure Interaction in Design of Structures

SEISMIC ASSESSMENT OF BRIDGES ACCOUNTING FOR NONLINEAR MATERIAL AND SOIL RESPONSE, AND VARYING BOUNDARY CONDITIONS

Andreas J. Kappos (ajkap@civil.auth.gr)* and Anastasios G. Sextos
Department of Civil Engineering, Aristotle University of Thessaloniki, 54124 Thessaloniki, Greece

Abstract. Seismic assessment of bridges using the pushover analysis technique often ignores the effect of some sources of nonlinearity such as those associated with the foundation soil and the boundary conditions, that may significantly modify the overall performance of a bridge. In this context, the seismic response of a typical overpass is assessed herein using lumped plasticity models to account for the inelastic behaviour of the critical cross-sections of piers and piles, and nonlinear springs to consider foundation-soil compliance; in addition, a detailed solid finite element model of the abutment-embankment-foundation soil system is set up and compared with the simpler models. The results of the analysis show a markedly different seismic behaviour when the abutment—soil system is included in the analysis, rather than simply considering a pinned support (in the transverse direction) as usually done in previous studies. Furthermore, for stronger excitations, it is seen that as inelastic mechanisms (of piers, piles, pile caps, and soil) are introduced and boundary conditions change (i.e., joint /gap closure), the assumptions made on the foundation and soil compliance play an increasingly important role that can potentially modify the anticipated failure hierarchy, as well as the ensuing pushover curves in both directions of the bridge.

Keywords: bridges, soil structure interaction, site effects, landfills, material nonlinearity, finite element analyses, permanent deformation analyses

1. Introduction

It is now well-established that elastic analysis of structures subjected to seismic actions, typically in the form of response spectrum analysis, cannot always predict the hierarchy of the failure mechanisms, nor is it able to quantify the energy absorption and force redistribution that result from the gradual plastic hinge development within the structure. For this reason, during the last 15 years or so, the development of analytical methods that would permit the quantification of the degree of global and/or local ductility (that depends on

*Corresponding author.

T. Schanz and R. Iankov (eds.), *Coupled Site and Soil-Structure Interaction Effects with Application to Seismic Risk Mitigation,* NATO Science for Peace and Security Series C: Environmental Security,

the level of earthquake excitation) has increasingly attracted the attention of both researchers and designers. Along these lines, nonlinear static (pushover) analysis has become a popular tool for the seismic assessment not only of buildings, but also of bridges (e.g., Isakovic et al., 2003), despite the fact that its main advantage of lower computational cost, compared to nonlinear dynamic time–history analysis, is counter-balanced by its inherent restriction to structures wherein the fundamental mode dominates the response. A recent contributor to this popularity is the extension of the pushover approach to consider higher mode effects for both buildings (Chopra and Goel, 2002; Goel and Chopra, 2004) and bridges (Paraskeva et al., 2006).

The aforementioned nonlinearity expected in bridges during strong ground motions, cannot be attributed solely to the structural system, e.g., yielding of reinforced concrete (R/C) sections, although structural members are often purposely designed to exhibit inelastic behaviour. Additional material nonlinearity mechanisms (of the foundation and/or backfill soil) as well as boundary nonlinearities (activation of control components such as stoppers or seismic joints) can also play a significant role in the overall system response. Moreover, despite the existence of specific guidelines in the US (AASHTO, 2008; Caltrans, 2006) and in Europe (CEN TC250, 2005) for the design of pile foundations and abutments and the decisions related to the above components, only minor guidance is provided for the modelling of the problem from a numerical point of view, or for the consideration and assessment (even statically) of the soil-foundation-pier-deck (CEN TC250, 2005; Kappos and Sextos, 2001), and soil-abutment-deck system (Karantzikis and Spyrakos, 2000; Mackie and Stojadinovic, 2002), interaction, which are associated with high level of uncertainty. As a result, in design, the abutment capacity and stiffness, as well as the effects of soil nonlinearity, are most commonly ignored, mainly for reasons of convenience.

The main goal of this paper is to focus on a real bridge structure that has been designed to resist seismic forces through both capacity-designed elements and control components (stoppers and seismic joints), in order to assess its performance using simple and refined finite element models of the abutment-embankment-pier foundation subsystems. It is noted that the dynamic interaction of the complete system is a far more complex and multi-parametric problem, especially in terms of the coupling between the geometry of the abutment and the backfill/embankment, the dynamic soil properties at high strains, and the dynamic characteristics of the structure (Zhang and Makris, 2002). Dynamic aspects of this behaviour are ignored herein, since the specific assessment presented here involves nonlinear static (pushover) analysis; these aspects are treated in a recent paper by Sextos et al. (2008).

2. Overview of the Bridge Studied

The bridge shown in Fig. 1 is an overpass (overcrossing) along the Egnatia motorway. It is a three-span, symmetric bridge (span lengths are 19, 32 and 19 m respectively) curved in elevation (max camber of 8%), that intersects the motorway axis at an angle of 75.3°. The deck is 11 m wide and 1.60 m high. The prestressed deck has a voided T-beam-like section (see Fig. 2) and is supported on two circular piers of 1.70 m diameter and 8.50 m height which are monolithically connected to the superstructure and the foundation. At the abutments (which have a 10.50 × 1.20 m wall section of 5.0 m height), the deck is connected through two pot bearings that permit sliding along the two principal bridge axes and a 120 mm longitudinal joint separates the deck from the backwall (Fig. 1). Transverse displacement at the abutments is blocked by a stopper system at the top of the backwall. The pier foundation consists of a 2 × 2 pile group of 28.0–32.0 m long piles, connected with a 1.60 × 5.0 × 5.0 m pile cap, while the abutments are supported on a 1 × 4 pile row 27–35.0 m long at 2.80 m axial spacing, all piles having equal diameter of 1.0 m.

The bridge was designed for normal loads according to the German Standards (DIN) while the seismic design was carried out according to the Greek standards EAK 2000 and E39/99, the first being the Greek Seismic Code for the design of structures (General and Buildings) and the latter the Code for the Seismic Design of Bridges. The bridge site is located in the Seismic Risk Zone I (peak ground acceleration of $a_g = 0.16$ g), and ground category is C (soft soil). The behaviour factors of the system were also adopted according to the E39/99 document and are: $q_x = 2.50$, $q_y = 3.50$, $q_z = 1.00$ for the response in the three principal directions, respectively.

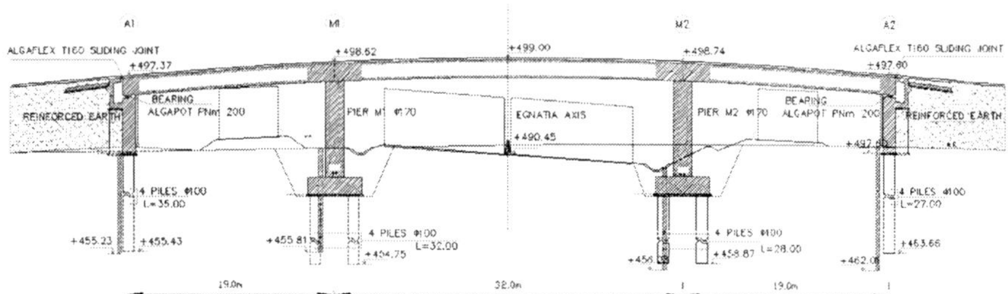

Figure 1. Longitudinal cross-section of the bridge.

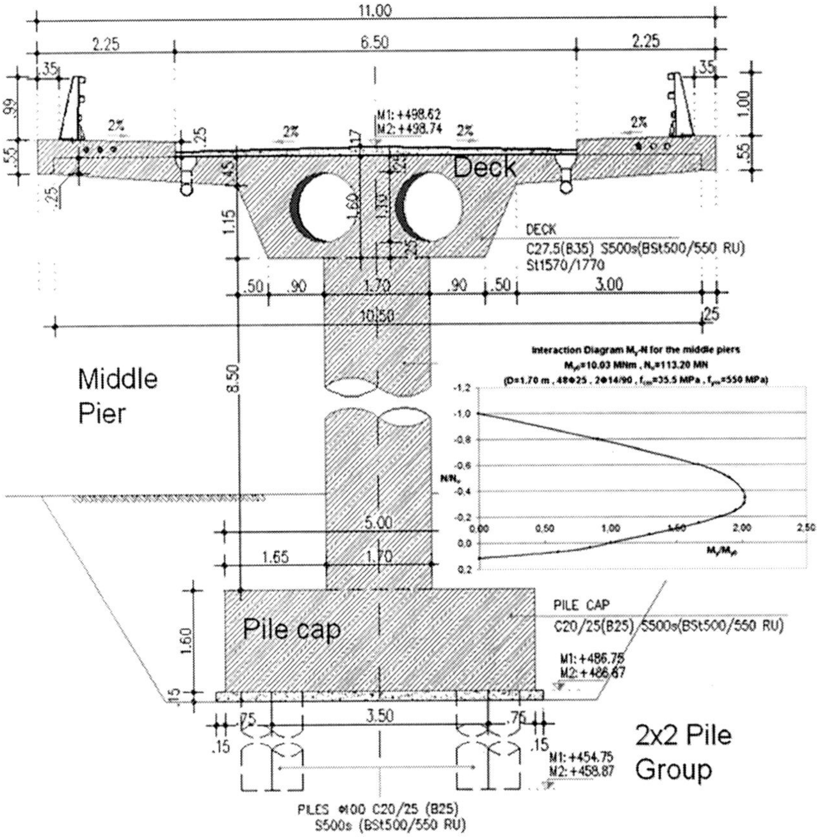

Figure 2. Overview of the foundation-pier-superstructure system cross-section.

3. Finite Element Modelling and Analysis of the Structure

3.1. MODELLING ASPECTS

The bridge was modelled using the 3D structural analysis program SAP 2000 (CSI, 2005). A three-dimensional model was created with linear Frame Elements to reflect the geometry, boundary conditions, and material behaviour of the bridge studied. The structure is adequately discretised to account for the bridge parabolic elevation shape and a continuous mass approach is used instead of lumped masses. The section stiffness was reduced as per the E39/99 guidelines; in particular, for the prestressed members that are designed to remain elastic during the seismic event (i.e., the deck), the uncracked stiffness is used, whereas for the piers, wherein development of plastic hinging is expected under the design earthquake, the secant stiffness at yield is adopted, based on the maximum expected axial load. For the two piers, the reduced

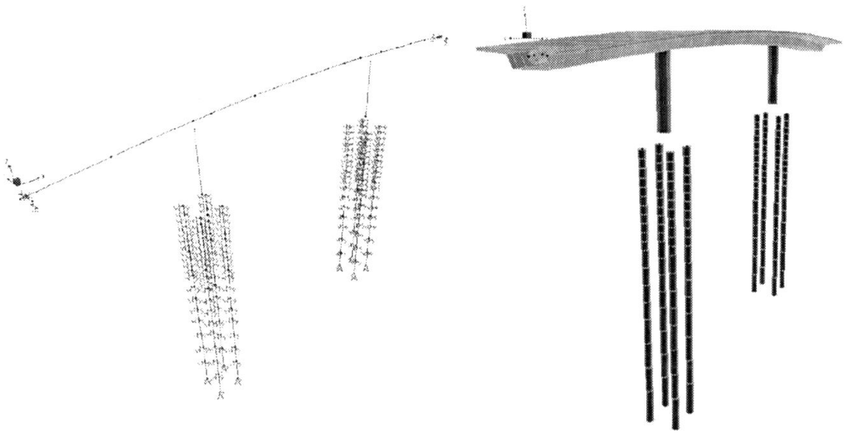

Figure 3. Finite element discretization of Model 4 of the bridge.

stiffness $EI_{eff} = 0.29 \div 0.37EI_g$ was calculated using two alternative cross-section analysis programs based on the fibre approach, FAGUS (Cubus, 2000) and RCCOLA (Kappos, 1993).

In order to investigate the effect of modelling assumptions on the bridge response, four different finite element models were developed for the bridge itself: in Model 1, the pier stiffness is taken uncracked and the pier supports are considered as completely fixed (apparently corresponding to the maximum stress and minimum displacements case). In Model 2, the pier stiffness is reduced according to the upper bound coefficient (0.37) described above, whereas Model 3 additionally accounts for the (linear elastic) pier foundation-soil system compliance. Finally, Model 4 is used to investigate the effect of adopting the lower bound of stiffness reduction coefficient (0.29), plus the pier foundation flexibility ('softest' model). The static pile-soil interaction is accounted for by attaching linear (at this stage) Winkler springs along the pile length. The corresponding stiffness is derived from the first branch stiffness of the P-y curve (i.e., lateral soil resistance vs. deflection relationship) proposed by Matlock (1970) after appropriate linearization. Model 3 was used for the seismic design of the piers, while seismic displacements of the deck, the design of the seismic joints and of the bearings at the abutments, are based on the analysis of Model 4 (Fig. 3).

3.2. MODAL AND RESPONSE SPECTRUM ANALYSIS OF THE SYSTEM

The longer periods of the four finite element models varied from 0.51 to 1.01 s in the longitudinal direction and from 0.46 to 0.51 s in the transverse direction. It is notable that although pier cracking and foundation flexibility lead to an increase of the fundamental period (longitudinal direction) by a factor of almost 2, the same conditions do not affect considerably the vibration along the transverse direction (mode 2); this is mainly due to blocking of transverse displacement at the abutments in this relatively short bridge.

3.3. NONLINEAR STATIC ANALYSIS OF THE ABUTMENT-BACKFILL SYSTEM

3.3.1. *Soil nonlinear behaviour*

The most commonly adopted engineering method for calculating the non-linear pseudo-static interaction between piles and the surrounding soil is bi-linear Winkler models, in which the soil reaction to pile movement is represented by independent unidirectional translational spring elements distributed along the pile shaft to account for the soil response in the elastic and inelastic range. Along these lines, the force-dependent stiffness of the foundation-soil system was estimated using the P-y curves proposed by Matlock (1970) for soft soils

$$\frac{p}{p_u} = 0.5\left(\frac{y}{y_{50}}\right)^n. \tag{1}$$

3.3.2. *Simplified modelling of abutment*

Various approaches for modelling abutments have been proposed in the literature (e.g., see Priestley et al., 1996); however, abutment compliance is only rarely accounted for in developing pushover curves for bridges. A more refined modelling of the abutment's actual stiffness was deemed necessary in order to obtain a realistic estimate of the nonlinear response of the bridge as a whole and especially of the base shear distribution among the structural elements. For this purpose, the foundation-abutment-backfill soil was studied separately and the appropriate pushover curves (i.e., seismic force vs. monitoring point displacement) were derived for the case of excitation along the two principal axes of the abutment (and the bridge).

The abutment wall is first modelled with 2D shell elements, while the piles with linear frame elements supported on the (depth-dependent) nonlinear springs as described above and shown in Fig. 4. For investigation purposes, the analysis is performed both for the (actual) soft soil conditions and for the case of a significantly stiffer supporting soil. The resulting four alternative finite element models of the abutment-foundation-soil system involve: a1) an abutment supported on soft soil conditions, where friction springs are used along the piles together and the appropriate vertical stiffness is introduced with the use of a (compression-only) spring at the tip of the piles, and a2) a similar FE system of practically infinite vertical stiffness (tip displacements are restrained). Two additional models (b1, b2) are also examined, with the same boundary conditions, but for the case of stiff soil.

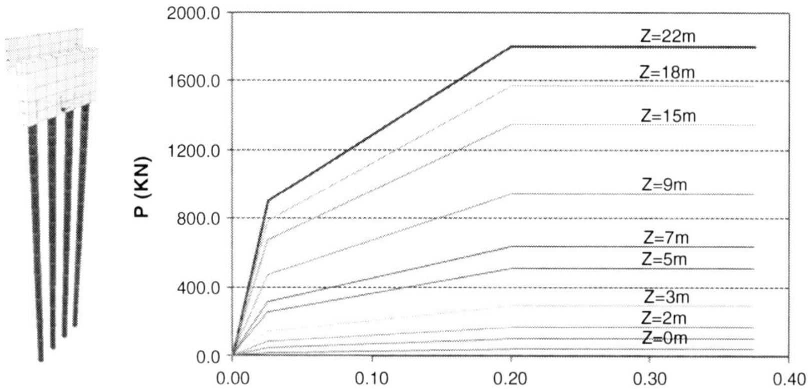

Figure 4. Depth-dependent multi-linear load-deflection curves used along the abutment piles to account for soil compliance (values for models a1 and a2—soft soil conditions).

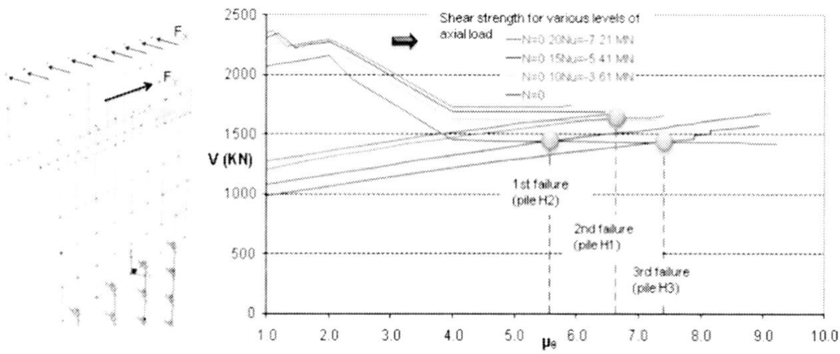

Figure 5. FE modelling (*left*) and shear strength vs. rotational ductility ($V_n - \mu_\theta$) and shear force vs. rotational ductility ($V_{sd} - \mu_\theta$) diagram (*right*) for soft soil conditions.

Apart from soil flexibility, pile nonlinearity is also considered for all models in terms of potential plastic hinge development and shear failure. The corresponding shear strength vs. rotational ductility ($V_n - \mu_\theta$) and shear force vs. rotational ductility ($V_{sd} - \mu_\theta$) diagrams are illustrated in Fig. 5 where it is seen that for the case of soft soil conditions, three piles fail in shear for large values of plastic rotation. Accordingly, four piles suffer shear failure right after their yield moment is exceeded. The fact that the foundation piles are able to resist the imposed bending moments and shear forces up to their flexural and shear capacity level affects the overall system stiffness. The resulting final pushover curve in the transverse direction of the complete abutment-foundation-soil system is illustrated in Fig. 6. As anticipated, for the case of soft soil conditions, the overall system stiffness is significantly lower. The pushover curve derived is used to control the transverse abutment stiffness of the final (most refined) finite element model of the complete bridge.

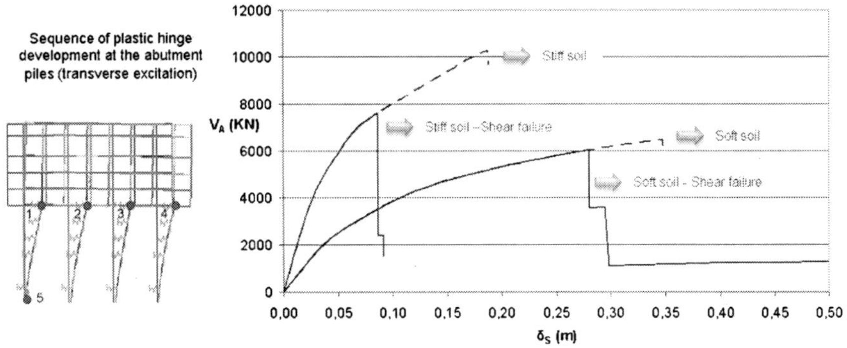

Figure 6. Sequence of plastic hinge development (*left*) and nonlinear response (in terms of pushover curve) of the abutment-foundation-soil system (*right*).

It is also noted that in the longitudinal direction, the flexibility of the backfill (attributed to the presence of reinforced soil at the back of the abutment as seen in Fig. 1) is also considered through an additional spring which, apparently, is independent of the overall foundation soil conditions. As a result, the stiffness of the backfill-abutment-foundation-soil system in the longitudinal direction is modelled as a series combination of two springs (i.e., backfill and abutment-foundation-soil); the assessment of the entire bridge under the prescribed seismic loads is discussed in the following.

3.3.3. *Detailed modelling of abutment-embankment-foundation soil system*
In addition to the relatively simple model described in Section 3.3.2, a 3D nonlinear finite element model of the full abutment-embankment system involving tetrahedral solid elements was set up in ABAQUS (2004), see inset in Fig. 7 and details in Sextos et al. (2008). The Mohr–Coulomb constitutive model implemented in ABAQUS was used to simulate the nonlinear soil behavior while a gradually increasing pressure was applied as distributed normal and shear forces on the abutment for the pushover analysis along the longitudinal and transverse directions, respectively. Soil parameters (modulus of elasticity E, cohesion c, and friction angle ϕ) for the various parts of the backfill-embankment-foundation soil system in the refined 3D model were directly estimated from the available geotechnical data, without specific attempt to match exactly the P-y curves used in the simpler model. It is observed with regard to the pushover curves resulting from the two models (Fig. 7) that, given the inherent differences between the nonlinear force-displacement curves of the springs used in the Winkler model and the stress-based yield criterion adopted for the solid of the 3D-FE model, the agreement between the two approaches is satisfactory, especially in the transverse direction (where the overall stiffness is controlled by the pile foundation).

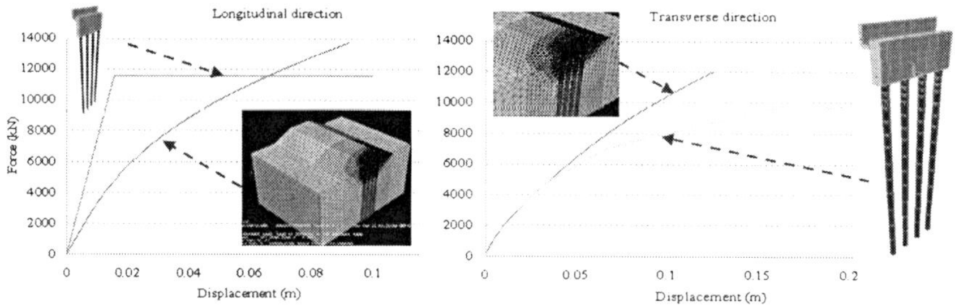

Figure 7. Comparison of pushover curves derived for the spring-supported abutment and 3D model of the bridge, for the longitudinal (*left*) and transverse direction (*right*).

4. Assessment of the Bridge Performance

4.1. LONGITUDINAL DIRECTION

Having modelled all sources of material nonlinearity (i.e., the effect of the backfill compliance, foundation soil yielding, pier and pile plastic hinge development and pile head failure in shear), as well as boundary nonlinearity (end gap closure), two nonlinear static analyses are performed along the two principal axes of the bridge (end springs are defined as described in Section 3.3.2). The pushover curve of the overall system for the longitudinal direction is illustrated in Fig. 8. It is observed that in the case of soft soil conditions (which is the case in the actual bridge), three plastic hinges develop (plastic rotations correspond to 2–10% of the ultimate rotation θ_{pu}) at the two piers before the system reaches the critical displacement $\delta = 12$ cm where the joint closes. At a displacement $\delta = 14$ cm the backfill soil yields and the system stiffness is considerably reduced until the ultimate displacement $\delta = 22$ cm, where the bridge abutments are considered unstable due to irrecoverable damage as the reinforced soil (which is not laterally restrained by wing walls) cannot be supported anymore. At this ultimate stage, four plastic hinges have developed at the two piers, reaching 35% and 49%, respectively, of the available plastic rotations. A similar picture is observed for the case of stiff soil. Clearly, differences in the bridge response are evident only for displacements that do not exceed $\delta = 12$ cm, that is, prior to the gap closure; for larger displacements the stiffness is controlled by the abutment and backfill system and is very similar to the soft soil case.

4.2. TRANSVERSE DIRECTION

As noted in Fig. 9, in the transverse direction, the nonlinear mechanism is indeed different, compared to the longitudinal excitation case. In particular, an abrupt stiffness reduction is observed for a mid-deck displacement equal

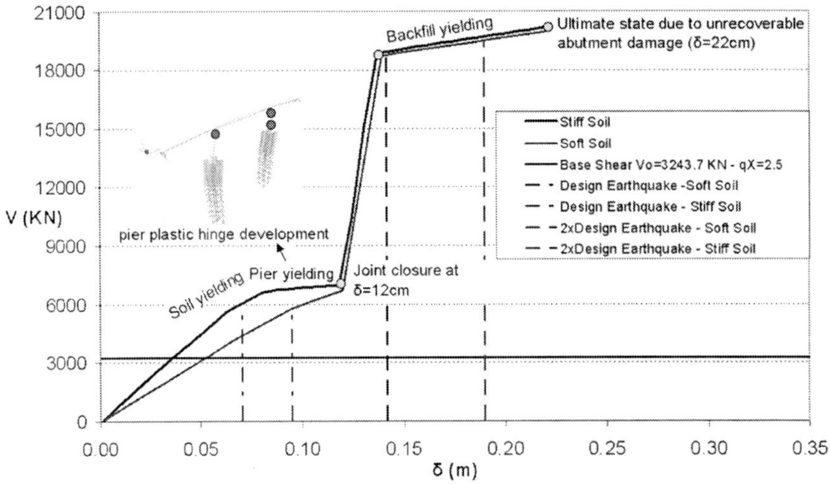

Figure 8. Pushover curves and seismic assessment of the bridge (longitudinal direction).

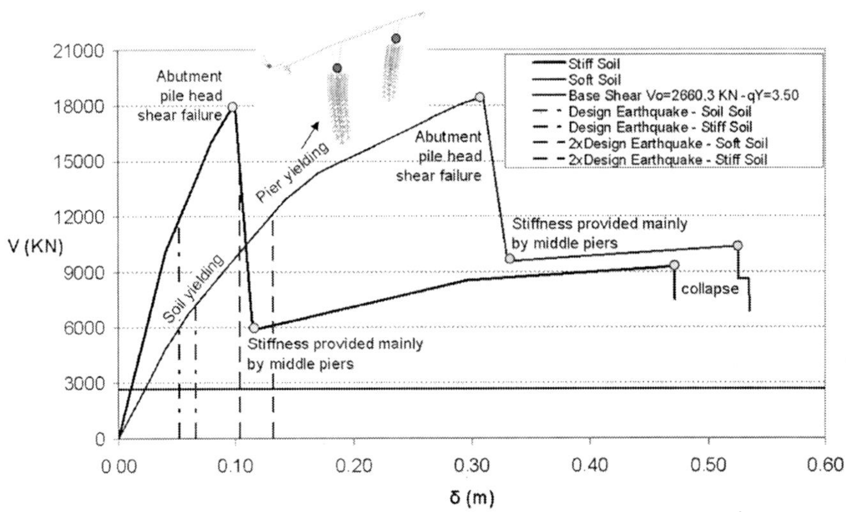

Figure 9. Pushover curves and seismic assessment of the bridge (transverse direction).

to $\delta = 31$ cm for the case of soft soil and $\delta = 10$ cm for the case of stiff soil, due to premature shear failure at the head of the abutment piles. At this stage, plastic hinges (with plastic rotation equal to 40–42% of the ultimate plastic rotation θ_{pu}) have developed at the base of the piers, but only for the case of soft foundation soil conditions. Once the abutment contribution to the system stiffness is eliminated due to damage to the piles supporting it, seismic forces are resisted mainly by the piers until bridge failure. It is notable that the particular design concept is not very commonly adopted, as the piers (that are monolithically connected to the deck) are typically designed as the primary lateral-load resisting elements. Therefore, it is necessary to evaluate the performance of the structure in the light of the (code-prescribed) anticipated level of seismic displacements, as done in the next section.

4.3. OVERALL ASSESSMENT

The target displacements of the bridge for the two directions, the two alternative soil conditions, and two earthquake levels (i.e., design earthquake and twice the design earthquake, which is a rather severe level of seismic action) are shown as vertical lines in Figs. 8 and 9. It is worth noting that whereas for medium and long-period structures and typical bilinear behaviour, the well-known equal displacement approximation is valid and target displacements can easily be calculated from elastic analysis (details are given in Potikas, 2006), this is not the case for nonlinear behaviour described by multilinear curves like those shown in Figs. 8 and 9 (and the associated hysteresis loops which can not be estimated from pushover analysis); hence the target displacements corresponding to stages subsequent to joint closure are subject to uncertainty. A more accurate estimation of the bridge response under forces significantly higher than those associated with the design earthquake intensity, should involve inelastic dynamic (time-history) analyses. Such analyses could also verify whether the *dynamic* soil-foundation-abutment-superstructure interaction, compared to the purely static approach adopted herein, essentially modifies the observed plastic hinge sequence and failure mechanisms.

It is observed from Figs. 8 and 9 that at least for the design earthquake, the bridge performance is very good, as no major damage is expected for both directions, independently of the soil type. This fact can be attributed up to a certain degree to the material overstrength, but, most importantly, to the rather conservative design approach adopted (i.e., cross-section stiffness was assumed for the piers). For twice the design earthquake, in the longitudinal direction, the joint is expected to close. Consequently, the stiffness of the overall bridge system in the longitudinal direction is significantly increased due to the activation of the backfill-abutment-foundation-soil subsystem. In the transverse direction, on the other hand, although damage is indeed minor for the case of soft foundation soil (even for displacements corresponding to twice the level of the design earthquake), the abutment piles are expected to suffer significant damage due to shear failure at their head when the supporting soil is stiff. This situation is clearly detrimental because the abutments can no longer resist even their own earth pressures, hence the bridge stability is jeopardized and the high ductility of the piers is never utilized.

5. Conclusions

The paper addressed a number of problems associated with nonlinearities encountered in the analysis of bridges interacting with the surrounding soil, with specific reference to an overpass bridge along the Egnatia Motorway in Greece. The performance of the bridge was assessed through nonlinear static (pushover) analysis accounting for various sources of material nonlinearity (backfill compliance, foundation soil yielding, pier and pile plastic hinge development, and pile failure in shear), as well as boundary nonlinearity (gap closure). Relatively simple, as well as refined (and expensive) models of the abutment-embankment-foundation soil subsystem were set up to define appropriate spring constants to apply at the ends of the deck of the bridge. Then, the inelastic behaviour of the bridge under seismic loads was analysed and finally a seismic assessment of the bridge was performed.

The analysis has shown that for a displacement corresponding to about twice the design earthquake, closure of the longitudinal joint is expected and the abutment-backfill-foundation-soil system is activated, a fact that drastically increases the resistance of the entire structure. The "full-range" pushover curves calculated from such an analysis are more complex than the typical bilinear ones found in the existing literature for both bridges and buildings; accommodating them in the pushover procedure poses a number of problems, for instance how the target displacement (for high earthquake actions) should be estimated. On the bright side, the analysis results show a very good behaviour of the bridge for the design earthquake and safety against collapse appears to be assured even for earthquakes much stronger than the design one.

In the transverse direction, yielding of piers leads to a full activation of the abutment-embankment-foundation soil system; failing to model this, completely distorts the actual picture of the bridge response. Moreover, the failure mechanism is different depending on whether the entire system is modelled or only a pin-supported bridge is considered, and also on whether the properties of the soil have been correctly estimated. For the extreme case of stiff soil (instead of the soft one considered in design), shear failure at the head of the abutment foundation piles was predicted by analysis, while piers were still well below their rotational capacity. Despite this being a severe worst-case scenario, this shear failure was predicted only for an earthquake intensity much higher than the design one. The encouraging implication from this observation is that modern code design ensures a very adequate safety margin against collapse, despite the code methods being unable to capture all salient features of the response of the bridge.

Based on the aforementioned observations, it is concluded that properly accounting for the abutment-soil stiffness is important, especially in relatively short bridges (such as the one studied here), hence refined modelling approaches are recommended, especially when a proper assessment of all possible failure modes is envisaged. Another notable finding is that the shape of the pushover curve calculated when gap closure is modelled in bridge analysis is different from the familiar bilinear one, which has several repercussions, as discussed previously. Finally, further investigation, especially towards the identification of the dynamic soil-abutment-embankment interaction effect is deemed necessary.

References

AASHTO Guide Specifications for LRFD Seismic Bridge Design (2008) American Association of State Motorway and Transportation Officials, Washington, DC.

ABAQUS/PRE (2004) Users Manual, Hibbit, Karlsson and Sorensen, Inc.

Caltrans (2006) Seismic Design Criteria, Ver. 1.4, California Department of Transportation.

CEN Techn. Comm. 250/SC8 (2005) Eurocode 8: Design provisions of structures for earthquake resistance, Part 2: Bridges (EN1998-2), CEN, Brussels.

Chopra, A. K. and Goel, R. K. (2002) A modal pushover analysis procedure for estimating seismic demands for buildings, *Earthquake Eng. Struct. Dynam.* **31**(3), 561–582.

Computers and Structures, Inc. (2005) SAP 2000 Nonlinear Ver. 10, User's Reference Manual, Berkeley, California.

Cubus (2000) FAGUS-4: Cross section analysis and check of concrete-, thin-walled, mixed- and post-tensionned cross-sections with biaxial bending, User's Manual, Zurich, Switzerland.

Goel, R. and Chopra, A. K. (2004) Extension of modal pushover analysis to compute member forces, *Earthquake Spectra* **21**(1), 125–139.

Isakovic, T., Fischinger, M., and Kante P. (2003) Bridges: when is single mode seismic analysis adequate? In *Proc. of the Institution of Civil Engineers—Structures and Buildings* **156**(2), pp. 165–173.

Kappos, A. J. (1993) RCCOLA-90: A Microcomputer Program for the Analysis of the Inelastic Response of Reinforced Concrete Sections, Department of Civil Engineering, Aristotle University of Thessaloniki, Greece.

Kappos, A. J. and Sextos, A. G. (2001) Effect of foundation type and compliance on seismic response of R/C bridges, *J. Bridge Eng., ASCE* **6**(2), 120–130.

Karantzikis, M. and Spyrakos, C. (2000) Seismic analysis of bridges including soil-abutment interaction. In *Proc. 12th World Conference on Earthquake Engineering*, No. 2471. New Zealand Earthquake, Auckland, New Zealand.

Mackie, K. and Stojadinovic, B. (2002) Bridge abutment model sensitivity for probabilistic seismic demand evaluation. In *Proc. 3rd Nat. Seismic Conference and Workshop on Bridges and Motorways*, Portland, USA.

Matlock, H. (1970) Correlations for design of laterally loaded piles in soft clay. In *Proc. 2nd Offshore Technology Conference*, Vol. 1, Houston, pp. 577–588.

Paraskeva, T. S., Kappos, A. J., and Sextos, A. G. (2006) Extension of modal pushover analysis to seismic assessment of bridges, *Earthquake Eng. Struct. Dynam.* **35**(10), 1269–1293.

Potikas, P. (2006) Seismic Design and Assessment of an Overpass Bridge, MSc Thesis, Aristotle University Thessaloniki, Greece (in Greek).

Priestley, M. J. N., Seible, F., and Calvi, G. M. (1996) *Seismic Design and Retrofit of Bridges*, Wiley, New York.

Sextos, A., Mackie, K., Stojadinovic, B., and Taskari, O. (2008) Simplified p-y relationships for modeling embankment-abutment systems of typical California bridges. In *Proc. 14th World Conference on Earthquake Engineering*, Beijing, China.

Zhang, J. and Makris, N. (2002) Kinematic response functions and dynamic stiffnesses of bridge embankments, *Earthquake Eng. Struct. Dynam.* **31**, 1933–1966.

STRUCTURAL RESPONSE TO COMPLEX SYNTHETIC GROUND MOTIONS

George D. Manolis (gdm@civil.auth.gr)*
and Asimina M. Athanatopoulou
Department of Civil Engineering, Aristotle University of Thessaloniki, 54124 Thessaloniki, Greece

Abstract. The purpose of this work is to study the response of 3D models of conventional multi-storey R/C buildings, in either bending or shearing mode behavior, in the presence or absence of asymmetries, as induced by artificial accelerations that take into account local site conditions. The first step is to model seismic waves propagating through complex geological profiles so as to recover artificial acceleration time histories at the surface of the ground. Next, we focus on the dynamic behavior of two common types of multi-storey buildings that comprise the modern building stock in Greece. These are modeled using finite elements and their dynamic response is computed for base motions in the form of artificial accelerograms and of recorded earthquake signals. The purpose is to compare the results of time-history analyses with the dynamic response spectrum method prescribed by the Greek National Earthquake Code, especially in the case where the buildings exhibit nonlinear material behavior.

Keywords: R/C buildings, time history analysis, artificial accelerograms, material nonlinearity, finite elements analyses

1. Introduction

Most modern American seismic codes (e.g., FEMA 356, 2000; NEHRP, 2003; UBC, 1997) suggest the use of time-stepping as the preferred method of analysis for residential and commercial structures in earthquake-prone zones past certain height and size standards. In contrast, the Greek seismic code (EAK, 2000) states that time-stepping should be used as a check on the results produced by the response spectrum analysis or by quasi-static analyses techniques. This, of course, pre-supposes the availability of ground motion time signals, be it recorded or synthetic in form. For instance (EAK, 2000) requires the use of at least five recorded signals that ought to be representative of the seismo-tectonic, geological and local site conditions of the geographical area where the structure is to be built. In case where actual

*Corresponding author.

T. Schanz and R. Iankov (eds.), *Coupled Site and Soil-Structure Interaction Effects with Application to Seismic Risk Mitigation,* NATO Science for Peace and Security Series C: Environmental Security,

recordings are absent, it is permissible to reproduce and use synthetic signals that not only account for seismicity particular to the broad geographical area, but also for local site conditions if possible.

Some open questions, however, remain. Specifically, do all these seismic signals, recorded or synthetic, produce a structural response that is consistent (in an overall sense) with what the seismic code-prescribed response spectrum analysis (RSA) method gives? Is this true in the linear elastic range as well as when material nonlinearities are induced through strong ground shaking? Do modeling details such as rigid floor diaphragms play a role, especially in the presence of directivity in the ground motions? What is the effect of soft soil layers in comparison with signals recovered at bedrock?

Thus, the main objective of this work is to try and answer some of these questions by studying the response of 3D models of conventional multi-storey reinforced concrete (R/C) buildings of mixed bending and shearing mode behaviour, in either the presence or absence of asymmetries (Athanatopoulou, 2005), as induced by both recorded seismic signals and artificial accelerations that take into account local site conditions. These buildings (see Figs. 1 and 2) are designed according to standard R/C design practice (EKOS, 2000) and are considered representative of the contemporary building stock for residential and office use in Greece.

2. The R/C Multi-Story Buildings

We consider two standard 3D residential building designs, consisting of a five-storey and an eight storey R/C frame. The first is an L-shaped, low-rise frame where all stories have the same floor plan (see Fig. 1). Total height is 16 m and the individual stories are 3.2 m high. Two variations are produced by changing the cross-section of the beam and columns proportionally so as to produce a stiff (A5) and a soft (B5) design.

Next, the medium-rise eight story frame has a rectangular floor plan and a total height of 28.4 m, with a soft first storey (h = 6 m) and the remaining floors at 3.2 m (see Fig. 2). Two variations are again produced by re-arranging the columns and the shear walls, thus producing designs with low (A8) and high (B8) eccentricities.

In all cases, the analysis tool used was the finite element method SAP 2000 (2005) computer program. The generalized beam element with six degrees-of-freedom (DOF) per node was used to model the beam and column spatial grid comprising the skeleton of the structure, while diaphragm action was assumed to hold for the flooring system. First, an eigenvalue analysis

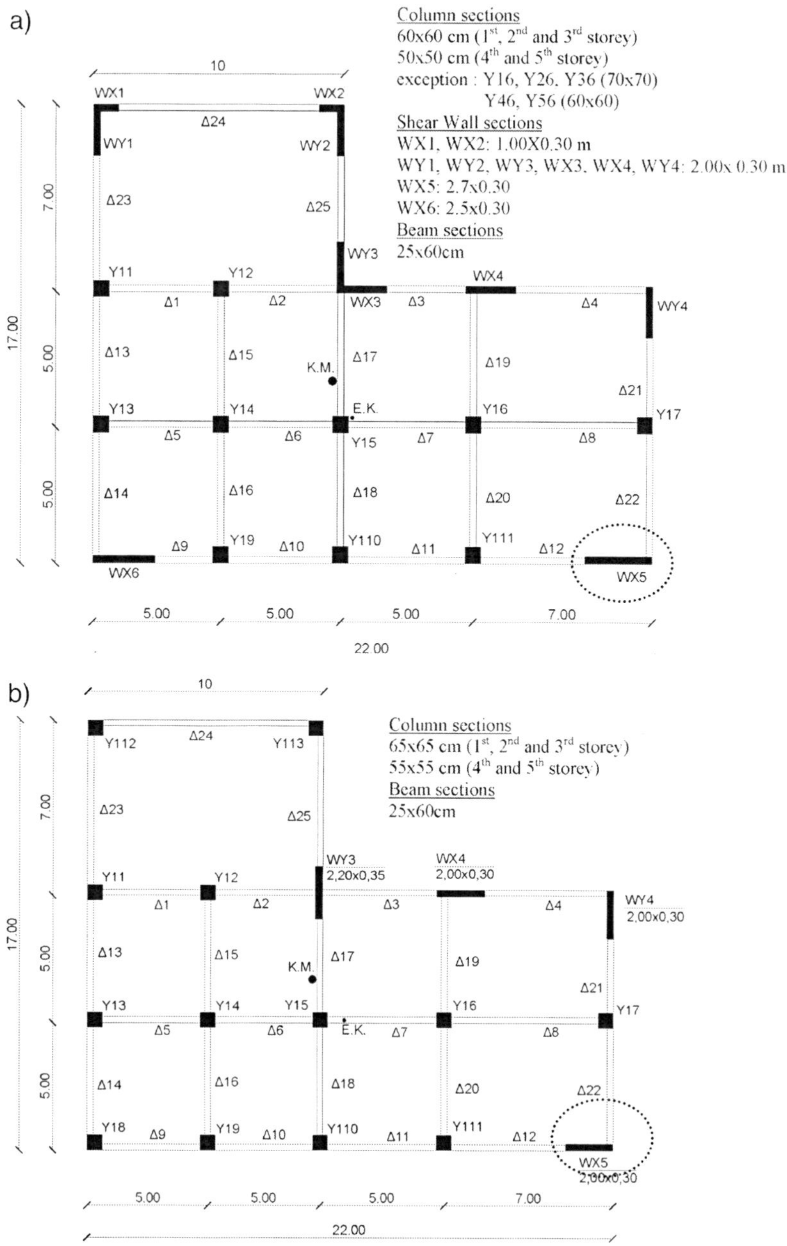

Figure 1. Floor plan of a five-story R/C building designed according to the National Greek Earthquake code (EAK, 2000): (a) version A5, with high stiffness and a first natural period T = 0.498 s; (b) version B5, with low stiffness and first natural period T = 0.645.

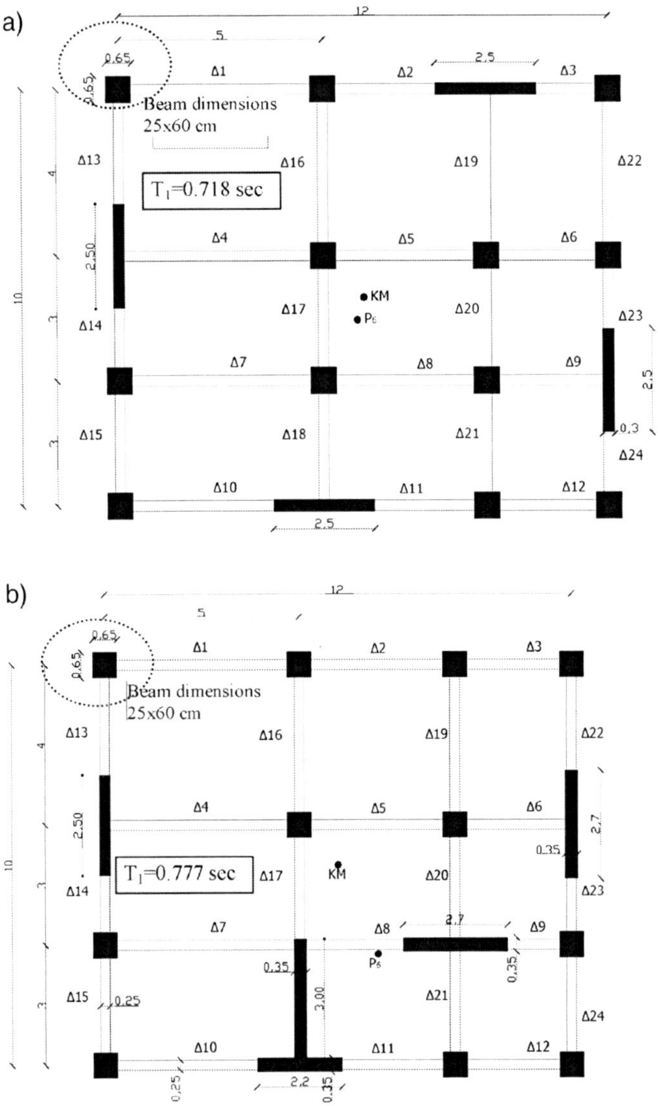

Figure 2. Floor plan of an eight story building designed according to EAK (2000): (a) version A8 with small eccentricity; (b) version B8 with large eccentricity. The first natural periods T_1 are roughly comparable in both versions.

was performed for all the building sub-cases to determine their fundamental natural frequencies and associated modal shapes. Next, regarding the ensuing time-history analyses, the Newmark-beta time integration algorithm was used in all cases with a time step gauges to be less than 1/20 of the smallest natural period of the structures for best quality results.

3. Synthetic and Recorded Ground Motions

In this work, we use on-line software that is accessible through the site `http://infoseismo.civil.auth.gr` (Manolis and Kapetas, 2007). Although the seismic wave propagation pattern used in the software is 1D, it is possible to account for soil layers over bedrock and (approximately) for the existence of tunnels or cavities in those layers. First, a family of five records is produced whose key parameters are seismic magnitude (M) in the Richter scale (6, 6.5, 7) and signal duration in seconds (20, 30). Next, a family of recently recorded earthquakes in the Greek domain is chosen as the second trial set, all of them public domain and downloadable from a national institute site (`www.itsak.gr`). In all cases, the records are normalized so their maximum spectral accelerations (SA) correspond to the Greek code (EAK, 2000) design spectrum (linear range, $q = 1$) for Zone II and for peak ground acceleration (PGA) = 0.24 g (2.354 m/s).

Figure 3 gives the spectra corresponding to the time histories for a damping ratio of $\zeta = 5\%$, along with the code-prescribed design spectrum. It should be noted here that the synthetic accelerations are defined for rock outcrop. In an ensuing section, we will introduce the presence of soft soil layers of horizontal structure and derive the corresponding signals.

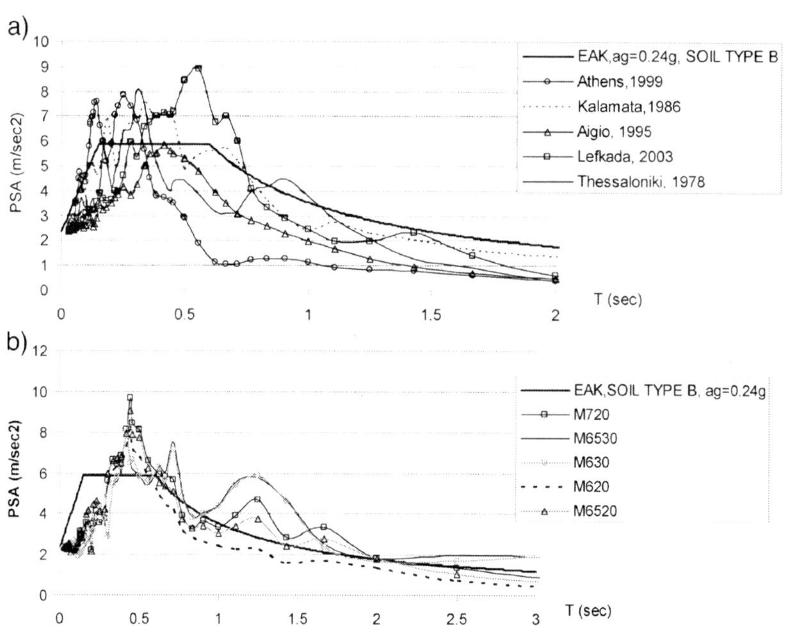

Figure 3. Earthquake acceleration spectra as functions of period T (s) corresponding to (a) recorded and (b) synthetic events.

4. Linear Elastic Analyses

The above four R/C building designs were allowed to behave in the linear elastic range for the given magnitude of the seismic signals that were shown in Fig. 3. Specifically, a given accelerogram (synthetic or recorded) was imposed simultaneously along two horizontal orthogonal directions. Two analyses were conducted for two different orientations of the angle of attack that allow for determination of maximum response values (Athanatopoulou, 2005) at any incidence angle using the SRSS rule.

Figure 4 collects maximum values for the displacements in the horizontal plane at the first floors of the two five story building designs, where shear wall WX5 ends. Also, Fig. 5 shows the same displacements at the top of column Y11 for the two eight story building sub-cases. A comparison of all the above results, and of the full set of information produced (Tsirikoglou and Batsiou, 2007), gives the following conclusions:

- All synthetic accelerograms induce a response in the buildings that is more pronounced compared to the code prescribed response spectrum analysis (RSA) response.
- For fixed earthquake magnitude, an increase in the duration of the seismic signal (e.g., going from M620 to M630) causes smaller displacements (and stresses) in the stiff five-story building, and in both eight-story ones. The only exception is the flexible five-story building. A similar situation was observed for the force response.
- As expected, an increase in the earthquake magnitude (e.g., going form M620 to M720) increases the response in all cases.
- Comparing synthetic and recorded signals of comparable magnitude and duration (e.g., Kalamata 1986 vs. M630, Thessaloniki 1978 vs. M6530,

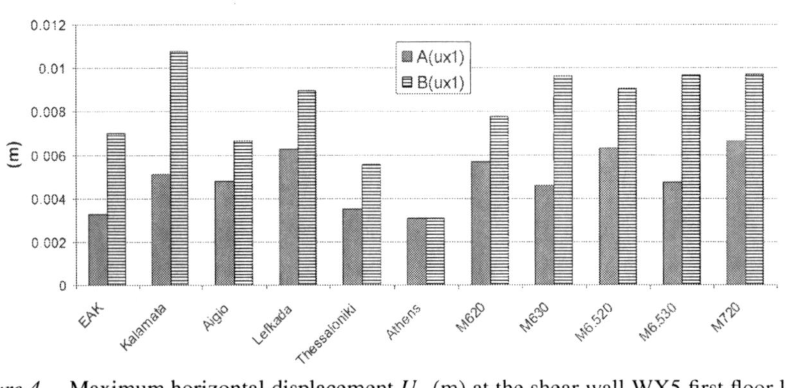

Figure 4. Maximum horizontal displacement U_x (m) at the shear wall WX5 first floor level for both versions (A5, B5) of the five story building.

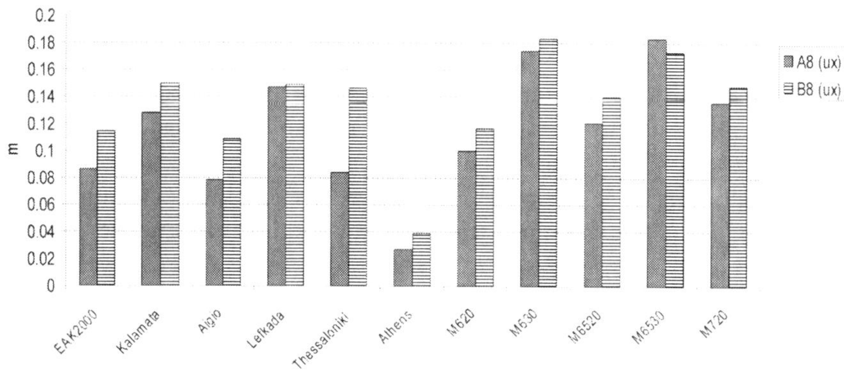

Figure 5. Maximum horizontal displacement U_x (m) at the top of column Υ11 for both versions (A8, B8) of the eight story building.

Aigion 1995 vs. M6530), the buildings' response is more pronounced for the former category, especially for the eight-story buildings. The only exception is again the flexible version of the five-story building.

- The largest displacements are observed in the eight-story building with the larger eccentricity (B8). Only two signals, namely Lefkada 2003 and M6530, produced greater displacements in the eight-story building with the smaller eccentricity (A8).

5. Non-Linear Analyses

In order to study the nonlinear response of the R/C buildings, it was first necessary to do a complete design based on the results of the previous linear analyses. The design was done according to the Greek R/C design code (EKOS, 2000) and member cross-section steel reinforcement was computed, which allowed evaluation of the maximum plastic moment capacity of that member. Next, the rather limited type of nonlinear capability of SAP 2000 (2005) was utilized by introducing 1D link elements (essentially a spring with three translational and three rotational DOF) capable of representing potential plastic hinge formation at select locations, primarily the column and beam intersections (joints). In the ensuing non-linear, direct time integration analyses, it was necessary to consider three directions for the mutually perpendicular seismic signals, namely $\theta = 0°$, $\theta = 45°$, $\theta = 90°$ with respect to the principal axes of the buildings, because the SRSS rule no longer holds.

We also note here that since the five-story R/C building A5 and B5 gave limited amount of yielding, all accelerograms in Fig. 3 that were used as input were scaled by a factor of 1.5 and the analyses were repeated.

The number of plastic hinges that forms in a building is widely accepted as a basic index for sustained damage (Chopra and Goel, 2002). Also quite important is the location of hinge formation (beam vs. column vs. shear wall), because it shows if repairs can be made or if the building is lost. Here, Figs. 6 and 7 list the total number of hinges that form in the five story buildings and similarly Figs. 8 and 9 in the eight story buildings, as functions of the incident angle of the incoming seismic signal with respect to the major X axis (i.e., $\theta = 0°, 45°$).

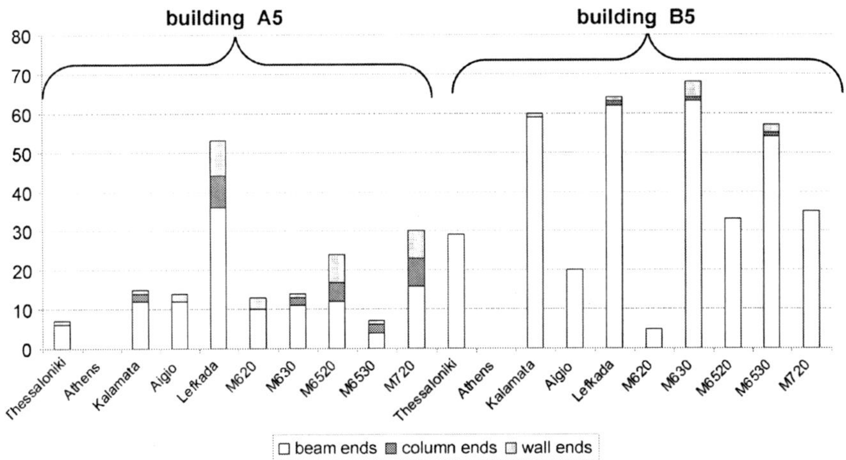

Figure 6. Five-story building: total number and distribution (beams; columns; shear walls) of plastic hinges for an incident angle of $\theta = 0°$.

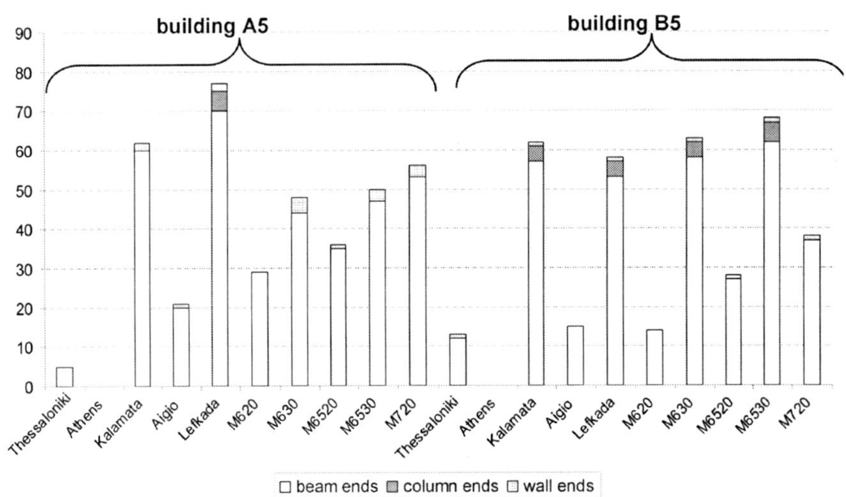

Figure 7. Five-story building: total number and distribution (beams; columns; shear walls) of plastic hinges for an incident angle of $\theta = 45°$.

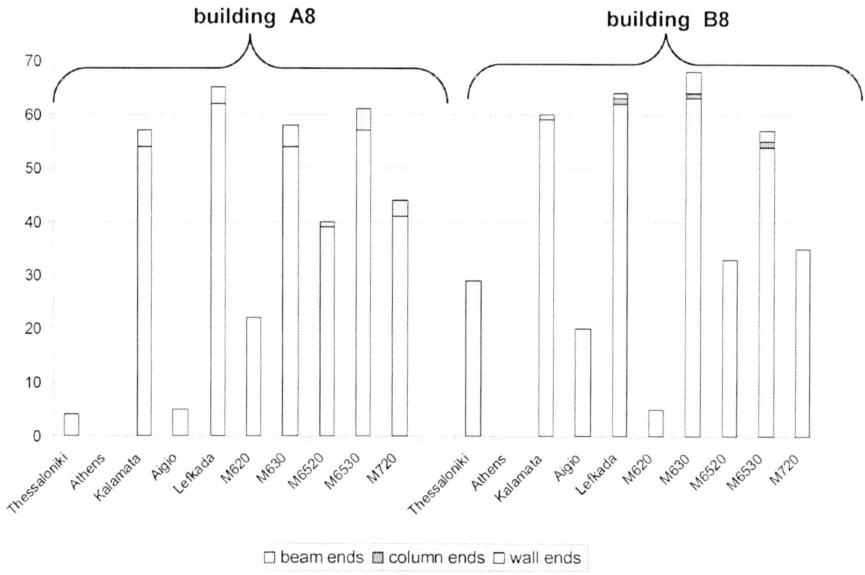

Figure 8. Eight-story building: total number and distribution (beams; columns; shear walls) of plastic hinges for an incident angle of $\theta = 0°$.

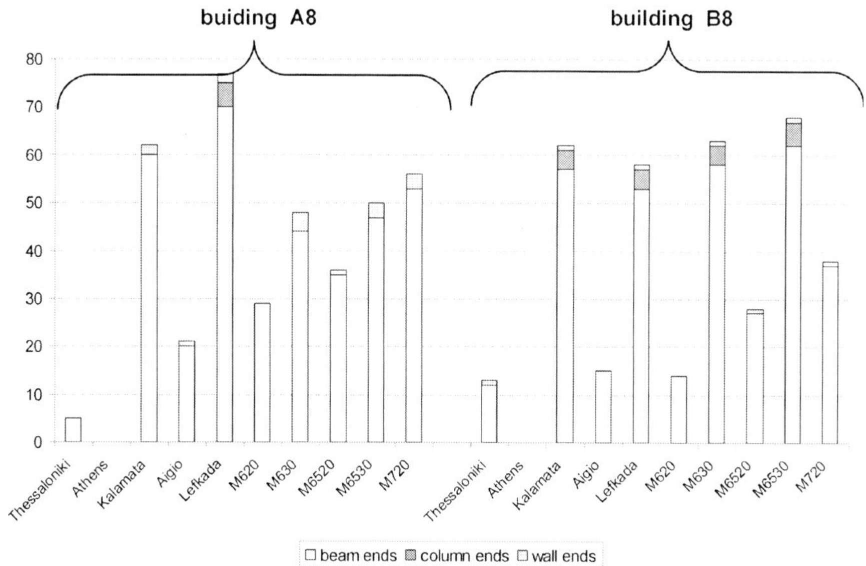

Figure 9. Eight-story building: total number and distribution (beams; columns; shear walls) of plastic hinges for an incident angle of $\theta = 45°$.

A synopsis of the results of the non-linear analyses is as follows:

- In all cases, the angle of orientation of the input seismic signal influences both the number of plastic hinges that eventually form as well as their location (e.g., on the beam or on the column side of the joint).
- As the magnitude of the earthquake that produces the synthetic signal increases (e.g., M620 to M720), so does the number of plastic hinges that form. This does not hold true for all recorded signals (e.g., the Thessaloniki 1978 vs. the Kalamata 1986 earthquakes).
- For fixed magnitude, synthetic signals of greater time duration (e.g., M620 vs. M630; M6520 vs. M6530) result in an increase of the number of plastic hinges. Again, this does not hold true for recorded signals (e.g., the Athens 1999 vs. the Kalamata 1986 earthquakes).
- Comparing synthetic and recorded input signals of comparable magnitude and duration (e.g, M6530 vs. Thessaloniki 1978 or Aigion 1995; M630 vs. Kalamata 1986), the former category always result in a much larger number of plastic hinge formations than the latter.
- Increased stiffness in a building (e.g., A5 vs. B5) greatly influences both the number and location of plastic hinge formation.
- Increased eccentricity in a building (e.g., A8 vs. B8) also influences both the number and location of plastic hinge formation, but the orientation angle of the input seismic signal plays a major role.

6. Layered Soil Profiles

In this section, we introduce soil layering on top of bedrock, where the synthetic signals were hitherto recovered. Three configurations, comprising three layers each, were considered (300/600/900, 300/900/600, 900/600/300), where the numbers correspond to the shear wave speed V_s (m/s) of the layer in question starting from the free surface and moving downwards. Also, the thickness d (m) of each layer was kept constant at 10 m. Next, the bedrock synthetic signal M6530 was introduced at the bottom of each layer formation and convolution-type calculations using software available at site `http://infoseismo.civil.auth.gr` were done to generate the free surface acceleration time histories. In all cases, the synthetic signals at the free surface were normalized by a value of PGA = 0.24 g. Finally, Fig. 10 plots the spectral accelerations of the synthetic signals for the three soil layer configurations against the spectral accelerations corresponding to the generating signal, namely M6530, and the code-prescribed RSA.

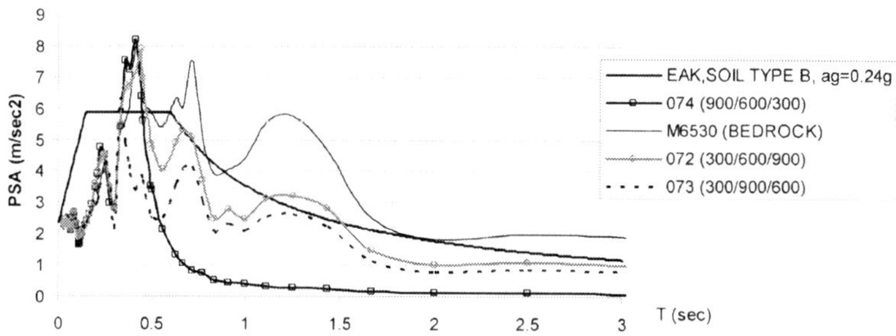

Figure 10. Spectral accelerations for the three layered profiles (codes 072, 073 and code 074) and for the generating M6530 bedrock synthetic signal.

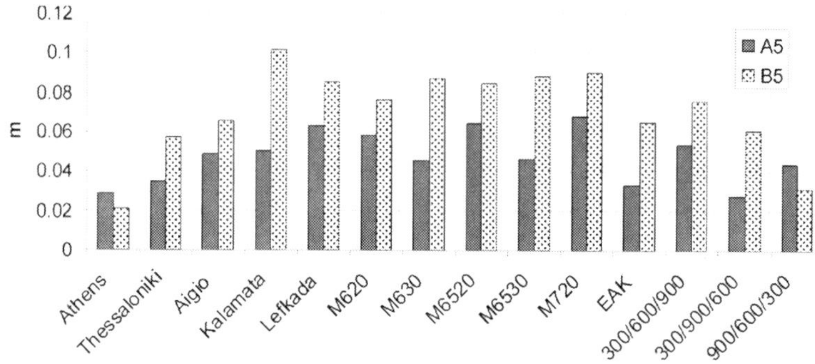

Figure 11. Maximum values for displacement component U_x (m) at the fifth floor (position WX5) for five recorded and eight synthetic earthquakes and the design spectrum of EAK (2000).

6.1. LINEAR ELASTIC ANALYSES

As before, the previous four building configurations were analyzed with these new signals as input in the linear elastic range. At first, Figs. 11 and 12 plot maximum values for displacement components U_x, U_y at the fifth floor, position WX5, of models A5 and B5.

Additional results for U_x and M_y are plotted in Figs. 13 and 14 for the eight story building versions with small (A8) and pronounced (B8) eccentricities, respectively, along the height of column Y11. More information can be found in Tsirikoglou and Batsiou (2007).

In all cases we see that the presence of layering does not change the response of the buildings by much; in most cases, it seems to be beneficial in that the maximum elastic response for the displacement and force fields is somewhat less pronounced. In particular, the 'reverse' profile where the layers progressively decrease in stiffness with depth seems to be the most effective filter for the seismic motions and results in a rather noticeable reduction of the response.

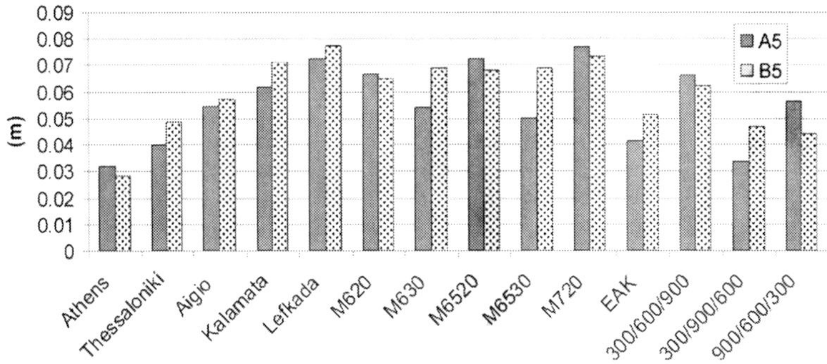

Figure 12. Maximum values for displacement component U_y (m) at the fifth floor (position WX5) for five recorded and eight synthetic earthquakes and the design spectrum of EAK (2000).

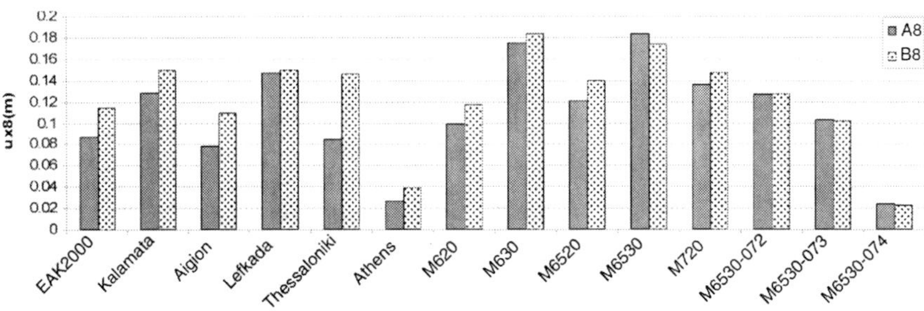

Figure 13. Maximum values for horizontal displacement U_x (m) at the eighth floor (column Y11) for five recorded and eight synthetic earthquakes and the design spectrum of EAK (2000).

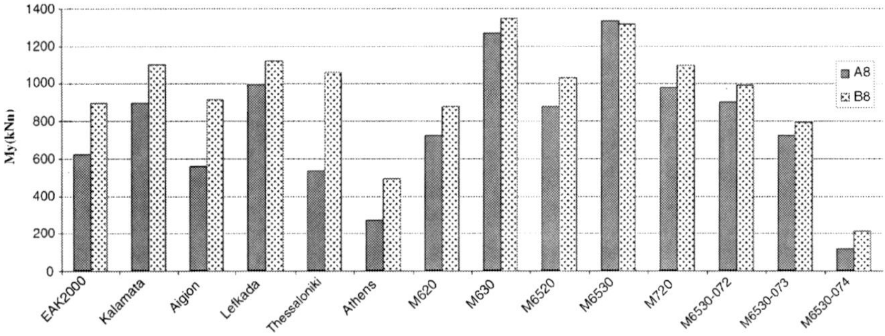

Figure 14. Maximum values for bending moment M_y (kN-m) at the base of column Y11 for five recorded and eight synthetic earthquakes and the design spectrum of EAK (2000).

6.2. NON-LINEAR ANALYSES

As before, the buildings were analyzed for nonlinear behavior by placing potential plastic hinges in all the joints and repeating the time history analysis. Listed below are total numbers of plastic hinges forming in the five story buildings A5 and B5 (see Fig. 15, while the same information for the eight story buildings A8 and B8 is given in Fig. 16). In both cases, the incident angle for the ground motion signal direction is $\theta = 0°$.

Consistent with the results of the preceding elastic analyses is the fact that relatively few plastic hinges form in the buildings when the seismic input is filtered through the soil layers, especially when a "reverse" profile with decreasing stiffness is present. One thing that should be mentioned, however,

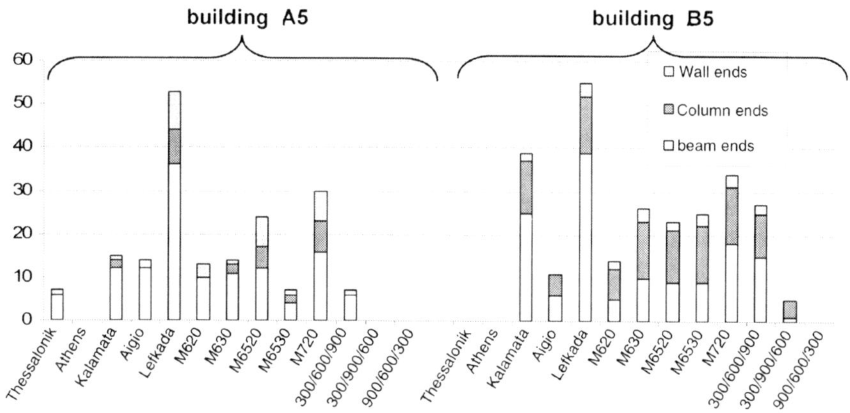

Figure 15. Five-story buildings (A5 and B5): total number and distribution of plastic hinges for an incident angle of $\theta = 0°$. Input is five recorded and eight synthetic signals

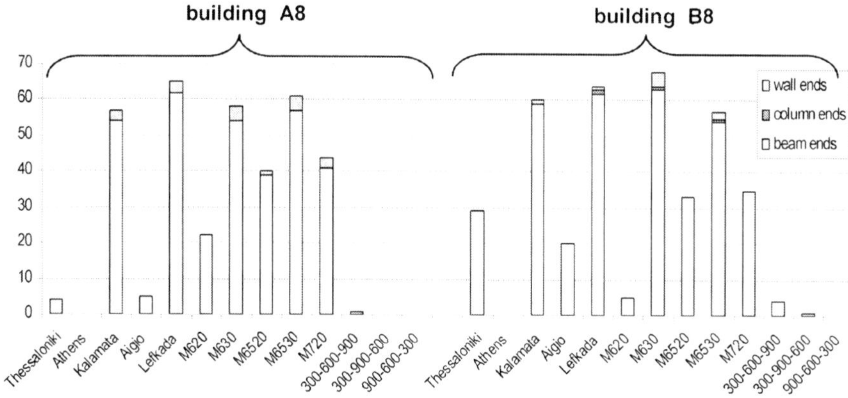

Figure 16. Eight-story buildings (A8 and B8): total number and distribution of plastic hinges for an incident angle of $\theta = 0°$. Input is five recorded and eight synthetic signals.

is the fact that soil layering produces large accelerations at certain instances that are lost in the scaling of the signals by the PGA concept.

7. Summary and Conclusions

In sum, the differences observed in the building response between the RSA and the time history analysis with seismic acceleration input are comparable in the linear elastic range as far as maximum values are concerned, but diverge when nonlinear response is manifested. A more careful observation of the corresponding spectra shows that the synthetic signals give relatively large acceleration values in three discrete ranges, namely at 0.5, 0.8 and 1.2 s. In contrast, the recorded signals show sizable spectral accelerations in the low (0.20–0.45 s) period range, and subsequently decrease. Given that the R/C buildings had fundamental periods of 0.50–0.64 s (the low-rise ones) and 0.72–0.78 s (the mid-rise ones) explains in part why the synthetic signals are more deleterious, since they impart more seismic energy to the structure and then decay. The introduction of soil layering over bedrock in the synthetic signals changes the picture completely. Soil layering produces a filtering effect in the higher period range, which in the seismic code is accounted for by using different drop curves for spectral accelerations in poor soil categories past the 0.56 s period range. Finally, it is possible to see large amplifications in the spectral accelerations, which are lost when the synthetic signals are normalized in accordance with the PGA concept. This last observation should be taken into account when examining local site effects.

References

Athanatopoulou, A. M. (2005) Critical orientation of three correlated seismic components, *Journal of Engineering Structures* **27**(2), 301–312.

Chopra, A. K. and Goel, R. K. (2002) A modal pushover analysis procedure for estimating seismic demands for buildings, *Earthquake Engineering and Structural Dynamics* **31**(3), 561–582.

EAK (2000) *National Greek Earthquake Design Code*, Athens, OASP Publication.

EKOS (2000) *National Greek Reinforced Concrete Design Code*, Athens, OASP Publication.

FEMA 356 (2000) Pre-standard and Commentary for the Seismic Rehabilitation of Buildings, Federal Emergency Management Agency, Washington, DC, USA.

Manolis, G. D. and Kapetas, M. S (2007) The effect of inhomogeneities on the dynamic response of layered soil with variable damping. In *Proc. 4th International Conference Earthquake Geotechnical Engineering*, Thessaloniki, Greece, Paper No. 1191.

NEHRP (2003) National Earthquake Hazard Reduction Program, Recommended Provisions for the Development of Seismic Regulations for New Buildings, Building Seismic Safety Council (BSSC), Washington, DC, USA.

SAP 2000 (2005) Structural Analysis Program, Integrated Finite Element Analysis and Design of Structures, Berkeley, California, USA, Computers and Structures, Inc.
Tsirikoglou, M. and Batsiou, E. (2007) Parametric analysis of buildings under both recorded and synthetic accelerograms, CE MSc Thesis, Aristotle University, Thessaloniki, Greece.
UBC (1997) Uniform Building Code, Vol. 2 on *Structural Engineering Design Provisions*, Int. Conf. Building Officials, Whittier, California, USA.

SINGLE AND MULTI-PLATFORM SIMULATION OF LINEAR AND NON-LINEAR BRIDGE-SOIL SYSTEMS

Anastasios G. Sextos (asextos@civil.auth.gr)* and Olympia Taskari
Department of Civil Engineering, Aristotle University of Thessaloniki, 54124 Thessaloniki, Greece

Abstract. Advanced computational tools are currently available for the dynamic analysis of bridge structures considering the complex phenomenon of embankment-foundation-abutment-superstructure interaction. Along these lines, analysis approaches of increasing complexity are applied for the study of a real, already built overcrossing along the Egnatia highway in Greece, each time making the appropriate assumptions and using purpose-specific software. Both single and multi-platform analysis is employed and the limitations and advantages of each approach are comparatively outlined and discussed in the linear and non-linear range. The results indicate that the last generation of computational tools available, is a promising alternative and certainly contribute towards the representation of the soil-foundation-bridge system as a whole and thus, the more accurate study of interaction problems that was not feasible to be examined in the past.

Keywords: abutment-embankment system, soil-structure interaction, multi-platform analysis

1. Introduction

The importance of soil-structure interaction for the assessment of the dynamic response of bridges has been widely recognized in numerous research studies. Consideration of the contribution of bridge lateral boundary conditions in the overall seismic response of bridges, has illustrated the significant role played by the embankment-foundation-abutment system not only in terms of the dynamic characteristics and response of the bridge (Goel and Chopra, 1997; Mackie and Stojadinovic, 2002; Dicleli, 2005; Kotsoglou and Pantazopoulou, 2007) but also regarding the modification of the incoming seismic motion (Zhang and Makris, 2002). Earthquake damage reports and laboratory tests have also indicated that abutment failure commonly caused by rotational and/or translational outward movement of the toe or even loss of subsoil bearing capacity is fairly common, hence refined analysis of the

*Corresponding author.

T. Schanz and R. Iankov (eds.), *Coupled Site and Soil-Structure Interaction Effects with Application to Seismic Risk Mitigation,* NATO Science for Peace and Security Series C: Environmental Security,

overall system is required. As a result, it is indeed a challenge to implement the computational tools and resources required to simulate the multi-parametric and complex nature of both the dynamic pier-foundation-subsoil and deck-abutment-embankment interaction as well as the shear deformation and failure of RC members (i.e., piers and piles), since coupled modelling of all these systems still requires extensive computation effort due to the model size and/or behavior complexity. It can be also argued that given the above complexity and computational demand, it is rather subjective whether a single software package exists that could possibly combine all the features required for advanced simulation of the non-linear response of bridge, foundations and abutments and their supporting soil.

Along these lines, this paper aims at investigating the application of distributed computational simulation as a means to comparatively assess the limitations and challenges of the most advanced modelling approaches currently available for the study of complex SSI systems. It is noted that, multi-platform simulation is one of the most promising approaches of this kind and was initially developed to accommodate multi-site hybrid simulation (Spencer et al., 2006). In particular, he dynamic response of full scale specimens that are physically separated is properly controlled with the use of purpose-specific coordination software that made feasible the incorporation of various numerical analysis platforms in the sub-structuring process. This concept has also been successfully applied (Kwon and Elnashai, 2008) for the coordination of purely numerical analysis modules (in contrast to the hybrid simulation application) for the case of real bridges in the U.S. for various soil conditions, as well as for the study of the potential impact of liquefaction susceptibility (Kwon et al., 2008). The advantage of this approach is that the appropriate selection and combination of different analysis packages, enables the concurrent use of the most sophisticated models and features of each package for each corresponding part of the system. In other words, different software can be used for different system components (i.e., abutments, superstructure and supporting pile groups) depending on the foreseen material constitutive laws and geometry.

In order to investigate the range of applicability of the advanced computational tools and methods currently available for simulating the embankment-abutment-bridge interaction, a typical, real and already built, overcrossing in Greece is chosen to serve as a benchmark and four different alternative modelling approaches are explored, namely: (1) a bridge frame model supported on complex dynamic impedance matrices that are specifically calculated for pile foundations and abutments; (2) a 3-dimensional spring-supported frame model consisting of the bridge, its abutment and its foundation, (3) a refined 3-dimensional solid model of the overall superstructure-abutment-embankment system and; (4) a multi-platform scheme (Kwon and Elnashai, 2008) using appropriate system sub-structuring. The analysis is performed both in the linear and the non-linear range. An overview of the bridge structure studied and the comparative assessment of the aforementioned approaches is presented in the following.

2. Overview of the Bridge Studied

The particular bridge adopted for study is an overpass (overcrossing) along the Egnatia highway, a large road network that has been constructed in northern Greece with more than 646 bridges built of a total of 40 km length most of which are structures of relatively small dimensions (i.e., $L < 100$ m). The particular bridge studied is a three-span, symmetric structure of 70 m length (span lengths are 19, 32 and 19 m, respectively) curved in elevation (maximum camber of 8%), that intersects the highway axis at an angle of 75.3° (Fig. 1). The deck is 11 m wide and 1.60 m high. The prestressed deck has a hollow T-beam-like section and is supported on two circular piers of 1.70 m diameter and 8.50 m height which are monolithically connected to the superstructure and the foundation. At the abutments (which have a 10.50 × 1.20 m wall section of 5.0 m height), the deck is connected through two pot bearings that permit sliding along the two principal bridge axes and a sliding joint

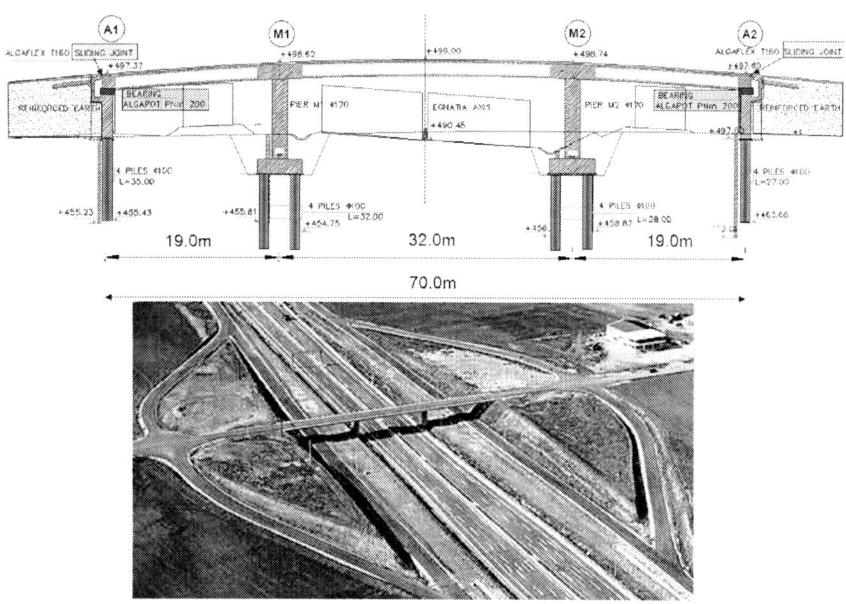

Figure 1. Longitudinal cross-section of the bridge (*above*) and indicative overview of a typical overcrossing along Egnatia Highway (*bottom*).

separates the deck from the backwall. Seismic forces are also resisted by the activation of stoppers (in the transverse direction) which are constructed at the seating of the abutments. The foundation on the other hand is deep, due to the soft clay formations characterizing the overall area. The pier foundation consists of a 2 × 2 pile group of 28.0–32.0 m long piles, connected with a 1.60 × 5.0 × 5.0 m pile cap, while the abutments are supported on a 1 × 4 pile row 27 to 35.0 m long at 2.80 m axial spacing, all piles having equal diameter of 1.0 m. The bridge was designed for normal loads according to the German Norms (i.e., DIN 1055, 1045, 1072, 1075, 1054, 4227, 4085, 4014) while the seismic design was carried out according to the Greek Seismic Code EAK 2000 (Ministry of Environment, physical planning, and public works, 2000) and the relevant Greek standards E39/99 (Ministry of Environment, physical planning, and public works,1999) for the seismic design of bridges. The bridge site is located in the Seismic Zone I which is equivalent to a peak ground acceleration of 0.16 g. The behaviour factors of the system adopted for design according to the E39/99 document were $q_x = 2.50$, $q_y = 3.50$ and $q_z = 1.00$ for the response in the three principal directions, respectively. The target displacements of the bridge under study for the two directions, the two alternative soil conditions and the two earthquake levels (i.e., design earthquake and twice the design earthquake) are also depicted in Fig. 2 (the complete calculation process can be found in Potikas, 2006). It is noted that for twice the design earthquake in the longitudinal direction, the joint is expected to close. Consequently, the overall bridge system stiffness in the longitudinal direction is significantly increased due to the activation of the backfill-abutment-foundation-soil subsystem. It is also noted that in the transverse direction, although damage is indeed minor for the case of soft foundation soil even for displacements corresponding to twice the level of the design earthquake, the abutment piles were found to suffer significant damage

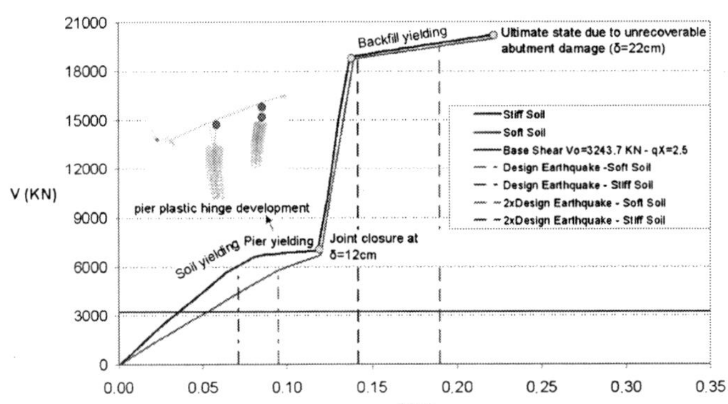

Figure 2. Pushover curve and seismic assessment of the overall system studied in the longitudinal direction for two different soil categories (after Kappos et al., 2007).

due to shear failure at their head when the supporting soil is stiff (Kappos et al., 2007). This situation is apparently detrimental because the abutments can no longer resist their own earth pressure, hence the bridge stability is jeopardized and the high ductility of the middle piers is never utilized. Given the above observations it is clear that for the particular bridge under study, the role played by the abutment is crucial and hence the appropriate modelling of the bridge lateral boundary conditions is necessary.

3. Computational Framework

In order to investigate and demonstrate the current capabilities of the various analysis approaches, four different models were developed. The assumptions made in each cases and the performance of all models is summarized in Table I, while a brief description of the overall concept is described in the following:

3.1. FRAME BRIDGE ON SPRING AND DASHPOT SYSTEMS (MODEL 1)

First, a bridge frame model supported on complex dynamic impedance matrices that are specifically calculated for pile foundations and abutments is developed. This consists of the superstructure (Fig. 3) whose pies are assumed to be connected to 6-DOF spring and dashpot systems with dynamic properties computed using the computer code ASING (Sextos et al., 2003) for coupled translational and rocking modes of vibration and given the foundation and soil properties described in Section 2. Pile-to-pile interaction was accounted through the formulation of the particular computer code. The abutment dynamic stiffness and damping is computed according to Zhang and Makris (2002). Kinematic interaction was ignored. The analysis is performed using the widely used FE program Sap2000 and represents the most refined approach that can be implemented in the design practice.

3.2. FRAME BRIDGE ON SPRING SUPPORTED ABUTMENT AND FOUNDATION (MODEL 2)

This approach involves the FE model illustrated in Fig. 4 inclusive of the superstructure, the abutments as modelled with 2D shell elements in 3D space, as well as the pile foundations modelled using beam-on-dynamic springs. Spring and dashpot values were computed as in Model 1 but distributed based on the area of influence of each particular spring. The analysis is performed using the widely used FE program Sap2000.

TABLE I. Summary of available features of the four approaches under study.

Model Features	Model 1	Model 2
Middle piers pile group foundation stiffness	Point Spring	Spring supported Frame
Middle piers pile group damping	✓	✓
Abutment stiffness	Spring	3D on shell
Embankment stiffness	Spring	Spring
Abutment foundation stiffness	✗	3D on shell
Embankment mass	Point mass	Point mass
Embankment soil stresses	✗	✗
Backfill soil stresses	✗	✗
Soil nonlinearity potential	Spring (p–y)	Spring (p–y)
R/C nonlinearity potential	Lamped plasticity	Lamped plasticity
Computational time*	2 min	4 min

Model Features	Model 3	Model 4
Middle piers pile group foundation stiffness	Point Spring	Fixed
Middle piers pile group damping	✗	✗
Abutment stiffness	3D	3D
Embankment stiffness	3D	3D
Abutment foundation stiffness	3D ✓	3D ✓
Embankment mass	3D ✓	3D ✓
Embankment soil stresses	✓	✓
Backfill soil stresses	✓	✓
Soil nonlinearity potential	Abaqus built-in models (Mohr–Coulomb Drucker–Prager, Cam–Clay)	Multiple software choices (Opensees, Abaqus)
R/C nonlinearity potential	Abaqus Built-in constitutive law	Multiple software choices (Opensees Zeus, FedeasLab)
Computational time*	2 ½ h	2 ¼ h

* Integration time refers to 10 s duration of uniform ground excitation at 0.01 s step executed on a Core 2 Duo processor, 2 GB RAM

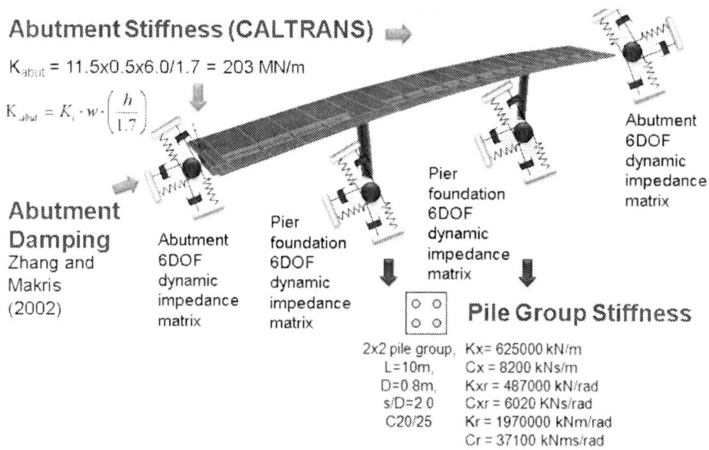

Figure 3. Overview of dynamic spring-supported Model 1.

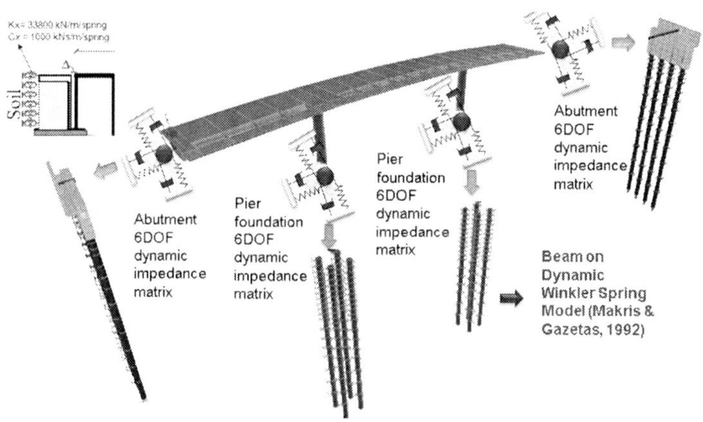

Figure 4. Overview of dynamic spring-supported Model 2.

3.3. FRAME BRIDGE ON 3D SOLID EMBANKMENT-FOUNDATION-ABUTMENT (MODEL 3)

Bridge superstructure is discretized using 3D frame elements which are then connected to a 3-D (solid) abutment-foundation-embankment system at both lateral supports of the deck (Fig. 5). The piers are assumed to be supported on the 6-DOF dynamic impedance matrices described above while retaining the same properties as previously. Soil is assumed as linear elastic for comparison purposes. The analysis is performed with the advanced FE software Abaqus.

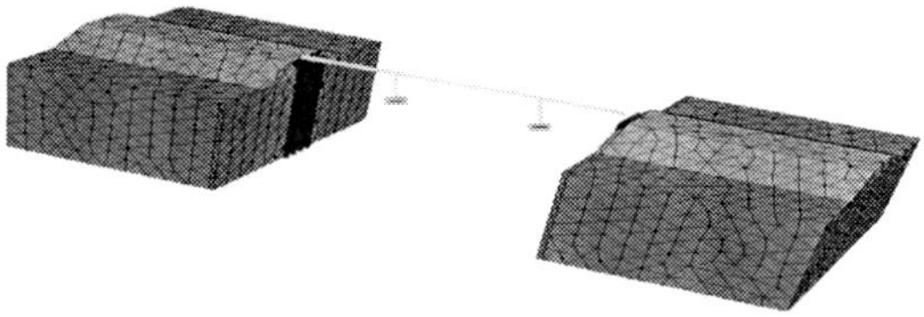

Figure 5. Overview of the 3-dimensional Model 3.

3.4. DISTRIBUTED SIMULATION: FRAME BRIDGE ON 3D SOIL EMBANKMENT-FOUNDATION-ABUTMENT SYSTEM (MODEL 4A)

In this approach, which is the most refined compared to the previously described ones, the structure is subdivided into several modules that are computationally simulated using different computer codes. The analysis of the distributed modules is coordinated with the aid of UI-SimCor (Spencer et al., 2006), an enhanced `Matlab` based script with its own GUI that was developed by University of Illinois in order to coordinate either software or hardware supporting NEESgrid Teleoperation Control Protocol (NTCP) as well as TCP-IP connections outside of the NEES system. The basic concept of the framework is that analytical models associated with various platforms or experimental specimens are considered as super-elements with many DOFs.

The main routine enforces static equilibrium during gravity load application and conducts dynamic time integration thereafter. Each of these elements are solved on a single computer or on different computers connected through the network. Interface programs for analytical platforms have been developed for Zeus-NL (Elnashai et al., 2002), OpenSees (McKenna and Fenves, 2001), FedeasLab (Filippou and Constantinides, 2004), and ABAQUS. In the particular analysis of the Egnatia Highway overcrossing, two different analysis packages were coordinated by UI-Simcor, corresponding to three distributed modules, namely (Figs. 6 and 7): (1) the bridge sub-system, which was modelled using the verified inelastic dynamic analysis program FedeasLab (Filippou and Constantinides, 2004), the left (2) and right (3), pile-supported, abutment-embankment system that was modelled using 3D solid elements and the commercial FE package Abaqus (Model 4a in Fig. 8). In order to minimize computational time, a relatively simpler 3D abutment-embankment system was adopted after appropriate calibration of damping and stiffness along the overall height with the refined 3-dimensional soil Model 3.

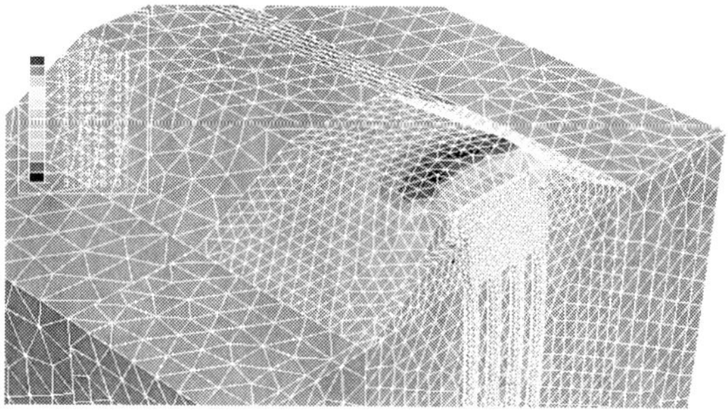

Figure 6. Plastic strains developed due to non-linear static (pushover) analysis of Model 3.

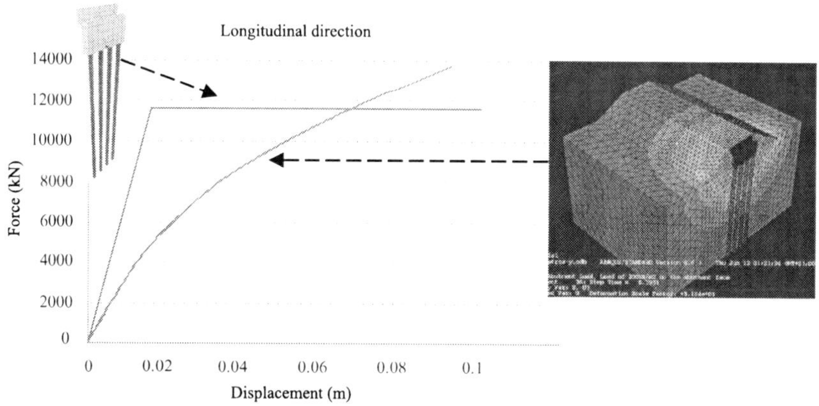

Figure 7. Comparison of the pushover curves derived for the spring supported abutment (Kappos et al., 2007) and 3D model of the Egnatia Highway bridge along the longitudinal direction (Sextos et al., 2008).

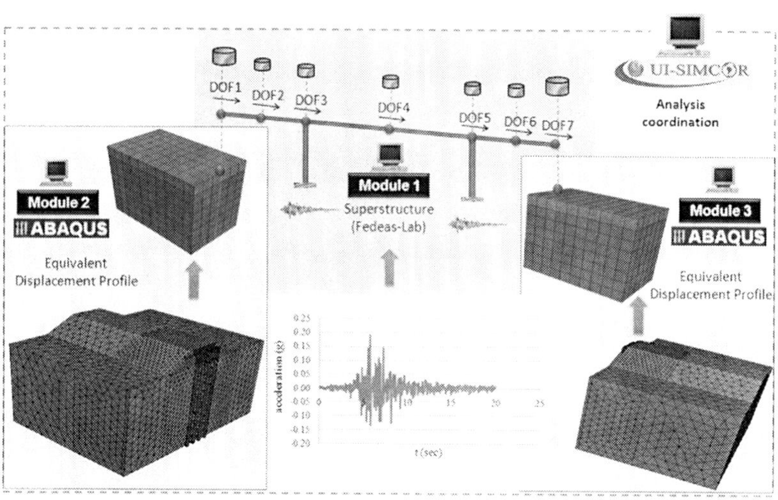

Figure 8. Distributed computational simulation (Model 4a)

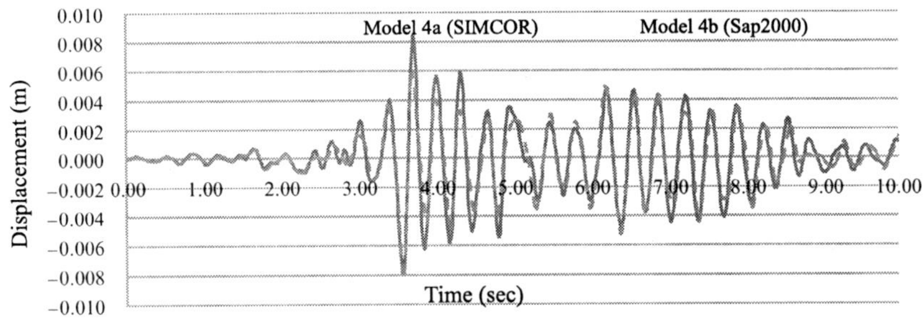

Figure 9. Dynamic response of the bridge deck using multi-platform and single-platform approaches of Models 4a and 4b for compatible material and geometry assumptions.

4. Validation

4.1. BENCHMARK MODEL: FRAME BRIDGE ON SPRING AND DASHPOT SYSTEMS WITH COMPARABLE PROPERTIES (MODEL 4B)

To validate the accuracy of multi-platform analysis, a firth frame model (i.e., Model 4b) was developed in addition to the aforementioned four, using comparable geometrical and material properties with those assumed in Model 4a. In particular, the deck cumber was ignored and a point mass was assumed at both lateral boundaries equal to the total abutment-embankment mass that was considered at the edge control points during the multi-platform analysis. Figure 9 illustrates that when compatible assumptions are made, the agreement between the single-platform (integral bridge model in Sap2000) and the multi-platform analysis (coordinating the response of Abaqus 3D volumes and FedeasLab bridge modules) was indeed very satisfactory.

5. Comparative Assessment of Single- and Multi-Platform Approaches in the Linear Elastic Range

Given the detailed data available for the particular case-study, an effort was made to use common assumptions regarding earthquake excitations and solution algorithms. Along these lines, the Kozani, Greece earthquake (PGA = 0.19 g) was uniformly applied in all cases, while the Hilbert–Hughes–Taylor integration method was used, with time step $\Delta t = 0.01$ s and a total of 1000 steps (10 s of input). A uniform damping value of 5% was assumed for the first and second modes of vibration, defined through the Rayleigh alpha and beta corresponding factors. Gaps and stoppers that have been designed for the particular structure were ignored to ensure maximum possible activation of the embankment-abutment system. Backfill and foundation soil properties were also taken identical between Models 3 and 4 based on the actual soil properties described in Section 2. All analyses were conducted in the elastic range and the excitation was performed in the longitudinal direction. Parametric analysis was also performed to investigate the relative influence of various assumptions that inevitably varied between the four approaches i.e., spring and dashpot constants of Models 1 and 2 in contrast to Poisson's ratio and modulus of elasticity for soils in Models 3 and 4, embankment finite element mesh dimensions and size, among many others. It was concluded that the parameter related to the maximum level of uncertainty was the critical embankment mass that was expected to be activated during the particular earthquake excitation and most importantly, the means to simulate its effect in the framework of the four different analysis strategies adopted. In contrast to the validation case (Section 4) the value of the single point mass that was used for Models 1 and 2 at the lateral boundaries of the bridge to represent the 'active' embankment-abutment system, was predicted independently (blindly) based on the concept of critical embankment length (Zhang and Makris, 2002) and without any calibration to 3D solid Models 3 and 4, where the activated embankment mass was inherently considered. Next, the dispersion in the dynamic response of the bridge due to the assumptions and modelling approach adopted is illustrated in Fig. 10. In particular, it is seen that following four different approaches to consider the effect of embankment-abutment-superstructure interaction, the maximum longitudinal displacement of the deck lies in the range of 0.6–1.0 cm whereas the fundamental period of the overall system may also differ by more than 100% despite the effort to use compatible properties where available. Further response measures (i.e., middle pier stresses) are not presented herein due to lack of available space; however, it is noted that the dispersion is of the same order. It is also seen (Table I) that multi-platform analysis is a very promising concept since it provides stable results within the envelope of the response produced by the other three approaches while enabling the consideration of 3-dimensional geometry without exceeding the computational time required for a conventional single-platform 3D modelling of the entire embankment-abutment-bridge system.

6. Comparative Assessment of Single- and Multi-Platform Approaches in the Nonlinear Range

Following the assessment of the four different approaches examined in the linear range, an effort was made to compare the response of the bridge in the non-linear range. The dynamic impedance matrices of Model 1 which represent the abutment-embankment as well as the dynamic pile group stiffness were properly modified based on intensity soil properties according to the

EC8-Part 5 provisions. The same procedure was applied for the case of the Model 2 through appropriate multi-linear springs and distinct dashpots that were also based on the aforementioned modified soil properties along the pile length.

On the other hand, the Mohr–Coulomb constitutive model implemented in ABAQUS was utilised to simulate the non-linear soil behaviour in the case of the, more refined Model 3. As for the Model 4, secant stiffness based on the detailed pushover analysis results along the longitudinal direction (Fig. 11, Sextos et al., 2008) was applied within the framework of an equivalent linear analysis. It is noted that the lateral assumption is not introducing any additional uncertainty to the problem since it essentially introduces an 'exact' (i.e., derived through refined pushover analysis) non-linear, force-displacement relationship at each abutment control point that is used through a multi-platform process that is also inherently pseudo-static. The dispersion of the results is presented in the Fig. 10 where it is seen that multi-platform analysis is not only feasible in the non-linear range, but it leads to comparable results with those derived with more conventional approaches both in terms of longitudinal displacement amplitude of the deck and of the fundamental period increase of the system as a whole.

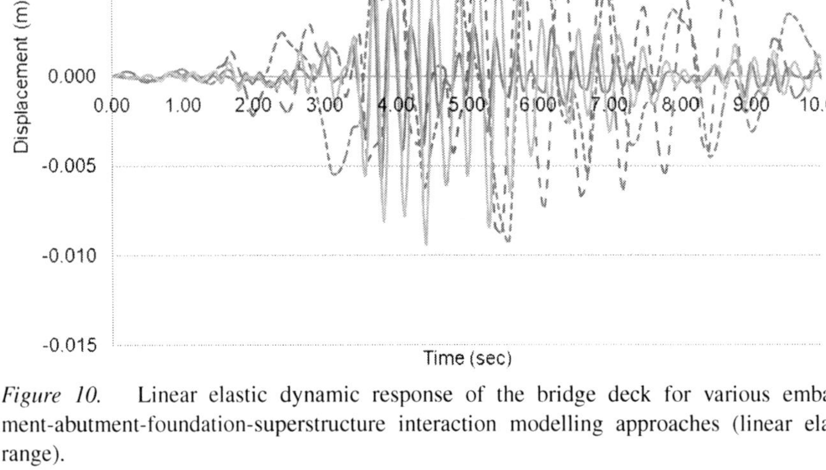

Figure 10. Linear elastic dynamic response of the bridge deck for various embankment-abutment-foundation-superstructure interaction modelling approaches (linear elastic range).

Moreover, it is noted that the multi-platform simulation coordinated by UI-Simcor, may also have significant advantages regarding for capturing the inelastic response of the R/C piers, as it can combine a number of specialised constitutive models and software currently available.

7. Conclusions

Four different analysis approaches to simulate the dynamic interaction between the embankment, the abutment, its foundation and the bridge were adopted and comparatively assessed for the study of a real overcrossing in Greece using a wide variety of software that involve both single and multi-platform analysis. It is concluded that the last generation computational tools available, provide a number of promising new capabilities, especially towards the concurrent use of different specialised software in the framework of multi-platform analysis. However, given the dispersion of the results observed through alternative analysis strategies which may well differ by a factor of 2 for the case of the particular overcrossing studied, it is deemed that research effort is needed in order to establish a uniform level of reliability among the various analysis approaches.

Acknowledgements

The authors would like to thank Prof. Amr Elnashai and Assist. Prof. Oh-Sung Kwon for their precious assistance regarding the application of the analysis coordinator UI-Simcor, developed at the University of Illinois, as

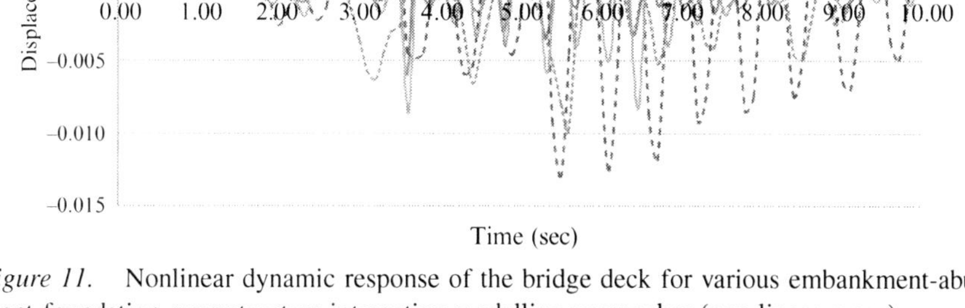

Figure 11. Nonlinear dynamic response of the bridge deck for various embankment-abutment-foundation-superstructure interaction modelling approaches (non-linear range).

well as the Earthquake Engineering Research Center (EERC) at University of California Berkeley and Prof. B. Stojadinovic for their overall support during the first author's visit in the framework of which part of this work was conducted. Acknowledgements are also due to Prof. A. Kappos and P. Potikas at Aristotle University Thessaloniki for kindly providing all the data and results derived during the non-linear static assessment of the bridge studied.

References

Dicleli, M. (2005) Integral abutment-backfill behavior on sand soil—pushover analysis approach, *J. Bridge Eng.* **10**(3), 354–364.

Elnashai, A. S., Papanikolaou, V., and Lee, D. (2002) *Zeus NL—A System for Inelastic Analysis of Structures*, Mid-America Earthquake Center, University of Illinois at Urbana-Champaign.

Filippou, F. C. and Constantinides, M. (2004) FEDEASLab Getting started guide and simulation examples, Technical Report NEESgrid-2004-22: www.nees-grid.org.

Goel, R. K. and Chopra, A. (1997) Evaluation of bridge abutment capacity and stiffness during earthquakes, *Earthquake Spectra* **13**(1), 1–23.

Kappos, A., Potikas, P., and Sextos, A. (2007) Seismic assessment of an overpass bridge accounting for non-linear material and soil response and varying boundary conditions. In *Conf. Comp. Meth. Struct. Dynam. Earthquake Eng.*, COMPDYN 2007, Rethymno, Greece, CD-ROM Volume.

Kotsoglou, A. and Pantazopoulou, S. (2007) Bridge-embankment interaction under transverse ground excitation, *Earthquake Eng. Struct. Dynam.* **36**, 1719–1740.

Kwon, O. S. and Elnashai, A. S. (2008) Seismic analysis of meloland road overcrossing using multiplatform simulation software including SSI, *J. Struct. Eng.* **134**(4), 651–660.

Kwon, O. S., Sextos, A., and Elnashai, A. (2008) Liquefaction-dependent fragility relationships of complex bridge–foundation–soil systems. In *Int. Conf. Earthquake Eng. Disaster Mitigation*, Jakarta, Indonesia, 14–15 April.

Mackie, K. and Stojadinovic, B. (2002) Bridge abutment model sensitivity for probabilistic seismic demand evaluation, *Proc. 3rd Nat. Seismic Conf. and Workshop on Bridges and Highways*, Portland, Oregon, USA, April 28–May 1.

McKenna, F. and Fenves, G. L. (2001) The OpenSees Command Language Manual, Version 1.2, Pacific Earthquake Engineering Research Center, University of California at Berkeley.

Ministry of Environment, physical planning, and public works (1999) Circular 39/99: Guidelines for the Seismic Design of Bridges, Athens (in Greek).

Ministry of Environment, physical planning, and public works (2000) Greek Seismic Code, EAK 2000, Athens (in Greek)

Potikas, P. (2006) Seismic design and assessment of an overpass bridge, MSc Thesis, Aristotle University, Thessaloniki, Greece (in Greek).

Sextos, A., Kappos, A., and Pitilakis, K. (2003) Inelastic dynamic analysis of RC bridges accounting for spatial variability of ground motion, site effects and soil-structure interaction phenomena. Part 2: Parametric analysis, *Earthquake Eng. Struct. Dynam.* **32**(4), 629–652.

Sextos, A., Mackie, K., Stojadinovic, B., and Taskari, O. (2008) Simplified P-y relationships for modelling embankment-abutment systems of typical California bridges. In *14th Word Conf. Earthquake Eng.*, Beijing, China, CD-ROM Volume.

Spencer Jr., B. F., Elnashai, A. S., Park, K., and Kwon, O. (2006) Hybrid test using UI-SimCor, three-site experiment, Final report to NEESit for Phase I project of hybrid simulation framework development, University of Illinois at Urbana-Champaigne.

Zhang, J. and Makris, N. (2002) Kinematic response functions and dynamic stiffnesses of bridge embankments, *Earthquake Eng. Struct. Dynam.* **31**, 1933–1966.

DESIGN SEISMIC RESPONSE EVALUATION OF WALL SYSTEMS INCLUDING FOUNDATION FLEXIBILITY

Zdravko Bonev (zbp_uacg@abv.bg)
University of Architecture, Civil Engineering and Geodesy, Hr. Smirnenski Blvd. 1, 1046 Sofia, Bulgaria

Golubka Necevska-Cvetanovska (golubka@pluto.iziis.ukim.edu.mk)
*Institute of Earthquake Engineering and Engineering Seismology, IZIIS, University "Ss. Cyril and Methodius" Salvador Aljende St. 73, 1000, Skopje, F.Y.R. of Macedonia**

Elena Vaseva
Central Laboratory of Seismic Mechanics and Seismic Engineering, Bulgarian Academy of Sciences, Acad. Georgi Bonchev St., Bl. 3, 1113 Sofia, Bulgaria

Roberta Apostolska (beti@pluto.iziis.ukim.edu.mk)
*Institute of Earthquake Engineering and Engineering Seismology, IZIIS, University "Ss. Cyril and Methodius" Salvador Aljende St. 73, 1000, Skopje, F.Y.R. of Macedonia**

Dilyan Blagov
University of Architecture, Civil Engineering and Geodesy, 1, Hristo Smirnenski Blvd., 1046 Sofia, Bulgaria

Abstract. The paper deals with application of capacity spectrum method which is extended to provide estimate for seismic demands for specific type of structure and variable boundary conditions. Flat slab—R/C wall systems are considered. Flexible shallow wall foundations are used because of soil deformations. Only elastic soil properties are considered. Foundation motion is implemented assuming new additional degrees of freedom instead of standard fixed end. The only rocking motion of the footing is accounted for as the most essential. The influence of foundation flexibility on the seismic demands is evaluated using capacity spectrum method. Numerical results for a single wall and for overall structure are graphically represented. It is concluded that soil conditions are very important for the design seismic response of wall systems. Some important conclusions are made.

Keywords: capacity spectrum method, effects of foundation flexibility, design seismic performance including soil effects

*Turkey recognises the Republic of Macedonia with its constitutional name.

T. Schanz and R. Iankov (eds.), *Coupled Site and Soil-Structure Interaction Effects with Application to Seismic Risk Mitigation*, NATO Science for Peace and Security Series C: Environmental Security, © Springer Science+Business Media B.V. 2009

1. Introduction

The Capacity Spectrum Method (CSM) is a tool to predict the design seismic performance of structures subjected to design seismic action. This method is recommended by the new generation of seismic resistant design codes such as EC8, 2004 for evaluation of seismic demands and for capacity assessment of newly designed or existing building structures. The theoretical background and application of CSM can be found in Chopra and Goel (2003) and Fajfar (1999). In Fajfar (1999) the seismic hazard data such as design ground acceleration and soil amplification effects are taken into account by the shape and scaling coefficient of the demand spectrum. The idea of the present paper is to account for the foundation flexibility in the capacity curve since soil deformations themselves influence overall stiffness of the structure. Thus the capacity curve is obtained for coupled foundation-R/C building system. Analyses in this direction are carried out in Bonev et al. (2005) and Schanz et al. (2007) for plane frame structures. In this paper the subject of interest is one 3D structure being designed as a wall system. It is shown in FEMA 450 (2003) the wall systems are much sensitive to soil deformations because the most stiff elements—walls dictate internal force distribution. The implementation of the soil conditions into numerical models of entire structure is discussed in FEMA 273 (1997), FEMA 274 (1998), FEMA 357 (2000), FEMA 450 (2003), Geotechnical Engineering (1997) and Kramer (1996). FEMA 357 (2000) provides foundation stiffness coefficients as dependent on the effective shear modulus, *G*, and Poisson's ratio. The model can be used for static and dynamic analyses. FEMA 273 (1997), FEMA 274 (1998) and FEMA 450 (2003) propose the constitutive relationship moment-rotation which is related to the rocking motion of the single shallow footing. Soil plasticity and uplifting phenomena can be taken into account in the resulting moment-rotation curve which can be used for static loads only. In this paper only linear soil properties are taken into account being represented by the unit foundation modulus (Winkler's constant). The paper deals with the changes in the capacity curves which are more or less steep depending on the elastic soil properties. Due to footing flexibility the target displacements are increased and the global ductility of the structure is reduced. It is shown that for soft soils the design performance of the structure may remain completely elastic. In contrary, for stiff soils the seismic demands of the structure are large enough to develop significant inelastic deformations. It is found that the influence of accidental eccentricity is much essential after collapse of the most loaded walls. The results are graphically illustrated and discussed.

2. Model Description

The general view of the model can be seen in Fig. 1. The numerical model used in calculations can be described starting with assumptions listed below:

1. Floor slab is treated as a rigid diaphragm in its own plane. The membrane stiffness of the floors is practically infinitely large and the slab may move horizontally as absolutely rigid body. On the other hand the slabs distribute the seismic loads between the walls.
2. The vertical loads are carried by shearwalls and columns. Lateral loads are carried by the shearwalls only. In case when the slabs are flat (no beams are used) the slab to column connection is not designed as moment resisting. It is assumed that columns are pinned at both ends and can resist to vertical loads only.
3. Shearwalls are modelled by vertical frame elements. The potential locations of plastic hinges are considered at each floor level.

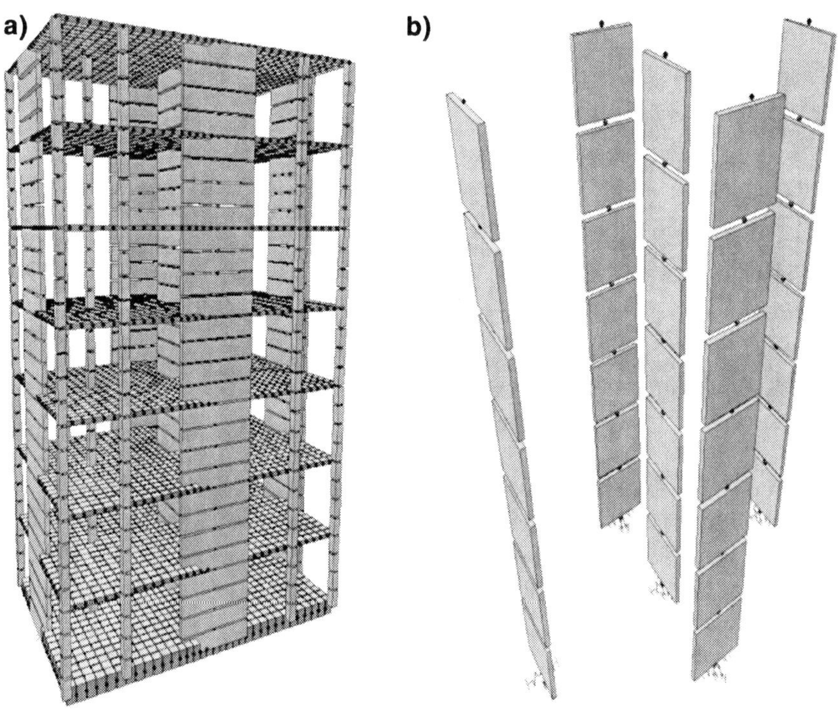

Figure 1. (a) General 3D view of the numerical model—walls, columns and slabs; (b) shearwall system only.

4. The structure is symmetric in plan with respect to X- and Y-axes. If the location of the centre of resistance (CR) is coincident with the centre of masses (CM) the structure is regular. The reference co-ordinate system is located in CR-point (level 0). If the centre of the masses CM has position different from the position of CR the structure is irregular with accidental eccentricity.
5. Bending stiffness of the slab is taken into account only to obtain the vertical loads distribution between the vertical elements—walls and columns.
6. Single footing under each wall is used. The foundation is supported by the soil with its vertical resistance. The elastic soil properties are implemented by the unit foundation modulus (Winkler's constant). Each foundation has 3 degrees of freedom in XZ-vertical plane and 3 degrees of freedom in YZ-vertical plane. The stiffness of the springs is determined using the unit foundation modulus. This approach is considered in FEMA 273 (1997), FEMA 274 (1998) and FEMA 450 (2003).
7. Loading pattern used for pushover analysis in both X- and Y-directions has the shape of inverted triangle and implies linear force distribution in elevation. Forces are applied in CM for each floor level.
8. The calculation of the spring stiffness implies that only rocking motion of the footing is considered.
9. Torsion effects due to different disposition of CM with respect to CR are taken into account. The influence of accidental eccentricities is accounted for as a source of torsion.
10. Axial forces are remaining constant during the lateral pushover analysis and plastic hinge properties once determined after application of the vertical loads are remaining the same.
11. The interface footing surface is unable to move horizontally because of the friction, passive pressure and embedment effect.

The general 3D view of the building model is shown in Fig. 1. The model is consisting of reinforced concrete walls, columns and slabs. Pin-joint connection is used and columns are unable to resist lateral loads. Foundations are modeled as single footings including the elastic support of the soil. Considering a single wall in X–Z plane for example, see Fig. 2a FEMA 274 (1998) the soil resistance is represented by equivalent soil springs whose stiffness can be calculated on the basis of the unit foundation modulus. This is shown in Fig. 3. The potential location of the first plastic hinge is at the base of the wall (in the centre of the plastic zone). It is indicated by the black point on

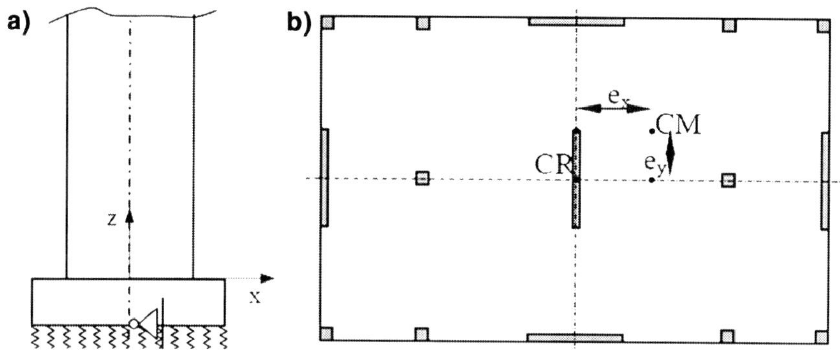

Figure 2. (a) Single footing under the wall in the plane of rocking; (b) structural layout.

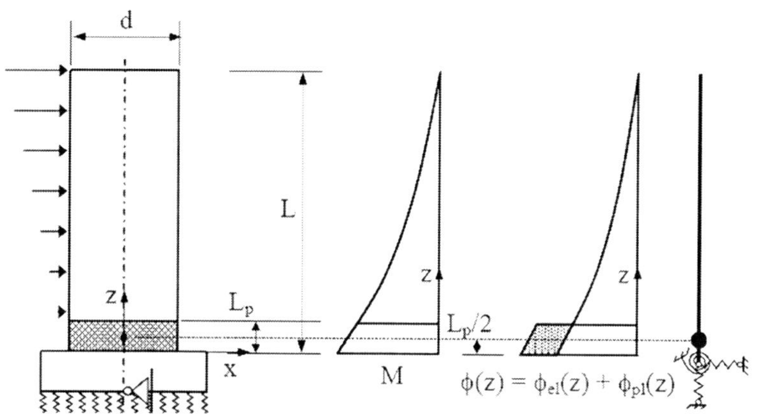

Figure 3. A single wall designed as dissipative wall—moment and curvature distribution. Plastic zone modelled by plastic hinge and spring elements including the soil resistance.

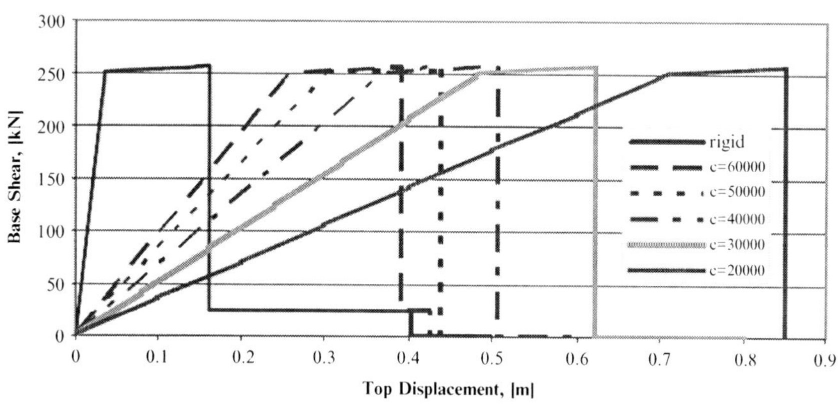

Figure 3a. Capacity curves for a single wall using different values for Winkler's modulus.

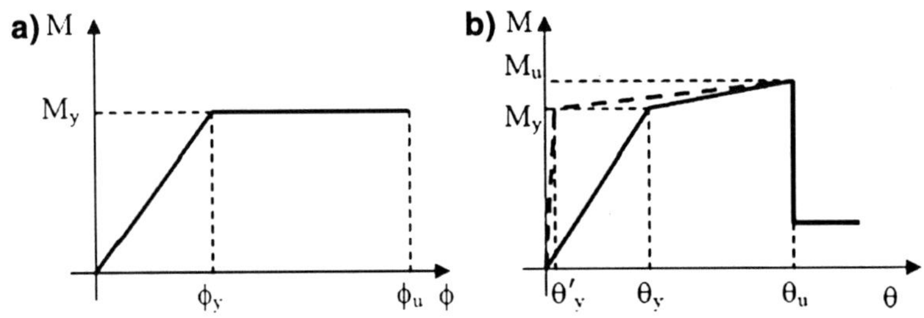

Figure 4. Typical constitutive relationships: (a) moment M—curvature ϕ for distributed plasticity model; (b) moment M—rotation θ for concentrated plasticity model.

the Z-axis.The plastic zone approach is based on distributed plasticity model. The typical constitutive relationship is moment-curvature. The distribution of elastic/inelastic curvature for a simple wall element is shown in Fig. 3.

The plastic zone length L_P is calculated following Paulay and Priestley (1992)

$$L_P = 0.08L + 0.022d_b f_y, \tag{1}$$

where L is the total wall length in (m), d_b is the bar diameter in (m), and f_y is the yield strength of steel in (MPa). For concentrated plasticity models (zero-length plastic hinges) used herein the basic constitutive relationship is moment-rotation, see Fig. 4. The parameters of the plastic hinges are calculated as follows:

$$\begin{aligned} \theta_y &= \frac{M_y L_P}{EI}, \\ \theta_P &= L_P(\varphi_u - \varphi_y)\left(1 - \frac{L_P}{2L}\right), \\ \theta_u &= \theta_y + \theta_P, \end{aligned} \tag{2}$$

where

ϕ_y and ϕ_u	the yielding and ultimate curvatures correspondingly,
M_y and M_u	the yielding and ultimate moments,
θ_y and θ_u	the yielding and ultimate rotations, see Fig. 4,
θ_P	the plastic rotation component.

For wall elements plastic hinges are implemented only in their own plane.

In order to simulate very large initial stiffness of rotation springs it is recommended to implement the "computational yield rotation" being defined as

$$\theta_y' = \frac{M_y L_P'}{EI}. \tag{3}$$

It is accepted in calculation of θ'_y that $L'_p = 0.10\,\text{m}$. The moment-rotation curve obtained by calculations is given in Fig. 4b as a solid line. The moment-rotation curve implemented in calculations is shown in the same figure with a dotted line.

The structural layout and wall/columns distribution is shown in Fig. 2b. The eccentricity is defined by the distances between CM center and CR centers. The behaviour of a single shearwall subjected to in-plane loading can be seen in Figs. 3 and 3a and is represented by the capacity curve. It is evident that the flexible foundations do not influence the base shear strength (capacity). An essential influence is found in target displacements and ductility. Target displacement is drastically increased for soft soils because elastic deformations are growing up. At the same time inelastic displacement demands are relatively constant. Ductility factor is significantly reduced and pure ductile behaviour of a single wall is practically not achieved.

The behaviour of single wall elements is in a large extent important for the behaviour of overall system (see Fig. 1).

3. Analysis Method

Capacity spectrum method is used as evaluation tool (Fajfar, 1999) where the use of design demand spectra is recommended. A version of the method based on elastic demand spectrum is available in Annex B of EC8, 2004. The basic relationship considering the global behaviour is "Base shear force V—Roof displacement u" derived numerically from pushover analysis. The normalized vector of lateral displacements distribution in elevation is denoted by $\{\Phi\}$, see Fig. 4a. For a single wall system it is reasonable to accept inverted triangle as a shape of the $\{\Phi\}$-vector.

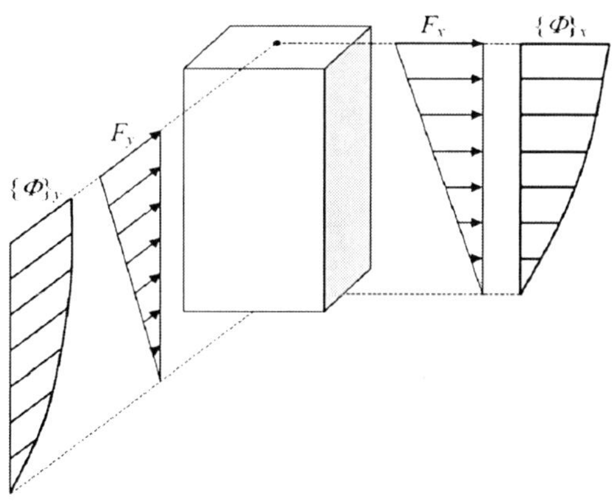

Figure 4a. Structural system subjected to two independent lateral loading patterns in main orthogonal directions. The direction is depicted by x and y subscripts.

The governing system of equations for dynamic equilibrium assuming that the effect of the damping takes part implicitly could be written as follows:

$$[m]\{\ddot{u}\} + \{f\} = -[m]\{1\}\ddot{u}_g, \tag{4}$$

where $[m]$ is the diagonal mass matrix of floor masses, $\{f\}$ is the vector of internal (restoring) forces, $\{\ddot{u}\}$ is the vector of level accelerations, $\ddot{u}_g$ is the ground acceleration. It is assumed further that the distribution of lateral displacements $\{u\}$ follows the shape of $\{\Phi\}$-vector with proportionality multiplier $u(t)$. This quantity is the roof displacement. Thus after replacement of $\{u(t)\} = \{\Phi\}u(t)$ in Eq. (4) and after multiplication of both sides of the new equations by $\{\Phi\}^T$ on the left the following scalar equation is obtained:

$$\{1\}^T[m]\{\Phi\}\ddot{u} + \{1\}^T\{f\} = -\{1\}^T[m]\{1\}\ddot{u}_g. \tag{5}$$

It is easy to implement the following quantities related to equivalent single degree of freedom system, namely: mass parameter $m^* = \{1\}^T[m]\{\Phi\} = \sum_i m_i\Phi_i$, total mass of the original system $\{1\}^T[m]\{1\} = \sum_i m_i = m$ and base shear force $V = \{1\}^T\{f\}$. Note that the mass of the original system is equal to the mass of the equivalent system. After taking into account the above notations Eq. (5) is transformed to give

$$\ddot{u}^* + \frac{V}{m} = -\ddot{u}_g. \tag{6}$$

The modified displacement u^* of equivalent single degree of freedom system is introduced by the following expression:

$$u^* = \frac{u}{\Gamma}, \tag{7}$$

where $\Gamma = m/m^*$ is modification factor used to transform the displacements in original system into displacement for equivalent system. Finally, the system of Eq. (4) provides the result:

$$\frac{V}{m} = -(\ddot{u}_g + \ddot{u}^*). \tag{8}$$

According to Eq. (8) it is implied in capacity spectrum method that capacity curve could be plotted following the rule: use the quantity V/m on the acceleration axis and use the quantity u^* on the displacement axis.

Two simplified and independent analyses in both X- and Y-directions are carried out. Two equivalent single degree of freedom systems are used (see Fig. 4a). After that the capacity spectrum method is applied in both orthogonal directions as well.

4. Numerical Results

The numerical model described in Section 2 is subjected consequently to monotonically increasing vertical and horizontal loads. Six values of unit foundation modulus are used in calculations: fixed base (infinitely large modulus), 60,000, 50,000, 40,000, 30,000 and 20,000 in kN/m^3 metric units. After completion of vertical loading procedure the horizontal loading pattern is applied. The effects of accidental torsion are studied considering eccentricity of 15% (large eccentricity and irregular structure) and 0% (regular structure). The modification factors Γ are dependent on the plastic mechanism reached during the loading. For structure with flexible foundation Γ is very close to 2.0. In case of fixed base structure this factor is slightly higher than 2.0. The total mass of the structure is 875 kN $\cdot$ s^2/m (125 per each storey level). The capacity spectrum method (Fajfar, 1999) is applied using the design demand spectrum in acceleration-displacement format. The analyses results are represented graphically in Figs. 5–14. Numerical calculations are based on the results obtained in Blagov et al. (2008).

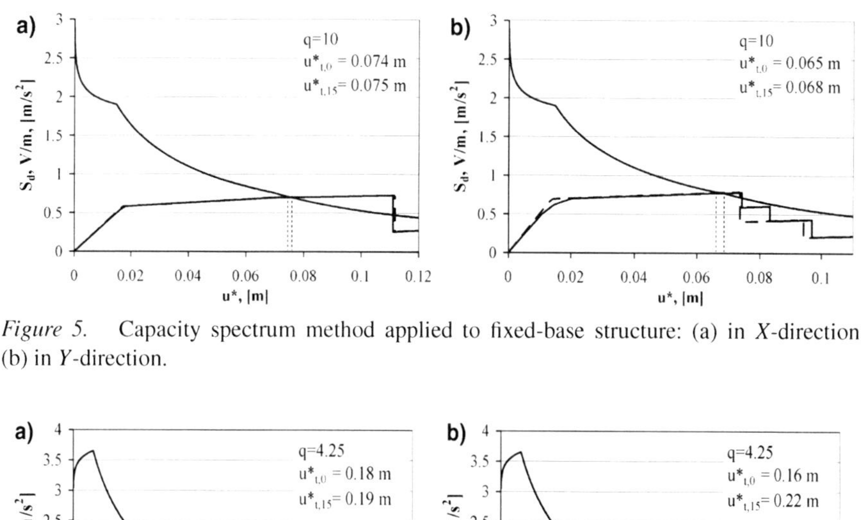

Figure 5. Capacity spectrum method applied to fixed-base structure: (a) in *X*-direction; (b) in *Y*-direction.

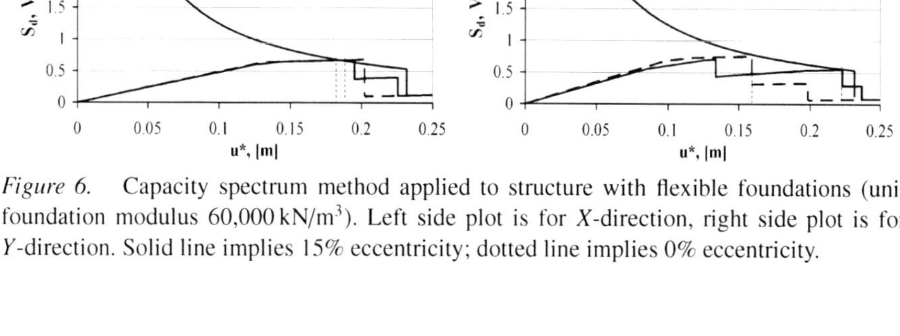

Figure 6. Capacity spectrum method applied to structure with flexible foundations (unit foundation modulus 60,000 kN/m^3). Left side plot is for *X*-direction, right side plot is for *Y*-direction. Solid line implies 15% eccentricity; dotted line implies 0% eccentricity.

Figures 5–10 illustrate the results obtained through capacity spectrum method. It is evident that the greatest inelastic seismic demands in the RC

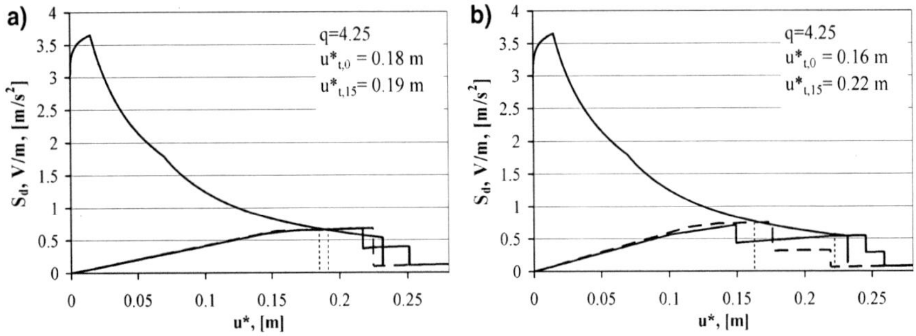

Figure 7. Capacity spectrum method applied to structure with flexible foundations (unit foundation modulus 50,000 kN/m³). Left side plot is for *X*-direction, right side plot is for *Y*-direction. Solid line implies 15% eccentricity; dotted line implies 0% eccentricity.

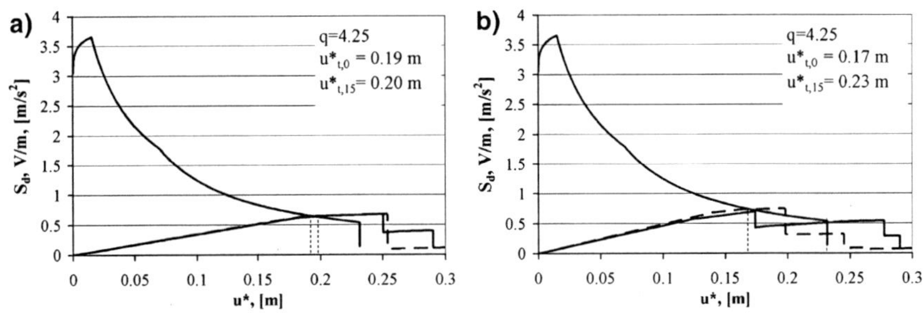

Figure 8. Capacity spectrum method applied to structure with flexible foundations (unit foundation modulus 40,000 kN/m³). Left side plot is for *X*-direction, right side plot is for *Y*-direction. Solid line implies 15% eccentricity; dotted line implies 0% eccentricity.

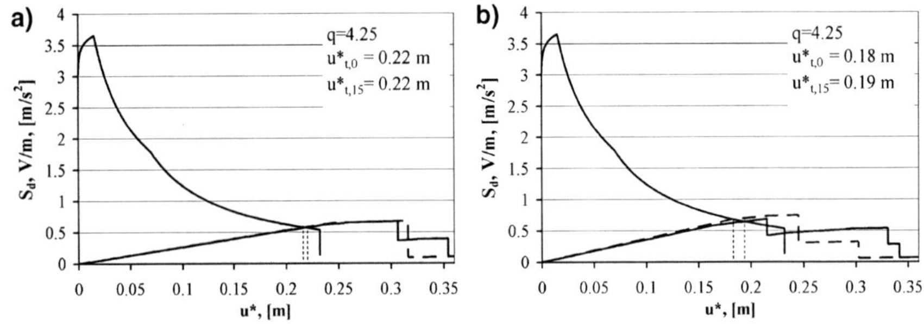

Figure 9. Capacity spectrum method applied to structure with flexible foundations (unit foundation modulus 30,000 kN/m³). Left side plot is for *X*-direction, right side plot is for *Y*-direction. Solid line implies 15% eccentricity; dotted line implies 0% eccentricity.

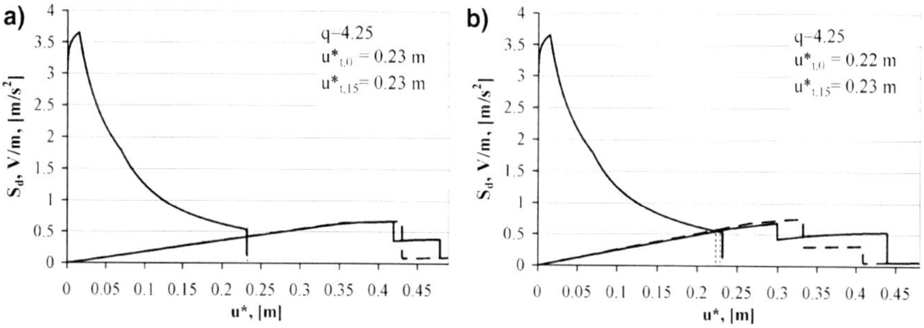

Figure 10. Capacity spectrum method applied to structure with flexible foundations (unit foundation modulus 20,000 kN/m^3). Left side plot is for X-direction, right side plot is for Y-direction. Solid line implies 15% eccentricity; dotted line implies 0% eccentricity.

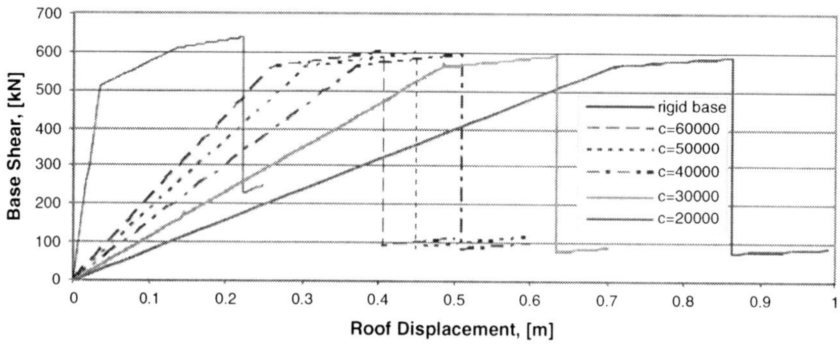

Figure 11. Capacity curves in X-direction (eccentricity 0%) obtained for different Winkler's constants.

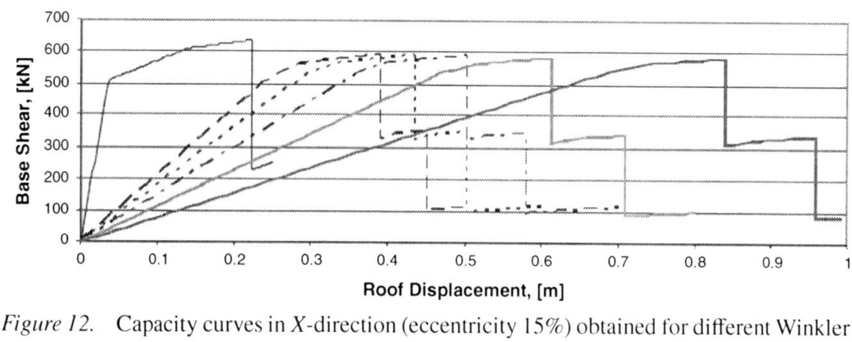

Figure 12. Capacity curves in X-direction (eccentricity 15%) obtained for different Winkler's constants.

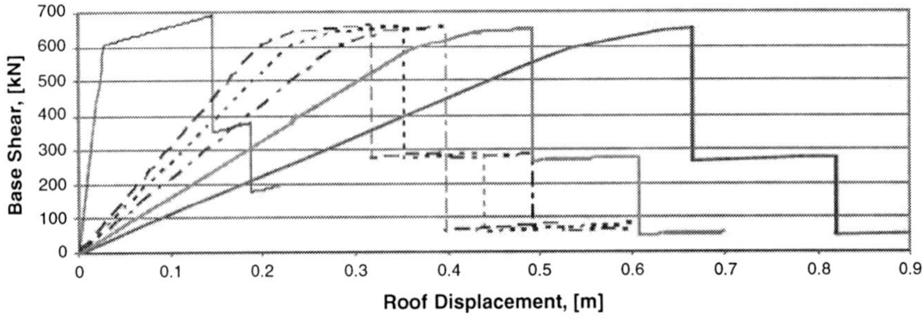

Figure 13. Capacity curves in Y-direction (eccentricity 0%) obtained for different Winkler's constants.

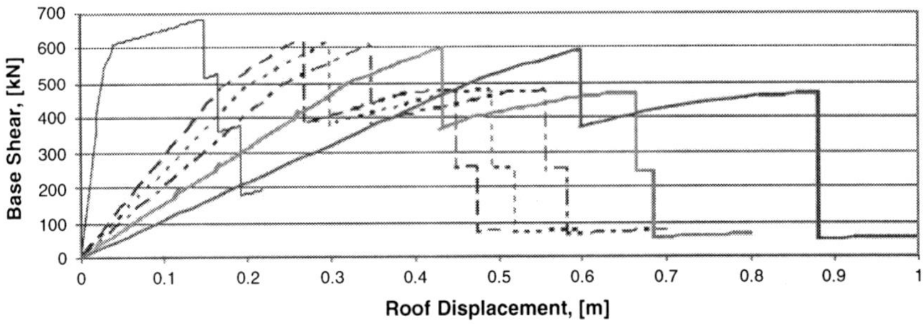

Figure 14. Capacity curves in Y-direction (eccentricity 15%) obtained for different Winkler's constants.

members is achieved when the Winkler's constant is infinitely large (fixed base). When foundation flexibility is increased the inelastic seismic demands are reduced. Elastic component of roof displacement is growing up. Figures 11–14 illustrate that the global ductility demand is reduced with increasing the footing flexibility. The same figures show that the initial (elastic) stiffness is also reduced due to flexibility. At the same time the base shear strength is relatively slightly influenced by the footing flexibility.

Figures 8–10 illustrate that when the soil is soft and structure reaches the target displacement the global behaviour of the structure may remain completely elastic. This mode of deformation implies that the soil failure precedes yielding of the structure. Figure 5 shows that the largest values for the behaviour factor could be achieved if the base is fixed and foundation flexibility does not exist. The smallest target displacements are observed in the same case. If the foundations are flexible (see Figs. 6–10) target displacement is increased but the behaviour factor decreases. The behaviour factor reduction is not on the side of safety. It is seen that the wall system is sensitive to soil deformations and the negative effect of flexibility can be avoided by making foundations less flexible.

5. Conclusions

Considering the analysis of the numerical results obtained in Section 4, the following conclusions can be made:

1. The foundation flexibility caused by soil deformations influence essentially the capacity curves. The global ductility factor is seriously reduced due to significant increase of elastic part of deformations. In general, the participation of soil deformation in overall structural deformations is essential. It is expected this participation to become more essential in case of plastic soil deformations and foundation uplift in new shearwalls in existing buildings, where the vertical load is not sufficient to prevent the footing from uplifting.
2. As a rule, target displacements are increased if the soil is softer. The behaviour factor however shows decreasing tendency. Safe design solutions could be expected if soil deformations are taken into account in capacity curves.
3. The influence of accidental torsion effects is small considering the elastic behaviour of the structure. More essential influence is observed when some plastic hinges yield and when some walls are collapsed. The capacity curves are sensitive to accidental torsion when wall elements yield or and plastic mechanism takes place.
4. The global structure strength is relatively independent of soil stiffness and accidental torsion effects influence slightly the strength.
5. It is concluded that wall systems are sensitive to flexible foundations in a large extent. It is demonstrated that overstrength factor of a single wall is 1.0 and no plastic redistribution mechanism is observed. This is easy to explain this by the fact that each single wall is not redundant system. Structural assemblages composed of separate walls and flat slabs in large degree follow this property (Figs. 11–14) and plastic force redistribution mechanism does not contribute significantly towards seismic loading reduction.

Acknowledgements

The project grant No. BM-6/2006 of the National Science Foundation at the Bulgarian Ministry of Education and Sciences is greatly acknowledged by the authors.

The project grant No. BN-84/2008 of the University of Architecture, Civil Engineering and Geodesy, Sofia, Bulgaria, is greatly acknowledged by the authors.

References

Blagov, D., Georgiev, V., and Bonev, Z. (2008) Influence of flexible foundations on the design response of buildings with accidental eccentricity. In *Proc. 5th European Workshop on Irregular and Complex Structures*, 16–17 September 2008, Catania, Italy.

Bonev, Z., Ganchev, S., Blagov, D., and Zerzour, A. (2005) Behaviour factor evaluation accounting for the elastic foundation, EE-21C, Topic 4: Structural Modelling, Analysis, Design and Seismic Safety, Technical report, Skopje, Republic of Macedonia.

Chopra, A. K. and Goel, R. K. (2003) Evaluation of the modal pushover analysis procedure using vertically "regular" and irregular generic frames, Technical Report No. EERC 2003–03, University of California, Berkeley.

Fajfar, P. (1999) Capacity spectrum method based on inelastic demand spectra, *Earthquake Engineering and Structural Dynamics* **28**, 979–993.

FEMA 273 (1997) NEHRP Guidelines for the Seismic Rehabilitation of Buildings, Issued by FEMA in Furtherance of the Decade for National Disaster Reduction.

FEMA 274 (1998) Seismic Rehabilitation Commentary C4: Foundations and Geotechnical Hazards.

FEMA 357 (2000) Global Topics Report on the Prestandard and Commentary for the Seismic Rehabilitation of Buildings.

FEMA 450 (2003) NEHRP Recommended Provisions for Seismic Regulations for New Buildings and Other Structures, Commentary C7A.

Geotechnical Engineering, Circular No. 3 (1997) Design Guidance: Geotechnical Earthquake Engineering for Highways, Vol. 1: Design Principles, US Department of Transportation, Federal Highway Administration.

Kramer, S. L. (1996) *Geotechnical Earthquake Engineering*, Prentice-Hall International Series in Civil Engineering and Engineering Mechanics, Upper Saddle River, New Jersey.

Paulay, T. and Priestley, M. J. N. (1992) *Seismic Design of Reinforced Concrete and Masonry Buildings*, John Wiley & Sons, New York.

Schanz, T., Bonev, Z., Georgiev, V., and Iankov, R. (2007) Application of capacity spectrum method to soil-foundation-structure interaction problems. In *Proc. of the Jubilee Scientific Conference, Devoted to 65 Years Anniversary of the University of Architecture, Civil Engineering and Geodesy*, 16–17 May 2007, Sofia, Bulgaria.

DESIGN SEISMIC PERFORMANCE OF R/C FRAME STRUCTURES TAKING INTO ACCOUNT FOUNDATION FLEXIBILITY

Tom Schanz (tom.schanz@rub.de)
Laboratory of Foundation Engineering, Soil and Rock Mechanics, Faculty of Civil Engineering, Ruhr-Universität Bochum, Universitätsstraße 150, D-44780 Bochum, Germany

Zdravko Bonev
University of Architecture, Civil Engineering and Geodesy, 1, Hristo Smirnenski Blvd., 1046 Sofia, Bulgaria

Frank Wuttke
Bauhaus-Universität Weimar, Coudraystraße 11C, D-99423 Weimar, Germany

Roumen Iankov
Institute of Mechanics, Bulgarian Academy of Sciences, Acad. Georgi Bonchev St., Bl. 4, 1113 Sofia, Bulgaria

Valeri Georgiev
University of Architecture, Civil Engineering and Geodesy, 1, Hristo Smirnenski Blvd., 1046 Sofia, Bulgaria

Abstract. This paper concerns the problem of design seismic performance of R/C frames with flexible foundations. Analysis is carried out by capacity spectrum method assuming two parametric nonlinear soil model and inelastic behaviour of structural R/C members. The overall structural model includes R/C part and flexible foundations. An additional rotational degree of freedom is implemented to capture foundation rocking motion. To do this "soil type" plastic hinge is proposed. Soil deformations are calculated through site-specific data such as unit foundation modulus and soil capacity. Capacity spectrum is then calculated considering coupled structure-foundation system. Two categories of results for target displacements are compared and analyzed—first for fixed base structure and second for structure with flexible foundation. The case of fixed base structure is used as reference case. Numerical results are graphically illustrated. An analytical estimate for target displacement—behaviour factor relationship is proposed. It allows for displacement-based control on seismic demands. Basing on the numerical results it is concluded that in general foundation flexibility influences the overall structural response. Due to soil deformations target displacements are larger than corresponding target displacements of the fixed-base structure. Foundation flexibility reduces the global ductility of structure. Displacement demands can effectively be controlled by suitable choice of behaviour factor. This procedure creates opportunity for displacement-based seismic design.

T. Schanz and R. Iankov (eds.), *Coupled Site and Soil-Structure Interaction Effects with Application to Seismic Risk Mitigation,* NATO Science for Peace and Security Series C: Environmental Security,

Keywords: flexible foundations, capacity spectrum method, design seismic response, displacement-based seismic design

1. Introduction

In the last two decades the idea to carry out performance-based seismic design of structures is dominant in a large number of research works. This concept is accepted in the new and modern generation of seismic resistant design codes. It is intended to bridge the existing gap between design practice and theory by development of approximate but reliable enough methods for assessment of structures subjected to design seismic loading. The theoretical background and application of Capacity spectrum method (CSM) is discussed by Chopra and Goel (1999, 2003), Fajfar (1999, 2000). Implementation of the method can be found in Eurocode 8, Annex B, 2004. Capacity spectrum method allows for relatively easy determination of seismic demands. The global behaviour of a building structure is studied using an equivalent single degree of freedom (ESDOF) system. Capacity curve and ESDOF parameters are derived through pushover analysis using inelastic deformations of structures. Capacity of the structure in resisting lateral loads is represented by the largest base shear force which can be carried by the structure. Design demand spectra are based on site-specific data such as design horizontal ground acceleration a_g, shear wave velocity V_s and soil amplification effect evident from the shape of the spectra. Calculation of target displacement is a subject of Annex B of Eurocode 8 (2004). Capacity spectrum method is extended by Fajfar (2000) and Kilar and Fajfar (2000) to some 3D applications for asymmetric spatial buildings with R/C frames and masonry infill.

It is a purpose of this paper to implement soil deformations into capacity curve assuming that reinforced concrete building and foundations are coupled in one structure. The structure is considered as flexibly supported by foundations. In this way various boundary conditions are imposed on the structure. Foundation flexibility influences overall frame-foundation response and capacity curves. The subject of the paper is to provide assessment of this influence numerically by adding the soil stiffness to overall stiffness. The shallow footing motion is simplified to have two components—vertical in case of vertical loading and rocking for horizontal loadings. Rocking component is the most important for lateral resistance, as is shown in FEMA 273 (1997), FEMA 274 (1998) and FEMA 450 (2003). In this study the only rocking additional degree of freedom of the footing is implemented into the model.

In a number of papers Blagov et al. (2008), Bonev et al. (2005) and Schanz et al. (2007) the influence of foundation flexibility is evaluated assuming elastic properties of the soil. It is found, Bonev et al. (2005) and Schanz et al. (2007) that for 2D structures, Blagov et al. (2008) for 3D structures soft soils reduce the global ductility and increase the target displacements. In the same time base shear capacity remains relatively unchanged and invariant with respect to the soil stiffness. Another purpose of the present paper is to implement soils with nonlinear behaviour in calculation the capacity spectra. The verification of aforementioned tendencies for nonlinear soils is important.

Two parametric model is implied in calculations. Model parameters are Winkler's modulus k (unit foundation modulus) and soil capacity σ_c (soil strength). The sources of nonlinearity are the soil plasticity (elastic-perfectly plastic stress law in compression) and soil uplift. This model is recommended by FEMA 273, 274 and 450 and discussed by Gazetas and Apostolou (2004) and by Kramer (1996). Closed-form solutions concerning moment-rotation relationship convenient for practical use are provided by El Naggar (2003). The numerical approach proposed in this reference offers rotation-controlled foundation moments which fits very well to existing finite element computer code. The present paper proposes a new type "shallow footing element" which allows for direct contribution of soil to overall stiffness of the structure. This element is applied to computer models directed to software SAP 2000 (1997) and PERFORM 3D (2006). Capacity curves are obtained assuming monotonically increasing lateral static loads.

The emphasis of the work is placed on opportunity to carry out displacement-based and performance-based approach for flexibly supported structures. Numerical analysis is based on comparison between structure under fixed-base conditions and flexibly supported structure. As a result following capacity spectrum method a new closed-form expression for flexibly supported structure is derived. It connects the target displacement with the behaviour factor. It allows to control the seismic demands by choosing appropriate value of the behaviour factor. It is concluded that smaller values (in comparison with fixed-base structure) got the behaviour factor should be used if the structure is flexibly supported. It is shown that plastic mechanism and global ductility can be influenced by the soil in a large extent. Finally, this influence can effectively be evaluated using capacity spectrum method.

2. Numerical Model

Numerical model is consistent with the following assumptions:

1. Foundations are stiff in horizontal direction because of passive resistance mobilization of the soil against the face of the footings and embedment and because of the friction between foundation and soil.

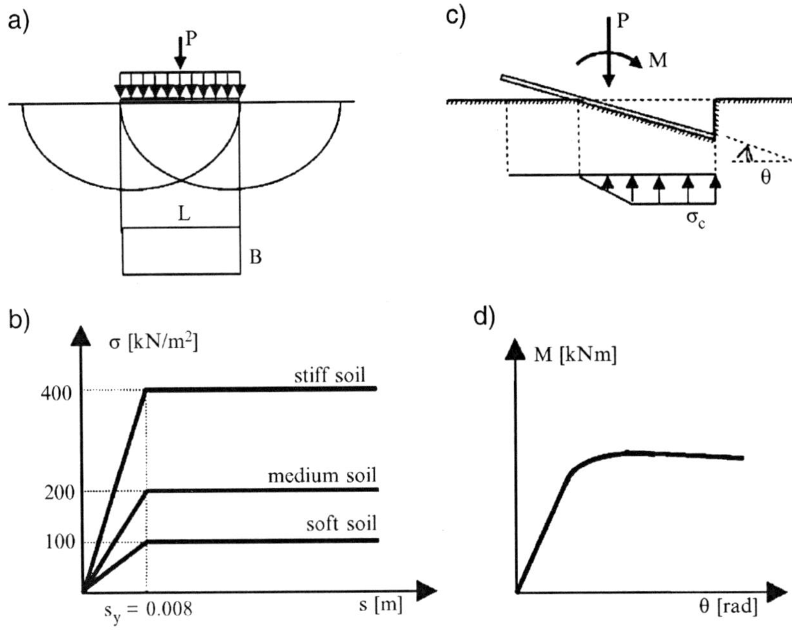

Figure 1. Two parameters soil model: (a) uniform vertical load acting on rigid foundation, (b) basic parameters of the model, (c) soil plasticity and soil uplift, (d) generalized moment-rotation relationship including both phenomena.

2. Foundations are assumed to be absolutely rigid.
3. Foundation rocking has the most essential influence on capacity spectrum diagram.
4. Two parameter constitutive model is used to represent the soil resistance against rocking. The model parameters are Winkler's modulus k and soil capacity σ_c, see Fig. 1.
5. Moment M—rotation θ constitutive relationship for shallow footing is obtained assuming that vertical load on foundation remains constant during the lateral pushover analysis. Because of this restriction the method is limited only to low- and medium-rise buildings (number of storeys less than 10).

The numerical model is consisting of reinforced concrete elements (beams, columns and walls) and foundation elements being all designed according to Eurocode 2 (2004), Eurocode 7 (2004) and Eurocode 8 (2004) provisions. Ductile local and global behaviour of the structure is ensured following the detailing rules and performance-based requirements provided by the code issues. Walls are designed to be "ductile walls" and brittle failure modes are avoided. Shear deformations are only elastic.

2.1. MODELING OF REINFORCED CONCRETE ELEMENTS

Reinforced concrete elements (beams, columns and shearwalls) are modeled by frame elements with zero-length plastic hinges at both ends. The rest part of the element remains completely elastic. The potential disposition of plastic hinges is near to beam-to-column joints where columns and beams elements are connected. Joint core is assumed to be absolutely rigid and plastic hinges are disposed on column/beam faces. Such numerical models are proposed in software codes SAP 2000 and PERFORM 3D. The basic hinge constitutive relationship is "moment-rotation". Plastic hinge parameters are determined according to FEMA 273. The potential places of plastic hinges for shearwalls is near the basement floor at the center of dissipative zone as discussed by Paulay and Priestley (1992).

2.2. IMPLEMENTATION OF SOIL PROPERTIES

Soil is modeled using a model which takes into account soil plasticity and soil uplift as shown in Fig. 1. The basic constitutive relationship for soil at a point level is "vertical stress-vertical settlement". Vertical stresses are implemented on interface surface of the rigid foundation loaded by vertical load. The "stress-settlement" curve is idealized and has two parameters—initial (elastic) modulus (unit foundation modulus or Winkler's modulus) k and soil strength. σ_c (see Fig. 1b). Loading pair consisting of vertical force P and rocking moment M may produce more general distribution of vertical stresses in conformity with uplifting and compression stress capacity area. The resulting constitutive relationship "moment-rotation" is then calculated by integration of stresses over the interface area as recommended by FEMA 273, 274, 450 and El Naggar (2003). P–M interaction curve given in Fig. 2 shows the shallow footing capacity when both loading components P and M are acting simultaneously.

Three different soils are used in further analyses. Their parameters are given in Table I. The plot of stress-settlement lines is presented in Fig. 1b).

Figure 3 shows the results of calculation moment-rotation constitutive relationship for a single shallow foundation which is subjected to constant vertical load and progressively increasing rotation.

Calculation of moment-rotation relationship is based on work of El Naggar (2003). Two dimensionless parameters $\chi = P/\sigma_c LB = \sigma/\sigma_c$ and $\psi = kL/\sigma_c$ are previously calculated. Considering the values of χ. The following two cases are possible considering the first parameter χ. For $0 \leq \chi \leq 0.5$ uplifting phenomenon takes place because compression

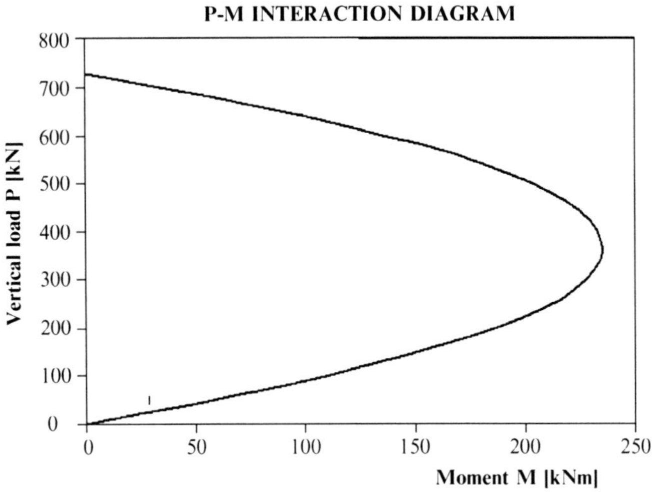

Figure 2. Vertical load P—rocking moment M interaction diagram for a single footing.

TABLE I. Soil properties used in analysis.

Soil No	Notation	Winkler's modulus k (kN/m^3)	Soil strength σ_c (kN/m^2)
1	Stiff soil	50,000	400
2	Medium soil	25,000	200
3	Soft soil	12,500	100

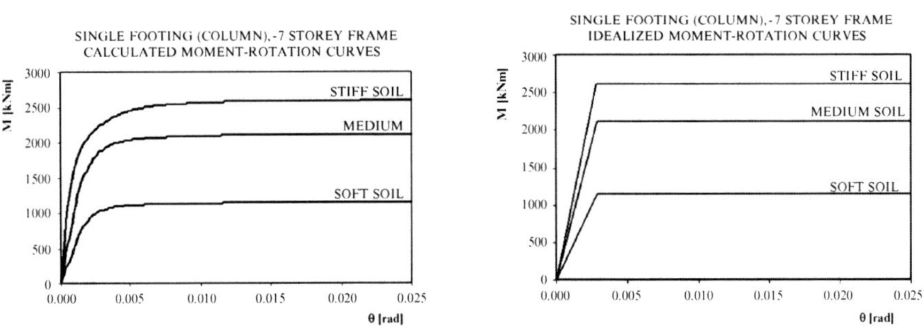

Figure 3. Calculated and idealized moment-rotation relationship for a shallow foundation under the column.

stresses are insufficient. Soil plasticity may also take place. For $\chi > 0.5$ compression stresses reach their capacity σ_c and uplift does not occur. In this case compressive stresses are close to σ_c and soil plasticity is taking part in calculation of footing capacity.

Figure 3 represents the result of moment-rotation calculations for a single footings used for 7-storey frame in Fig. 5. Two plots are presented—one obtained from calculations and one idealized relationship which is easy to implement in the existing computer software as "plastic hinge". Each curve is obtained for soils specified in Table I. First graphics are the result of direct calculation of M when P and angle of rocking rotation are given. It is evident from Fig. 3 that the largest capacity of the rocking moment is reached for stiff soil. The largest initial (elastic) Winkler's modulus is obtained again for stiff soil. Thus for stiff soils larger moment capacity and Winkler's modulus are anticipated.

2.3. NUMERICAL MODEL OF SOIL IMPLEMENTED BY "SOIL ELEMENTS"

Soil properties are implemented through "soil plastic hinge element" being defined as rotational spring with inelastic stiffness, see Fig. 4. The element is the same as "R/C plastic hinge element" but the properties are different. It is evident that foundation element has two hinges—one of reinforced concrete type and one of soil type. The foundation element is assumed to be rigid because its deformations are negligibly small compared to deformations of other elements.

2.4. VARIETY OF FRAME MODELS STUDIED

Three reinforced concrete plane frame systems are studied. They are shown in Fig. 5. The frames have two bays with and 3, 5 and 7 storey correspondingly. Interstorey height is $3m$ for each frame, bay length is 6 m. Seven storey frame is a dual system. Loading patterns for lateral pushover analysis are given on the left, distribution of lateral floor displacements in elevation is presented on the right of each frame figure.

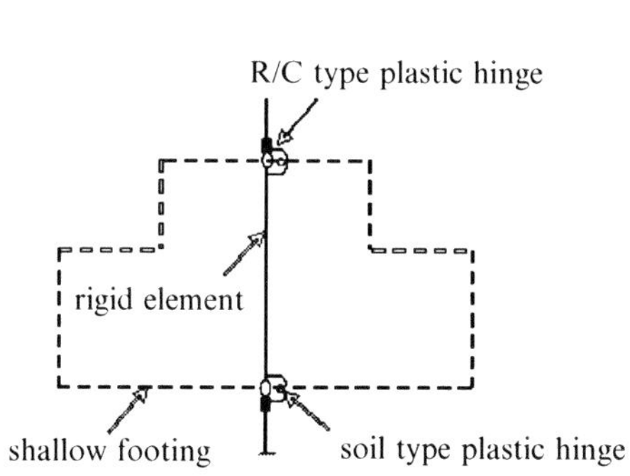

Figure 4. Soil element having two potential plastic hinges—lower hinge is of "soil" type, upper hinge is of "reinforced concrete" type.

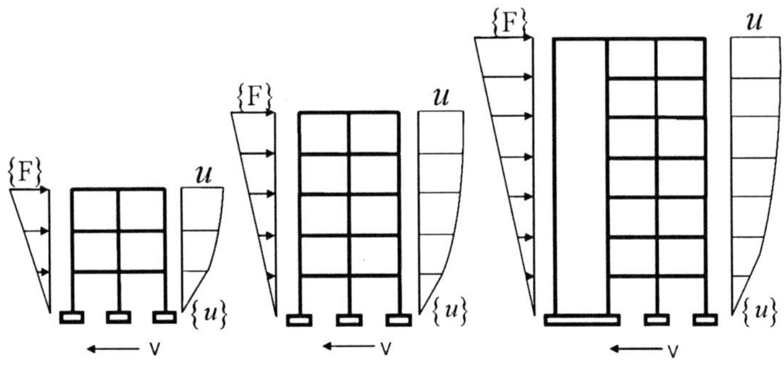

Figure 5. Three, five and seven storey frames studied numerically in the paper.

3. Analysis Methods

Capacity spectrum method is used as evaluation tool as recommended in Eurocode 8, Annex B, 2004. One variety of capacity spectrum method demonstrated by Blagov et al. (2008) is used herein. In general, the method is approximate but proposes relatively easy performance-based assessment of seismic resistance. Very useful developments and applications of the method are proposed by Fajfar (1999, 2000) and Fajfar and Marusic (2005).

3.1. EQUIVALENT SINGLE DEGREE OF FREEDOM SYSTEM AND CAPACITY SPECTRUM METHOD

The basic relationship considering the global behaviour is "*Base shear force V—Roof displacement u*" derived numerically from pushover analysis. The normalized vector of lateral displacements distribution in elevation is denoted by $\{\Phi\}$, see Fig. 5. Normalization is carried out with respect to roof displacement of corresponding frame considering fixed base boundary conditions.

The governing system of equations for dynamic equilibrium assuming that the effect of damping takes part implicitly, can be written as follows:

$$[m]\{\ddot{u}\} + \{f\} = -[m]\{1\}\ddot{u}_g, \tag{1}$$

where $[m]$ is the diagonal mass matrix of floor masses, $\{1\}^T$ is a unity vector defining the transfer of the ground motion in each floor level, $\{f\}$ is the vector of internal (restoring) forces, $\{\ddot{u}\}$ is the vector of floor level accelerations, $\ddot{u}_g$ is the ground acceleration record. It is assumed further that the distribution of lateral displacements $\{u\}$ follows the shape of $\{\Phi\}$-vector with proportionality multiplier $u(t)$ identified as the roof displacement. Thus after replacement of $\{u(t)\} = \{\Phi\}u(t)$ in Eq. (1) and after multiplication of both sides of the new equations by $\{1\}^T$ on the left the following scalar equation is obtained:

$$\{1\}^T[m]\{\Phi\}\ddot{u} + \{1\}^T\{f\} = -\{1\}^T[m]\{1\}\ddot{u}_g. \tag{2}$$

It is relevant to implement the following quantities related to equivalent single degree of freedom system (ESDOF system), namely: mass parameter $m^* = \{1\}^T[m]\{\Phi\} = \sum_i m_i\Phi_i$, total mass of the original system $\{1\}^T[m]\{1\} = \sum_i m_i = m$ and base shear force $V = \{1\}^T\{f\}$. It is worth noting that m is also the mass of the ESDOF system. After taking into account the above notations Eq. (2) is transformed to give

$$\ddot{u}^* + \frac{V}{m} = -\ddot{u}_g. \tag{3}$$

The modified displacement u^* of ESDOF system is introduced with the following expression:

$$u^* = \frac{u}{\Gamma}, \tag{4}$$

where $\Gamma = m/m^*$ is modification factor used to transform the displacement of original system into displacement in the equivalent system. Finally, Eq. (3) provides the result:

$$\frac{V}{m} = -(\ddot{u}_g + \ddot{u}^*). \tag{5}$$

The ratio V/m has acceleration metric units (m/s^2). It can be treated as "*base shear force per unit mass*". This ratio is the capacity spectrum. It is dependent on structural design and ability of structure to dissipate energy through development of inelastic deformations. In the present study capacity spectrum is dependent also on elastic and inelastic soil properties.

The right hand side term in Eq. (5) represents the total acceleration being composed of two counterparts—ground acceleration and relative acceleration of ESDOF system. The maximum values of total accelerations considering ESDOF system are given by the design response spectrum. The intersection point defined by V/m curve (capacity spectrum curve) and spectral design acceleration curve (demand curve) is called "performance point". This intersection point provides the target displacement.

3.2. TAKING INTO ACCOUNT THE SOIL CONDITIONS

Soil conditions are taken into account using "soil plastic hinge element" (Fig. 4) defined by its nonlinear "moment-rotation" relationship. Elastic and inelastic soil properties influence capacity curve of overall structure. For soft and medium soils this influence may become essential considering the position of performance point and the amount of target displacement. Sources of nonlinearity are soil plasticity and soil uplift, see Fig. 1. After reaching the moment capacity the soil becomes ideally plastic (Fig. 3) and soil spring yields. As a matter of fact, elastic-plastic boundary conditions influence seriously the process of re-distribution of internal forces. Soil conditions may significantly change the plastic mechanism of overall structure.

3.3. EXPLICIT ESTIMATE OF SOIL INFLUENCE

It is relevant to search for an explicit expression for soil influence on the structure. To do this two different stages of the structure are considered—"fixed base" structure (denoted by fb subscript) and flexibly supported structure (no subscript is added). The natural frequency of ESDOF system is then defined as

$$(\omega^*)^2 = \frac{V_Y \Gamma}{u_Y m}, \tag{6}$$

where V_Y is the shear base capacity, u_Y is the yielding roof displacement, Γ is participation factor, defined by $\Gamma = m/m^*$ and m is the mass of ESDOF system. Equation (6) defines the natural frequencies of both aforementioned stages. Natural periods of ESDOF systems related to flexible foundations and "fixed base" structure are T^* and T^*_{fb} correspondingly, q and q_{fb} are the behaviour factors for flexibly supported and "fixed base" structures. The behaviour factor ratio can be found using Fig. 6:

$$\frac{q}{q_{fb}} = \frac{S_e(T^*)}{S_e(T^*_{fb})}, \tag{7}$$

where S_e is the elastic design spectrum for accelerations proposed by EC8 provisions. Finally, using again Fig. 6 the target displacement ratio is found to be

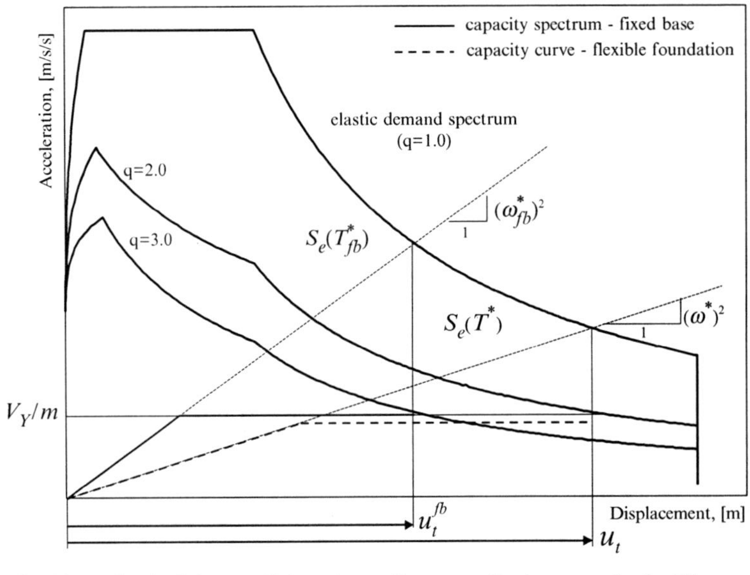

Figure 6. Towards obtaining explicit estimate for target displacement in flexibly supported structure.

$$\frac{u_t}{u_t^{fb}} = \frac{q}{q_{fb}} \left(\frac{T^*}{T^*_{fb}} \right)^2 . \tag{8}$$

The expression (8) creates opportunity to carry out displacement-based control on the displacement demand of flexibly supported structure by appropriate choice of the behaviour factor q. The control on u_t can be performed by q and by Eq. (8), assuming that u_t^{fb} and q_{fb} for "fixed base" structure are already known. The latter parameters are used as reference values. It is proven numerically (Schanz et al., 2007) that $q < q_{fb}$ if soil with elastic properties is implemented into analysis. This paper proposes an explicit expression in closed form which allows for efficient control on target displacement assuming to be carried out considering elastic or inelastic soil behaviour, see Eq. (8). The advantage of this expression is that it can be applied to structures with various types of foundation. Application of Eq. (8) can be extended to structures with deep foundations using more complex and sophisticated mathematical foundation model.

The most essential advantage of Eq. (8) is found to be in opportunity to to determine "site dependent behaviour factor q" depending on soil deformation properties. Usually the behaviour factor depends on structural properties (structural type and ductility class) and does not depend on soil deformability. It is possible due to Eq. (8) to predict the change in q just to ensure the required seismic displacement demands. The application of Eq. (8) is shortly discussed in the next paragraph.

It is known that as a rule flexible foundations lead to greater lateral displacement demands (Schanz et al., 2007). The main question arising herein is how to control the growing up displacements using displacement-based design procedure. Following the result (8) this is possible by selection of "site dependent q factor" (less than q_{fb}).

4. Numerical Results, Observations and Comments

Numerical results are presented in Figs. 7, 8 and 9.

Figure 7 represents the pushover curves when inelastic soil deformations take place. In three studied cases "soil plastic hinge" appears due to large vertical load and lateral pushover loading pattern. In all of the three cases $\chi > 0.5$ which means that uplifting phenomenon does not occur. The plots in Fig. 7 are representative curves for the global structural behaviour. Three main observations are evident from the plots:

1. Initial (elastic) stiffness is reduced due to foundation flexibility. This reduction is greater for soft soils and smaller for stiff soils.

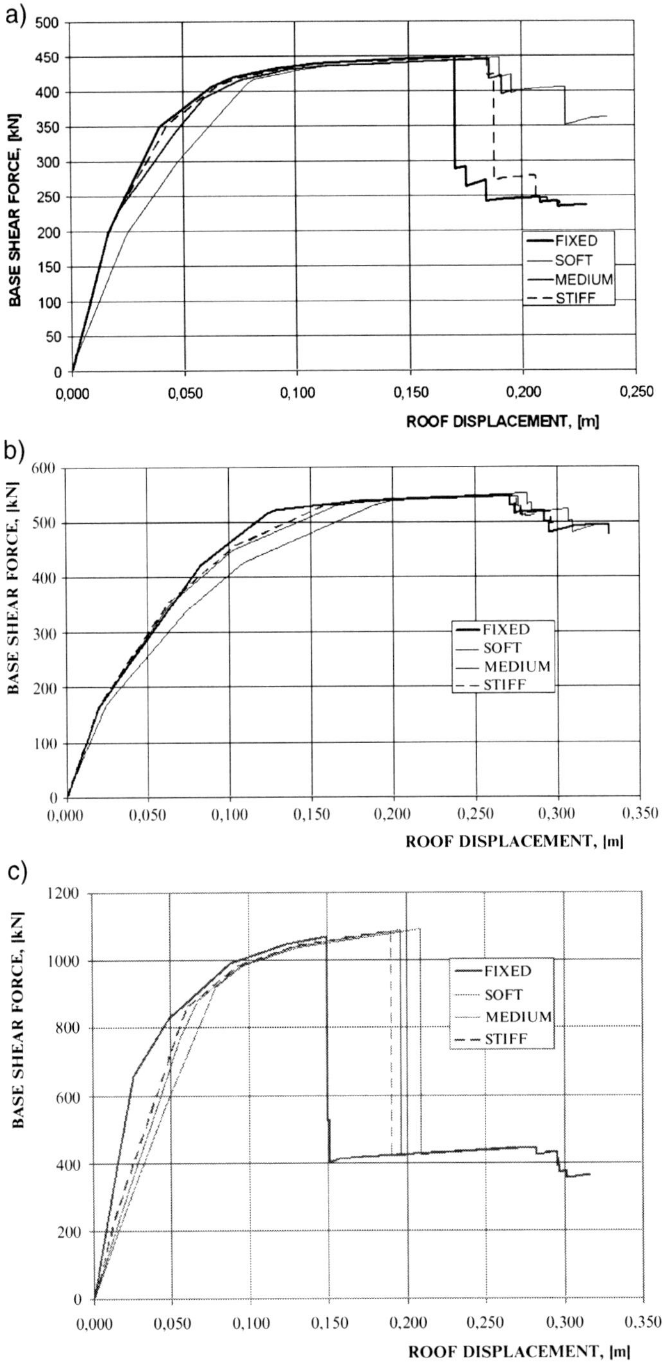

Figure 7. Capacity curves for frames shown in Fig. 6: (a) three storey frame, (b) five storey frame, (c) seven storey frame. Different family of lines are determined for corresponding soil or fixed base conditions.

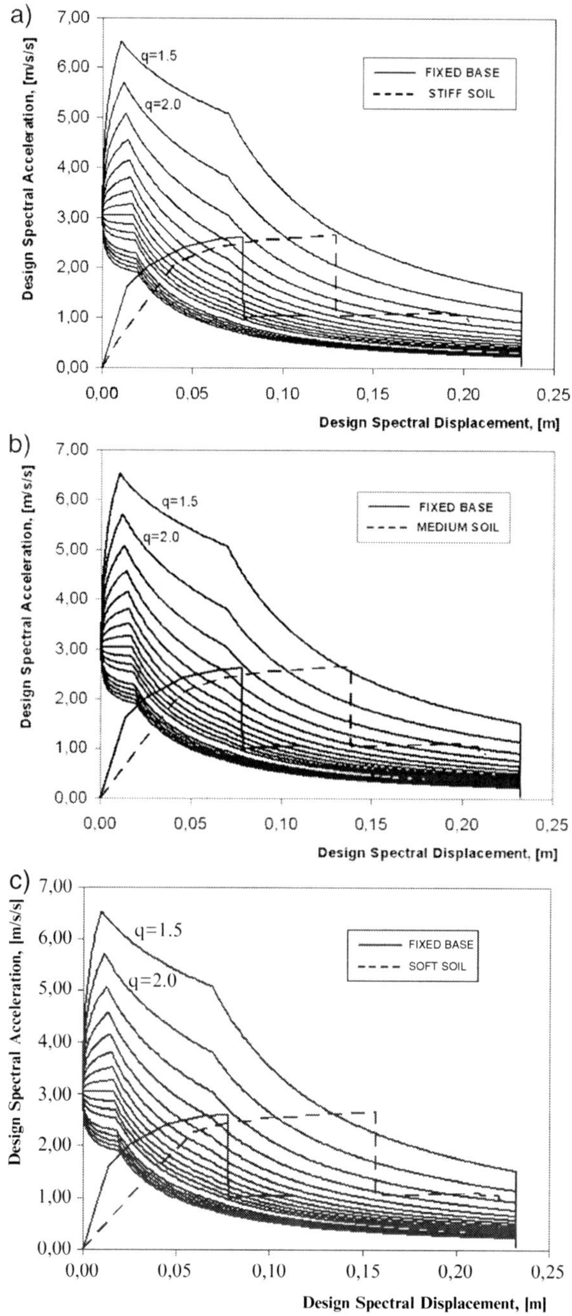

Figure 8. Demand–Capacity diagrams for 7 storey frame from Fig. 5. The family of design demand spectra are obtained for q-factor within the range 1.5–10 (step 0.5).

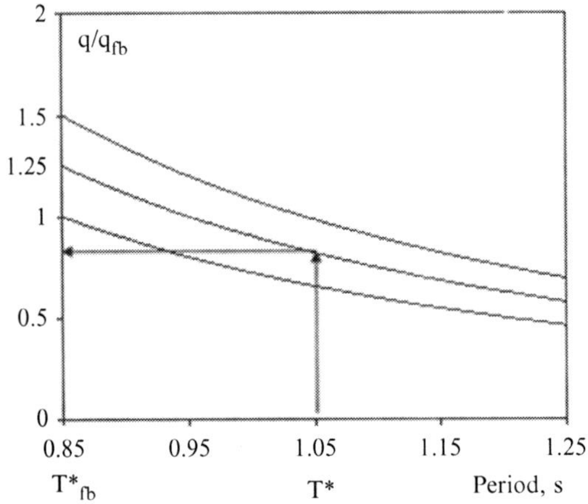

Figure 9. Displacement-based control by the behaviour factor.

2. Plastic mechanism is characterized by the fact that first plastic hinges are opened in the soil spring elements. Plastic hinges in R/C elements are activated later. Soil plastic hinges influence the redistribution of internal forces.
3. Due to soil flexibility target displacements are increased. This increase is more essential for soft soils.
4. The global ductility is reduced. The most essential reduction is observed for soft soils.
5. Structure base shear capacity is relatively independent on the soil stiffness properties.

Figure 8 shows the demand-capacity diagrams for 7 storey dual structure using different soil conditions. Design demand curves are defined and plotted as “equal q curves” where each point on the curve has the same value of the behaviour factor. Intersection point between each capacity diagram and demand curve defines the displacement demand. Target displacement results are also dependent on structural resistance (represented by capacity diagram) and on the behaviour factor (represented by the demand diagram). It is evident from Fig. 8 that for flexibly supported structure the increase of target displacement is achieved (compared to “fixed-base” structure) for smaller value of behaviour factor. Thus the increase of displacements demands can be reached by decreasing of q-factor. It is also seen that elastic component of displacements is increased and inelastic displacement demand remains approximately the same. The frame is designed using $q = 3.5$.

The opportunity to control the displacement demand of flexibly supported structure is demonstrated in Fig. 9. The plot of relationship

$$\frac{q}{q_{fb}} = \frac{u_t}{u_t^{fb}} \left(\frac{T_{fb}^*}{T^*} \right)^2$$

based on Eq. (8) shows decreasing tendency of the ratio q/q_{fb} with respect to period T^* increase. This result is also evident from Fig. 8 and attached comments to it. The ratio u_t/u_t^{fb} defines a family of lines. Each line is proportional to the line corresponding to the line of "fixed-base" case. Proportionality coefficient is the displacement ratio which shows how much the displacement u_t^{fb} is exceeded by u_t. The displacement ratio is reached if behaviour factor ratio is used for determination of q.

If the period of the "fixed-base" structure is 0.85s and the period of flexible supported structure is 1.05s, and if the displacement ratio is 1.25 (displacement enlargement 25%), the behaviour factor of flexibly supported structure should be obtained by multiplying q_{fb} with 0.75. Thus the idea for displacement-based control on seismic displacement demands can be carried out in practice.

5. Conclusions

Conclusions can briefly be summarized in the following items:

1. As general rule foundation flexibility can influence the overall designresponse of structure. As a result target displacements are increased for soft soils.
2. Initial (elastic) stiffness is reduced due to soil deformations. Base shear capacity of the structure remains relatively independent of soil stiffness properties.
3. Elastic component of the global displacement is increased progressively when the soil is soft. Inelastic displacement demand is remaining relatively unchanged. However the global ductility is reduced due to soft soils.
4. Global displacement demands can be controlled by appropriate choice of the behaviour factor taking into account site specific data.
5. The values of the behaviour factor can be derived using displacement-based criterion.

Acknowledgements

This work is supported by the NATO grant CLG 982064. The grant is greatly acknowledged by the authors.

The project grant No. BN-84/2008 of the University of Architecture, Civil Engineering and Geodesy, Sofia, Bulgaria, is greatly acknowledged by the authors.

References

Blagov, D., Georgiev, V., and Bonev, Z. (2008) Influence of flexible foundations on the design response of buildings with accidental eccentricity. *Proceeding of the Fifth European Workshop on the Seismic Behaviour of Irregular and Complex Structures*, pp. 375–386, 16–17 September 2008, Catania, Italy.

Bonev, Z., Ganchev, S., Blagov, D., and Zerzour, A. (2005) Factor evaluation accounting for the elastic foundation. In *Proc. EE-21C, Topic 4: Structural Modelling, Analysis, Design and Seismic Safety*, Skopje, Republic of Macedonia.

Chopra, A. and Goel, R. (1999) Capacity-demand diagram methods for estimating seismic deformation of inelastic structures: SDOF systems, A Report on Research Conducted under Grant No. CMS-9812531, US-Japan Cooperative Research in Urban Earthquake Disaster Mitigation, from the National Science Foundation, Report No. PEER-1999/02 Pacific Earthquake Engineering Research Center, College of Engineering University of California, Berkeley.

Chopra, A. K. and Goel, R. (2003) Evaluation of the modal pushover analysis procedure using vertically "regular" and irregular generic frames, Report No. EERC 2003-03, University of California, Berkeley.

EN 1992-1-1 (2004) Eurocode 2: Design of Concrete Structures, Part 1-1: General Rules and Rules for Buildings, December.

EN 1997-1 (2004) Eurocode 7: Geotechnical Design, Part 1: General Rules, November.

EN 1998 (2004) Parts 1 and 5, Eurocode 8: Design of Structures for Earthquake Resistance, Part 1: General Rules, Seismic Actions and Rules for Buildings (EN 1998-1), December.

Fajfar, P. (1999) Capacity spectrum method based on inelastic demand spectra, *Earthquake Engineering and Structural Dynamics* **28**, 979–993.

Fajfar, P. (2000) A nonlinear analysis method for performance-based seismic design, *Earthquake Spectra* **16**(3), 573–592.

Fajfar, P. and Marusic, D. (2005) The extension of N2-method to asymmetric buildings. In *Proc. 4th EWICS* (*European Workshop on the Seismic Behaviour of Irregular and Complex Structures*), Thessaloniki, Greece.

FEMA 273 (1997) NEHRP Guidelines for the Seismic Rehabilitation of Buildings, Issued by FEMA in Furtherance of the Decade for National Disaster Reduction, October.

FEMA 274 (1998) Seismic Rehabilitation Commentary C4: Foundations and Geotechnical Hazards.

FEMA 357 (2000) Global Topics Report on the Prestandard and Commentary for the Seismic Rehabilitation of Buildings, Prepared by American Society of Civil Engineers, Reston, Virginia.

FEMA 450 (2003) NEHRP Recommended Provisions for Seismic Regulations for New Buildings and Other Structures, Commentary C7A.

Gazetas, G. and Apostolou, M. (2004) Nonlinear soil-structure interaction: Foundation uplifting and soil yielding. In *Proc. 3rd UJNR Workshop on Soil-Structure Interaction*, Menlo Park, California, USA.

Geotechnical Engineering, Circular No. 3 (1997) Design Guidance: Geotechnical Earthquake Engineering for Highways, Publication No. FHWA-SA-97-076, Vol. 1: Design Principles, US Department of Transportation, Federal Highway Administration, Office of Engineering, Office of Technology Applications, 400 Seventh street,, SW, Washington, DC 20 590, May, 1997.

El Naggar, H. M. (2003) Seismic response of structures with underground storeys, JCLR Research Paper Series, No. 26, Institute for Catastrophic Loss Prediction, University of Western Ontario.

Kilar, V. and Fajfar, P. (2000) Simplified nonlinear seismic analysis of asymmetric multistorey R/C building. In *Proc. 12th European Conference on Earthquake Engineering*, Pap. Ref. 033.

Kramer, S. L. (1996) *Geotechnical Earthquake Engineering*, Prentice-Hall International Series in Civil Engineering and Engineering Mechanics, Upper Saddle River, New Jersey.

Paulay, T. and Priestley, M. J. N. (1992) *Seismic Design of Reinforced Concrete and Masonry Buildings*, John Wiley & Sons, New York.

PERFORM 3D v. 4 (2006) *Nonlinear Analysis and Performance Assessment of Structures*, CSI, Berkeley, California, August.

SAP 2000 (1997) *Integrated Finite Element Analysis and Design of Structures*, CSI, Berkeley, California.

Schanz, T., Bonev, Z., Georgiev, V., and Iankov, R. (2007) Application of capacity spectrum method to soil-foundation-structure interaction problems. In *Proc. of the Jubilee Scientific Conference, Devoted to 65 years Anniversary of the University of Architecture, Civil Engineering and Geodesy*, Sofia, Bulgaria.

EFFECT OF DEPTH OF GROUND WATER ON THE SEISMIC RESPONSE OF FRAME TYPE BUILDINGS ON SAND DEPOSITS. PART I: SOIL RESPONSE

S. Umit Dikmen (`u.dikmen@iku.edu.tr`)*, A. Murat Turk, and Guven Kiymaz
Department of Civil Engineering, Istanbul Kultur University, Bakirkoy, 34156, Istanbul, Turkey

Abstract. Earthquake codes often times classify the subsoil based on their shear wave velocities which are mostly calculated based on the situ static conditions. However, it has been shown by various researchers that saturated sands exhibit a softening behavior under cyclic conditions. As a result of this behavior the surface response may be significantly affected by the existence of ground water. In this study, the effect of depth of water table on the seismic response of sand deposits is investigated.

Keywords: seismic, sand, ground water, ground response

1. Introduction

Current versions of the seismic codes in general classify the subsoil based on their shear wave velocities (EUROCODE-8, 2004; ASCE Standard 7-05, 2006; TEC-2007, 2007). In this respect, both Eurocode 8 and TEC-2007 define the shape and the amplitudes of the response spectrum by applying the local soil conditions and the expected maximum ground acceleration on a predefined raw response spectrum skeleton. Then the response spectrum amplitudes are further multiplied by building importance and load reduction factors to account for the post earthquake importance of the structure and for the ductility respectively.

The raw response spectrum skeleton used in both codes is based on the earlier research done by Seed et al. (1976a,b) and by Newmark and Hall (1982). In both of those studies, actual field data recorded in various earthquakes at different site conditions were classified to obtain a design spectrum of base motion to be used in structural calculations. Thus the design spectrum obtained is from an ensemble of records containing different water table depths. In other words, water table depth was not a classification parameter.

*Corresponding author.

T. Schanz and R. Iankov (eds.), *Coupled Site and Soil-Structure Interaction Effects with Application to Seismic Risk Mitigation,* NATO Science for Peace and Security Series C: Environmental Security,

However on the other hand, under seismic conditions, especially in case of loose and medium density saturated sands, increasing pore water pressure will cause a decrease in the effective stress which in turn will cause a decrease of the effective shear modulus apart from the decrease due to the nonlinear material behavior of sands. Whereas, the degradation of shear modulus for dry sands will be purely due to the nonlinear material behavior of sands. This property of the saturated sands may lead to considerable differences on the amplification at the surface when compared with those of the deposits with same soil characteristics but no ground water. This characteristic of saturated sands has been studied by various researchers and documented (Ishihara and Towhata, 1980; Dikmen and Ghaboussi, 1984; Ghaboussi and Dikmen, 1984; Yoshida and Iai, 1998; Youd and Carter, 2005; Todorovska and Al Rjoub, 2006; Zorapapel and Vucetic, 1994).

2. Method of Analysis

A two stage analysis approach has been selected to investigate the effect of ground water depth on the seismic response of frame type buildings.

In the first stage, amplification of the seismic motion at the base rock level to the ground surface response of sand deposits with varying depths of water table has been investigated. For this purpose, LASS-IV (Dikmen and Ghaboussi, 1984; Ghaboussi and Dikmen, 1984) software is utilized. The main reason for selection of the LASS-IV code at this stage is its fully non-linear two phase medium solution capability, namely effective stress analysis.

In the second stage, the ground surface motion obtained in the first stage were applied as base motion to steel frames and reinforced concrete frames with differing number of storeys designed per the current codes. SAP2000 software is used for the seismic structural analysis of the frames. The details of this second stage will be presented in a companion paper.

2.1. LASS-IV SOFTWARE

As mentioned above, LASS-IV software is capable of performing fully non-linear effective stress analysis of sands under seismic conditions in time domain. The theoretical background used in LASS-IV has been presented in detail in a publication by (Dikmen and Ghaboussi, 1984). However, a short summary of the method will only be presented here.

The method models the horizontally layered ground by a number of "layer elements". Thus the response of the system is described in terms of the nodal plane displacement degrees of freedom. Each nodal plane has three degrees of freedom, two components of the solid displacement and the displacement of the pore water relative to solid. Respectively, the stresses considered are the vertical normal stress, the horizontal shear stress and the pore water pressure. The nodal planes are assumed to remain horizontal and undergo parallel displacements. Thus, the corresponding stresses considered in a layered ground are; the vertical normal stress σ, the horizontal shear stress τ, and the pore water pressure π.

The method uses a plasticity based material model for analyzing the behavior of sands under cyclic loading. A modified Masing type material model is used to define the stress–strain relationship. In this respect, the increments of shear strain are assumed to be the sum of the elastic and plastic shear strain increments

$$d\gamma = d\gamma^e + d\gamma^p$$

and the relationship between the shear stress and strain increments are related through the following equation

$$d\tau = \frac{GH}{G + H} d\gamma,$$

where G is the shear modulus and H is the plastic modulus. The shear modulus G is assumed to vary with the effective stress and the plastic modulus is the slope of the shear stress—plastic strain curve which is defined by a hyperbolic curve having an asymptote M as defined in the next paragraph. Hence, within the computer program the stiffness matrix is continuously updated to reflect the current values of shear and plastic moduli per changes in the stress conditions.

As for the failure criteria, shear strength is assumed to be isotropic with respect to shear stresses and assumed to vary linearly with the effective pressure, implying a circular cone shaped failure surface in the stress space as expressed with the following equation

$$F(\sigma) = \tau - M\sigma' = 0$$

in which $M = \tan(\phi)$, where ϕ is the internal angle of friction. Thus, M defines the failure condition τ/σ'.

The undrained effective stress path is assumed to be a quarter of an ellipse in $\tau - \sigma'$ plane, equation of which is given as follows

$$f_e = \tau^2 + \lambda^2 \left[\sigma'^2 - \left(\frac{2\lambda}{\lambda + M}\right)\sigma'\sigma_0 + \left(\frac{\lambda - M}{\lambda + M}\right)\sigma'^2_0\right] = 0.$$

Two parameters, namely σ_0 and λ completely define the effective stress path. The material parameter λ is the ratio of vertical to horizontal axis of the effective stress path which can be correlated to the density of sand, generally increasing with relative density. The effective stress path is followed only during the elasto-plastic shear stress increments when yielding takes place. The criterion for the onset of liquefaction is stated as the effective stresses reaching a residual value.

In the method, the nonlinear material model is considered to inherently comprise of two damping mechanisms, namely hysteretic damping and dissipative damping of the pore water. Thus no additional viscous damping ratio is used.

2.2. ANALYSIS MODEL

To show the effect of water table depth on the surface response of sand deposits subject to seismic loading a 30.0 m deep sand layer laying over fractured rock has been selected.

Furthermore, to account for the effects of the relative density of the sand deposit two different soil profiles are used per TEC-2007. The first soil profile has the characteristics of Z3 type subsoil per TEC-2007 or approximately type C subsoil per Eurocode 8. Per TEC-2007, Z3 type subsoils should have shear wave velocities ranging 200 m/s between 400 m/s for sand deposits. Whereas the type C subsoil per Eurocode 8 is defined as dense or medium dense sand deposits having an average of 180–360 m/s shear wave velocity at the top 30 m of the deposit. The second soil profile has the characteristics of Z4 type subsoil per TEC-2007 or approximately type D subsoil per Eurocode 8. Per TEC-2007, Z4 type subsoils should have shear wave velocities less then 200 m/s for sand deposits. Whereas the type C subsoil per Eurocode 8 is defined as loose to medium dense sand deposits having an average shear wave velocity less then 180 m/s for the top 30 m of the deposit.

The material characteristics of both profiles are summarized in Table I. In both cases the shear modulus is assumed to vary linearly between the top and bottom layers.

TABLE I. Properties of the selected soil profiles

	Soil Type (per TEC-2007)	
	Z3	Z4
Initial Shear Modulus at top layer (kPa)	70,000	35,000
Shear Modulus increase by depth (kPa)	4,000	2,000
Mass density of soil	1.70	1.70
Coefficient of Permeability (m/s)	0.000003	0.000003

From the data presented in Table I, the shear wave velocities for top and the bottom of the soil profile for Z3 type soil can be calculated as approximately 200.0 and 330.0 m/s respectively. The values for Z4 type soil will be approximately 140.0–240.0 m/sec for top and bottom respectively. The rock layer is assumed to have a shear wave velocity of 750.0 m/sec which closely represents the fractured rock layers per TEC-2007 where fractured rock layers are defined to have shear wave velocity between 700 and 1,000 m/s.

2.3. BASE MOTION

As for the base motion, May 1940 El Centro Earthquake record has been selected for its rich frequency content. The record is digitized to 0.005 s and 31.0 s of the data has been used in calculations. However, the record has been scaled down to a peak acceleration of 0.20 g.

3. Analysis of the Results

Analyses have been performed for both selected profiles for five different water table depths of 1.0, 2.0, 5.0, 10.0, and 15.0 m. In all cases analyzed, no liquefaction was observed. For comparison, the peak ground surface accelerations and the response spectrum of the ground surface motion values that are deemed to be significant for the structural calculations will be presented here.

The peak ground surface accelerations are presented in Table II. As can be seen from the results tabulated, almost in all cases there is an increase in the peak ground surface accelerations by increasing water table depths.

As mentioned above, the ground surface acceleration response spectra, for 5% damping, for different levels water table depth is calculated for

TABLE II. Peak ground surface accelerations

	Z3-soil		Z4-soil	
Water Table Depth (m)	Peak Acc. (g)	Time (s)	Peak Acc. (g)	Time (s)
1.0	0.46	2.27	0.38	2.57
2.0	0.33	2.22	0.41	2.30
5.0	0.38	2.54	0.43	2.51
10.0	0.45	2.48	0.52	2.38
15.0	0.47	2.50	0.54	2.41

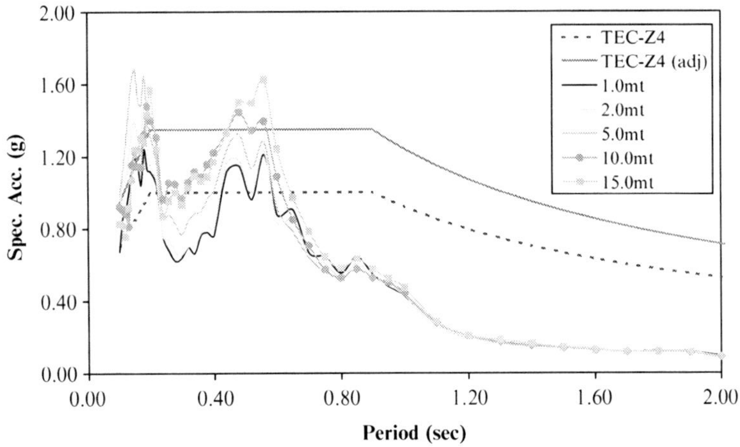

Figure 1. Ground surface response spectra for Z3 type soil.

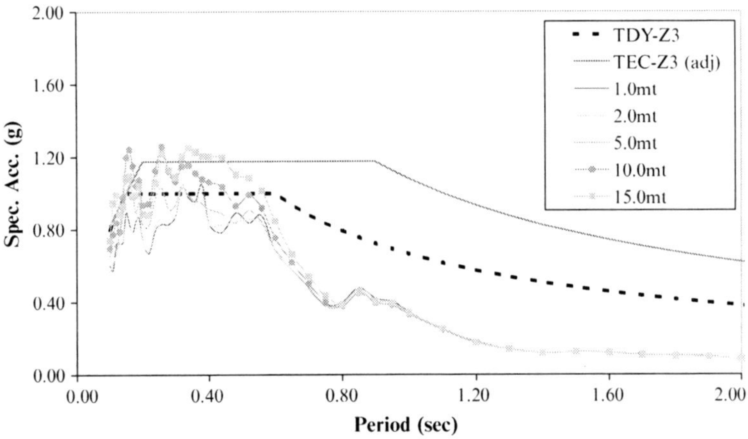

Figure 2. Ground surface response spectra for Z4 type soil.

comparison. The results are presented in Figs. 1 and 2 for Z3 and Z4 type soils, respectively. As can be seen from both figures, the spectral values for all the water table depths remain practically the same for periods above 1.0 s. However, for lower periods especially in the range of 0.2–0.8 s significant differences are observed. The response spectra in this range show an increasing trend with the increasing water table depth.

The ground surface response spectra obtained are also compared and presented in Figs. 1 and 2. In both figures the design spectra shown with lower spectral acceleration values represent the spectra for earthquake region 1 in Turkey, namely a peak ground acceleration of 0.4 g. Whereas the design spectrum with higher values is by using the actual peak acceleration obtained from the analysis, namely 0.47 and 0.54 g for Z3 and Z4 soils respectively. In the case of Z3 type soil in all cases the code spectra seemed to be reasonably conservative for all periods. But whereas for the Z4 type soil, code specified spectra is not conservative for all periods especially in the range of 0.2–0.8 s periods which is significant for the structural analysis.

The reduction of both the spectral accelerations and the spectral values by decreasing water table depth can be attributed to the reduction of effective stress under seismic conditions and an associated reduction in the shear modulus. This behavior observed closely agrees with the findings by various researchers (Youd and Carter, 2005; Trifunac and Todorovska, 1996; Trifunac and Todorovska, 1998; Trifunac et al., 1999).

4. Conclusions

A numerical simulation has been conducted and presented about the effects of the ground water table depth on the ground surface motion under seismic conditions. An effective stress time history analysis method is used via a software code named LASS-IV. From the results of this study, following conclusions can be drawn,

- Under seismic conditions soil softening occurs due to decreasing effective stress as a result peak ground surface accelerations may reduce significantly.
- The code specified response spectrum may not be conservative at all periods.

References

ASCE Standard 7-05 (2006) Minimum design loads for buildings and other structures, ASCE.

Dikmen, S. U. and Ghaboussi, J. (1984) Effective stress analysis of seismic response and liquefaction: Theory, *J. Geotech. Eng.*, ASCE **110**(5), 628–644.

EUROCODE-8 (2004) Design of structures for earthquake resistance, 1998-1.

Ghaboussi, J. and Dikmen, S. U. (1984) Effective stress analysis of seismic response and liquefaction: Case studies, *J. Geotech. Eng.*, ASCE **110**(5), 645–658.

Ishihara, K. and Towhata, I. (1980) One-dimensional soil response analysis during earthquake based on effective stress method, *J. Fac. Eng., Tokyo* **XXXV**(4), 656–700.

Newmark, N. M. and Hall, W. J. (1982) *Earthquake Spectra and Design*, EERI Monograph, Earthquake Engineering Research Institute, Berkeley, California, USA, pp. 103.

SAP2000 Integrated Software Structural Analysis and Design, Computers and Structures, Inc., Berkeley, California, USA.

Seed, H. B., Murarka, R., Lysmer, J., and Idriss, I. M. (1976a) Relationships of maximum acceleration, maximum velocity, distance from source and local site conditions for moderately strong earthquakes, *B. Seismol. Soc. Am.* **66**(4), 1323–1342.

Seed, H. B., Ugas, C., and Lysmer, J. (1976b) Site dependent spectra for earthquake resistant design, *B. Seismol. Soc. Am.* **66**(1), 221–243.

TEC-2007 (2007) Turkish earthquake resistant design code, Ministry of Public Works, Turkey.

Todorovska, M. I. and Al Rjoub, Y. (2006) Effects of rainfall on soil-structure system frequency: Examples based on poroelasticity and a comparison with full-scale measurements, *Soil Dyn. Earthq. Eng.* **26**(6), 708–717.

Trifunac, M. D. and Todorovska, M. I. (1996) Nonlinear soil response—1994 Northridge California earthquake, *J. Geotech. Eng.*, ASCE **122**(9), 725–735.

Trifunac, M. D. and Todorovska, M. I. (1998) Nonlinear soil response as a natural passive isolation mechanism—The 1994 Northridge California earthquake, *Soil Dyn. Earthq. Eng.* **17**(1), 41–51.

Trifunac, M. D., Hao, T. Y., and Todorovska, M. I. (1999) On recurrence of site specific response, *Soil Dyn. Earthq. Eng.* **18**(8), 569–592.

Yoshida, N. and Iai, S. (1998) Nonlinear site response and its evaluation and prediction. In *Proc. 2nd Int. Symp. on the Effect of Surface Geology on Seismic Motion*, Yokusuka, Japan, pp. 71–90.

Youd, T. L. and Carter, B. L. (2005) Influence on soil softening and liquefaction on spectral acceleration, *J. Geotech. Geoenviron. Eng.*, ASCE **131**(7), 811–825.

Zorapapel, G. T. and Vucetic, M. (1994) The effects of seismic pore water pressure on ground surface motion, *Earthquake Spectra*, EERI **10**(2), 403–437.

Part IV

General and Related Subjects

EXTRACTING THE TIME-DOMAIN BUILDING RESPONSE FROM RANDOM VIBRATIONS

Roel Snieder (rsnieder@mines.edu)
Center for Wave Phenomena and Dept. of Geophysics,
Colorado School of Mines, Illinois Street1500, CO 80401-1887 Golden, USA

Abstract. The extraction of the response from field fluctuations excited by random sources has received considerable attention in a variety of different fields. I show application of this principle to the motion recorded after an earthquake in the Millikan Library at the California Institute of Technology in Pasadena, California. Deconvolution of the recorded motion at different floors unravels the building response from the complicated excitation and from the unknown soil-structure interaction. I give arguments why analyzing the response function in the time domain is more informative than only using the amplitude spectrum of the transfer function. I provide examples showing that it is possible to extract the building response that satisfies the same dynamic equations as does the real building, but that may satisfy different boundary conditions at the base. This means one can obtain from the data the building response with different soil-structure interaction than that of the real building.

Keywords: building response, time domain

1. Introduction

The extraction of information from random field fluctuations is a rapidly growing field in physics, acoustics, engineering, and geophysics. The widespread application of this idea has resulted in a variety of different names for the method, including Green's function extraction, daylight imaging, the virtual source method, and seismic interferometry (Curtis et al., 2006; Larose et al., 2006a).

Aki (1957) pioneered the use of microseismic noise to extract the properties of the near surface; a technique later called microtremor analysis, having numerous applications (Louie, 2001; Chávez-Garcia and Luzón, 2005). Lobkis and Weaver (2001) gave the field new momentum with their derivation of Green's function extraction based on normal modes, showing how useful the Green's function extraction can be in practical applications. A flurry of applications appeared in different fields, including ultrasound (Weaver and Lobkis 2001, 2003; Malcolm et al., 2004; van Wijk 2006; Larose et al., 2006b), helioseismology (Rickett and Claerbout, 1999, 2000),

T. Schanz and R. Iankov (eds.), *Coupled Site and Soil-Structure Interaction Effects with Application to Seismic Risk Mitigation,* NATO Science for Peace and Security Series C: Environmental Security,

ocean acoustics (Roux and Fink, 2003; Roux et al., 2004; Sabra et al., 2005b), structural engineering (Snieder and Şafak, 2006; Snieder et al., 2006; Thompson and Snieder, 2006; Kohler et al., 2007; Sabra et al., 2008), medical diagnostics (Sabra et al., 2007), exploration seismology (Schuster et al., 2004; Bakulin and Calvert, 2006; Hornby and Yu, 2007; Draganov et al., 2007; Bakulin et al., 2007), crustal seismology (Campillo and Paul, 2003; Shapiro and Campillo, 2004; Shapiro et al., 2005; Sabra et al., 2005a; Roux et al., 2005; Mehta et al., 2007b), and hazard monitoring (Sabra et al., 2006; Sens-Schönfelder and Wegler, 2006; Wegler and Sens-Schönfelder, 2007). In most of the applications, the system response was extracted by cross-correlating field fluctuations, but the alternative data-processing technique of deconvolution has been applied in some of these studies.

The fields obtained from correlation or deconvolution satisfy the same equation as does the original system; these methods are thus guaranteed to give valid states of the field in the real medium (Snieder et al., 2006). However, the field extracted with any of the three approaches may satisfy different boundary conditions than the real physical system satisfies. This makes it possible to determine the wave state as if the system satisfied different boundary conditions than does the real system.

I illustrate the extraction of the system response by showing in Section 2 the response extracted from the recorded motion of a building after an earthquake. In Section 3, I show advantages of extracting the transfer function in the time domain, rather than only considering the amplitude spectrum of the impulse response. Section 4 contains examples of the extracted wave state of the building as if it was placed on the reflectionless subsurface, or on a subsurface that gives total reflection.

2. Extracting the Building Response

Using recordings of vibrations of the Millikan Library at Caltech that occurred during an earthquake, I illustrate extracting the time-domain response of a building from random vibrations (Snieder and Şafak, 2006). This building is shown in the left panel of Fig. 1. The location of accelerometers in the basement and the ten floors is marked with solid circles. In the right panel, the north–south component of the acceleration after the Yorba Lina earthquake is shown. The motion increases with height in the building because of the increased sway of the building with height.

The recorded motion is a consequence of the combination of (i) the waves excited by the earthquake that strike the building from below, (ii) the coupling of the building to the subsurface, and (iii) the mechanical properties of the

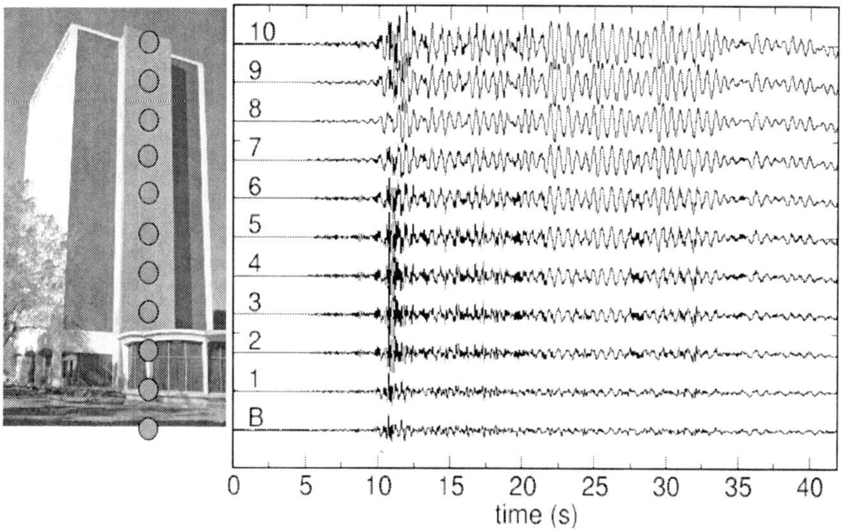

Figure 1. Left panel: the Millikan Library at Caltech with the location of accelerometers. Right panel: north–south component of acceleration after the Yorba Linda earthquake of 03 Sep 2002 (ML = 4.8, Time: 02:08:51 PDT, 33.917N 117.776W, Depth 3.9 km). Traces are labeled with the floor number (B indicates basement).

building. It is the goal to extract the mechanical properties of the building from the motion shown in Fig. 1. This can be achieved by deconvolving the motion at every level with the motion recorded at one of the floors. Deconvolution of the motion $u(z, t)$ at height z, with the motion $u(z_0, t)$ at reference level z_0, in the frequency domain, is given by:

$$D(z, z_0, \omega) = \frac{u(z, \omega)}{u(z_0, \omega)} , \tag{1}$$

where $u(z, \omega)$ is the temporal Fourier transform of $u(z, t)$. At notches in the spectrum $u(z_0, \omega)$ this spectral division is unstable. For this reason I replace the deconvolution in expression (1) by

$$D(z, z_0, \omega) = \frac{u(z, \omega)u^*(z_0, \omega)}{|u(z_0, \omega)|^2 + \varepsilon} . \tag{2}$$

Figure 2 shows the motion of the Millikan Library after deconvolution with the motion at the top floor. The deconvolved motion in the top of the building is a bandpass-filtered delta function in the basement, because any function deconvolved with itself gives a delta function. The deconvolved waves are extremely simple and consist of an upgoing wave that propagates up the building, is reflected by the top of the building, and then propagates downward. The velocity of this wave can easily be measured and is equal

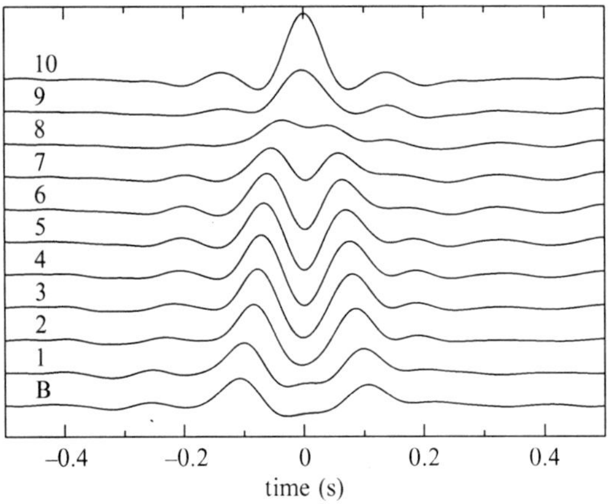

Figure 2. Waveforms of Fig. 1 after deconvolution with the motion at the top floor.

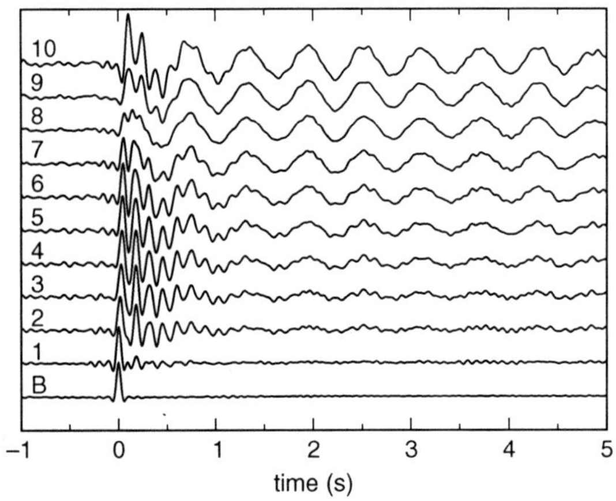

Figure 3. Waveforms of Fig. 1 after deconvolution with the motion in the basement. Note that the time scale is different than in Fig. 2.

to 322 m/s (Snieder and Şafak, 2006). During their propagation through the building, the deconvolved waves attenuate, and the quality factor of the building can be estimated from the attenuation (Snieder and Şafak, 2006).

The response extracted by deconvolving the motion at every floor with the motion in the basement is shown in Fig. 3. Now the motion in the basement is a bandpass-filtered delta function. Note that these deconvolved waveforms are more complicated than those in Fig. 2. The motion in Fig. 3 is for early time ($t < 1$ s) given by a superposition upward and downward propagating waves. For later time the motion has a harmonic character and is dominated by the motion of the fundamental mode of the building.

Note that the motion in Fig. 3 is causal, while the motion in Fig. 2 is not. The Millikan Library is excited at its base, and the response shown in Fig. 3 shows the response to this excitation. This response is causal because the motion of the building follows the excitation in time. In contrast, the deconvolved motion in Fig. 2 is nonzero for negative time. This is due to the absence of a source at the top of the building. By definition, the deconvolved motion of this wave state is a bandpass filtered delta function centered at $t = 0$ at the top of the building. The only way to generate this impulsive motion at the top of the building in the absence of a source at that location is to send a wave upward in the building that arrives at the top at $t = 0$. In order to achieve this, the wave must be launched from the base of the building at negative time, and this is exactly what can be seen in Fig. 2.

3. Time Domain Versus Frequency Domain Response

As an alternative to the deconvolved waveforms in the time domain, one could analyze the amplitude spectrum of the motion recorded in Fig. 1. The amplitude spectrum thus obtained is the product of the amplitude spectra of the waves striking the building from below, the ground coupling, and the building response. In this approach, the contribution of these three physical factors is still mixed. In order to eliminate the contribution of the unknown excitation, the amplitude of the spectral ratio of expression (1) in the frequency domain is often considered.

As an example, let us consider what this amplitude ratio gives for the deconvolved waves shown in Fig. 2. At any level, the deconvolved motion is approximately given by

$$D(t) = s(t + \tau) + s(t - \tau) , \tag{3}$$

where $\pm\tau$ are the arrival times of the upgoing and downgoing waves, respectively. For simplicity I ignore attenuation and assume that the amplitudes of the upgoing and downgoing waves are identical. In the frequency domain, this expression corresponds to

$$D(\omega) = \left(e^{-i\omega\tau} + e^{i\omega\tau}\right) s(\omega) = 2\cos(\omega\tau)S(\omega) , \tag{4}$$

with $S(\omega)$ the Fourier transform of the band-limited delta function $s(t)$. The amplitude of this spectral ratio is given by

$$|D(\omega)| = 2|\cos(\omega\tau)|\,|S(\omega)| , \tag{5}$$

while the phase φ satisfies

$$\tan(\varphi) = \frac{\mathrm{Im}(S(\omega))}{\mathrm{Re}(S(\omega))} . \tag{6}$$

The time-domain deconvolved waves of expression (3) are trivial to interpret; they consist of two waves arriving at opposite arrival times. In contrast, if one only knew the amplitude spectrum (4), it would not be trivial to infer that the deconvolved motion consists of just two waves with opposite arrival times because the information of the arrival time, $|\cos(\omega\tau)|$, is multiplied with the amplitude spectrum $|S(\omega)|$ of the band-limited delta function. Information of the travel time τ is encoded in the notches of the deconvolved spectra, which may be difficult to interpret.

Note that the phase φ in expression (6) does not depend on the travel time τ at all; hence, the phase is useless to infer that waves consist of upgoing and downgoing waves with opposite travel time τ. If one arbitrarily changes the origin of the time axis in Fig. 2 over a time shift t_0, the phase is given by

$$\tan(\varphi) = \frac{\mathrm{Im}(S(\omega))}{\mathrm{Re}(S(\omega))} - \omega t_0 . \tag{7}$$

The measured phase is restricted to between 0 and 2π, and following the phase wrapping, the measured phase is a complicated function of frequency, which carries no information at all about the travel time τ.

This analysis of the waves in Fig. 2 illustrates the advantages of analyzing the deconvolved response in the time domain rather than in the frequency domain. When the motion in a system consists of the superposition of a limited number of resonances it is, of course, informative to study the amplitude spectrum of the deconvolved motion because this provides direct information of the frequency of these resonances. In the example of the motion deconvolved with the motion in the basement (Fig. 3), this procedure gives the frequency of the normal mode of the building. There is, of course, no reason why one could not analyze both the time dependence and the amplitude spectrum of the deconvolved response. This can be useful because these different quantities provide complementary information of the building response.

4. Changing the Boundary Conditions

It is not widely known that in the Green's function extraction one can alter the boundary conditions of the system. For the correlation and deconvolution approaches the extracted response satisfies the same field equation as does the physical system (Snieder et al., 2006).

I illustrate the freedom to change the boundary conditions in seismic interferometry with the response of the Millikan Library at Caltech extracted from recorded vibrations of the building after an earthquake, shown in Fig. 1. Figure 2 shows the motion of the Millikan Library after deconvolution with the motion at the top floor. The motion at the top floor is collapsed into a bandpass-filtered delta function because the recorded motion at the top floor is deconvolved with itself, and the deconvolution of any function with itself yields a delta function. The response in Fig. 2 is a-causal, but it still is a valid wave state of the building that consists of one upgoing wave that is reflected by the top of the building into a downgoing wave. Note that this downgoing wave is not reflected at the base of the building, this wave state thus corresponds to a fictitious building that has reflection coefficient $R = 0$ at its base. This example shows that the response extracted by this deconvolution makes it possible to determine the building response for the hypothetical situation that the building would have been placed on a reflection-free subsurface.

The response extracted by deconvolving the motion at every floor with the motion in the basement is shown in Fig. 3. The extracted response in the basement is a bandpass-filtered delta function because it consists of the motion in the basement deconvolved with itself. Since the delta function is equal to zero for $t \neq 0$, the extracted motion in the basement vanishes for non-zero time. In the real building, the motion at the base does, of course, not vanish for $t \neq 0$. In fact, one can see in the bottom trace of the original data in Fig. 1 that the building is being shaken at its base throughout the arrival of the body-wave coda and the surface waves that excite the building. In contrast, the extracted response in Fig. 3 is for a fictitious building whose base is excited by a bandpass filtered delta pulse and then remains fixed. Such a fictitious building has reflection coefficient $R = -1$ at the base (Snieder et al., 2006), which precludes the transmission of energy from the subsurface into the building!

The examples of Figs. 2 and 3 show that from one data set one can retrieve wave-states that satisfy different boundary conditions. The real building has neither reflection coefficient $R = 0$ nor $R = -1$ at its base. A reflection coefficient $R = 0$ precludes the resonance that is clearly visible in Fig. 1 because, for a reflectionless ground coupling, all wave energy is radiated downward at the base. If the reflection coefficient at the base of the real building would be given by $R = -1$, energy would not be able to be transmitted into the building. Additional examples of wave-states of the building that have reflection coefficient $R = 0$, but that are either purely causal or a-causal can be found in Snieder et al. (2006). Note that the extracted response in Figs. 2 and 3 is solely based on data processing of the recorded motion in Fig. 1.

5. Conclusion

The example of the motion of the Millikan Library shows that deconvolution is an effective method to extract the building response from recorded motion after a complicated excitation by an earthquake. The retrieved time-domain response shows distinct waves arriving. One can measure the velocity and attenuation of these waves and retrieve the velocity of shear waves and the associated attenuation of the building. Extracting such information from the amplitude spectrum of the impulse response is much more difficult because arrival time information is coded in a complicated way into the notches of the extracted spectral amplitude. The examples in this work show that it is possible to extract the time-domain response of structures under different boundary conditions than those of the real structure. This can be a valuable tool for obtaining independent estimates of the building response from the soil-structure interaction. Note that the extraction of the building response under different boundary conditions at the base only involves data processing; it does not entail numerical modeling of the building, and the mechanical properties of the building need not be known. We have also applied interferometric techniques that change the boundary conditions in marine exploration seismology with the goal of suppressing waves that are reflected from the ocean's free surface (Mehta et al., 2007a, 2008).

Acknowledgements

This work was supported by the NSF (grant EAS-0609595), and by the GameChanger program of Shell.

References

Aki, K. (1957) Space and time spectra of stationary stochastic waves with special reference to microtremors, *B. Earthq. Res. Inst.* **35**, 415–456.

Bakulin, A. and Calvert, R. (2006) The virtual source method: Theory and case study, *Geophysics* **71**, SI139–SI150.

Bakulin, A., Mateeva, A., Mehta, K., Jorgensen, P., Ferrandis, J., Sinha Herhold, I., and Lopez, J. (2007) Virtual source applications to imaging and reservoir monitoring, *The Leading Edge* **26**, 732–740.

Campillo, M. and Paul, A. (2003) Long-range correlations in the diffuse seismic coda, *Science* **299**, 547–549.

Chávez-Garcia, F. and Luzón, F. (2005) On the correlation of seismic microtremors, *J. Geophys. Res.* **110**, B11313, doi:10.1029/2005JB003686.

Curtis, A., Gerstoft, P., Sato, H., Snieder, R., and Wapenaar, K. (2006) Seismic interferometry – turning noise into signal, *The Leading Edge* **25**, 1082–1092.

Draganov, D., Wapenaar, K., Mulder, W., Singer, J., and Verdel, A. (2007) Retrieval of reflections from seismic background-noise measurements, *Geophys. Res. Lett.* **34**, L04305.

Hornby, B. and Yu, J. (2007) Interferometric imaging of a salt flank using walkaway VSP data, *The Leading Edge* **26**, 760–763.

Kohler, M., Heaton, T., and Bradford, S. (2007) Propagating waves in the steel, moment-frame factor building recorded during earthquakes, *B. Seismol. Soc Am.* **97**, 1334–1345.

Larose, E., Margerin, L., Derode, A., van Tiggelen, B., Campillo, M., Shapiro, N., Paul, A., Stehly, L., and Tanter, M. (2006a) Correlation of random wavefields: An interdisciplinary review, *Geophysics* **71**, SI11–SI21.

Larose, E., Montaldo, G., Derode, A., and Campillo, M. (2006b) Passive imaging of localized reflectors and interfaces in open media, *Appl. Phys. Lett.* **88**, 104103.

Lobkis, O. and Weaver, R. (2001) On the emergence of the Green's function in the correlations of a diffuse field, *J. Acoust. Soc. Am.* **110**, 3011–3017.

Louie, J. (2001) Faster, better: Shear-wave velocity to 100 meters depth from refraction microtremor analysis, *B. Seismol. Soc. Am.* **91**, 347–364.

Malcolm, A., Scales, J., and van Tiggelen, B. (2004) Extracting the Green's function from diffuse, equipartitioned waves, *Phys. Rev. E* **70**, 015601.

Mehta, K., Bakulin, A., Sheiman, J., Calvert, R., and Snieder, R. (2007a) Improving the virtual source method by wavefield separation, *Geophysics* **72**, V79–V86.

Mehta, K., Snieder, R., and Graizer, V. (2007b) Downhole receiver function: A case study, *Bull. Seismol. Soc. Am.* **97**, 1396–1403.

Mehta, K., Sheiman, J., Snieder, R., and Calvert, R. (2008) Strengthening the virtual-source method for time-lapse monitoring, *Geophysics* **73**, S73–S80.

Rickett, J. and Claerbout, J. (1999) Acoustic daylight imaging via spectral factorization: Helioseismology and reservoir monitoring, *The Leading Edge* **18**, 957–960.

Rickett, J. and Claerbout, J. (2000) Calculation of the sun's acoustic impulse response by multidimensional spectral factorization, *Sol. Phys.* **192**, 203–210.

Roux, P. and Fink, M. (2003) Green's function estimation using secondary sources in a shallow water environment, *J. Acoust. Soc. Am.* **113**, 1406–1416.

Roux, P., Kuperman, W., and Group, N. (2004) Extracting coherent wave fronts from acoustic ambient noise in the ocean, *J. Acoust. Soc. Am.* **116**, 1995–2003.

Roux, P., Sabra, K., Gerstoft, P., and Kuperman, W. (2005) P-waves from cross correlation of seismic noise, *Geophys. Res. Lett.* **32**, L19303, doi:10.1029/2005GL023803.

Sabra, K., Conti, S., Roux, P., and Kuperman, W. (2007) Passive in-vivo elastography from skeletal muscle noise, *Appl. Phys. Lett.* **90**, 194101.

Sabra, K., Gerstoft, P., Roux, P., Kuperman, W., and Fehler, M. (2005a) Surface wave tomography from microseisms in Southern California, *Geophys. Res. Lett.* **32**, L14311, doi:10.1029/2005GL023155.

Sabra, K., Roux, P., Thode, A., D'Spain, G., and Hodgkiss, W. (2005b) Using ocean ambient noise for array self-localization and self-synchronization, *IEEE J. of Oceanic Eng.* **30**, 338–347.

Sabra, K., Roux, P., Gerstoft, P., Kuperman, W., and Fehler, M. (2006) Extracting coherent coda arrivals from cross-correlations of long period seismic waves during the Mount St. Helens 2004 eruption, *J. Geophys. Res.* **33**, L06313, doi:1029.2005GL025563.

Sabra, K., Srivastava, A., Lanza di Scalea, F., Bartoli, I., Rizzo, P., and Conti, S. (2008) Structural health monitoring by extraction of coherent guided waves from diffuse fields, *J. Acoust. Soc. Am.* **123**(1), EL8–EL13.

Schuster, G., Yu, J., Sheng, J., and Rickett, J. (2004) Interferometric/daylight seismic imaging, *Geophys. J. Int.* **157**, 838–852.

Sens-Schönfelder, C. and Wegler, U. (2006) Passive image interferometry and seasonal variations at Merapi volcano, Indonesia, *Geophys. Res. Lett.* **33**, L21302, doi:10.1029/2006GL027797.

Shapiro, N. and Campillo, M. (2004) Emergence of broadband Rayleigh waves from correlations of the ambient seismic noise, *Geophys. Res. Lett.* **31**, L07614, doi10.1029/2004GL019491.

Shapiro, N., Campillo, M., Stehly, L., and Ritzwoller, M. (2005) High-resolution surface-wave tomography from ambient seismic noise, *Science* **307**, 1615–1618.

Snieder, R. and Şafak, E. (2006) Extracting the building response using seismic interferometry: Theory and application to the Millikan Library in Pasadena, California, *Bull. Seismol. Soc. Am.* **96**, 586–598.

Snieder, R., Sheiman, J., and Calvert, R. (2006) Equivalence of the virtual source method and wavefield deconvolution in seismic interferometry, *Phys. Rev. E* **73**, 066620.

Thompson, D. and Snieder, R. (2006) Seismic anisotropy of a building, *The Leading Edge* **25**, 1093.

van Wijk, K. (2006) On estimating the impulse response between receivers in a controlled ultrasonic experiment, *Geophysics* **71**, SI79–SI84.

Weaver, R. and Lobkis, O. (2001) Ultrasonics without a source: Thermal fluctuation correlations at MHz frequencies, *Phys. Rev. Lett.* **87**, 134301.

Weaver, R. and Lobkis, O. (2003) On the emergence of the Green's function in the correlations of a diffuse field: Pulse-echo using thermal phonons, *Ultrasonics* **40**, 435–439.

Wegler, U. and Sens-Schönfelder, C. (2007) Fault zone monitoring with passive image interferometry, *Geophys. J. Int.* **168**, 1029–1033.

RECENT MEASUREMENTS OF AMBIENT VIBRATIONS IN FREE-FIELD AND IN BUILDINGS IN CROATIA *

Marijan Herak (`herak@irb.hr`)[†] and Davorka Herak
Department of Geophysics, Faculty of Science and Mathematics, University of Zagreb, Horvatovac 95, 10000 Zagreb, Croatia (`herak@irb.hr`)

Abstract. Recent measurements of ambient vibrations in Croatia include free-field measurements in Zagreb and Ston, as well as those done within buildings. Free-field data at both localities are consistent with the properties of shallow geological structures there. In Zagreb, horizontal-to-vertical spectral ratios (HVSR) of microtremors point to existence of thick alluvial cover (over 100 m) that gradually gets thinner as it reaches the southern slopes of the Mt. Medvednica. In Ston HVSR profiles reveal several tens of meters thick sedimentary cover over the bedrock which gets exposed at the Stoviš hill. Over 100 measurements of ambient vibrations in buildings that have recently been performed across Croatia, were analysed by a suite of `Matlab` routines yielding frequencies and damping for the fundamental and higher modes of oscillation. Applications to real buildings proved that analyses of shaking induced by ambient vibrations in most cases lead to well constrained, reliable, and time independent estimates of frequencies and damping of the building vibrational modes. The measurements done so far form the initial nucleus of the Croatian building inventory, a collection of observed fundamental periods, damping, and spectral shapes, which can prove important in documenting building structural integrity and in assessing the degree of possible damage in future earthquakes.

Keywords: building frequency, soil frequency, soil-structure resonance, height-period relationship

1. Introduction

Measurements of microtremors have been used in Croatia for assessing local site effects since 1960s. Based on works of Kanai (1957a,b), the measurements were used to constrain and verify available geotechnical models by assessing the fundamental soil period, but there have also been attempts to compare the spectrum of recorded noise with theoretical amplification spectra. Limited mostly by the need to manually digitize analogue recordings

*A substantial part of material presented herein is taken over from a chapter in the book: Mucciarelli M., Herak M., Cassidy J. (Eds) (2009) *Increasing Seismic Safety by Combining Engineering Technologies and Seismological Data*, NATO Science for Peace and Security Series – C: Environmental Security, Springer.

[†]Corresponding author.

T. Schanz and R. Iankov (eds.), *Coupled Site and Soil-Structure Interaction Effects with Application to Seismic Risk Mitigation*, NATO Science for Peace and Security Series C: Environmental Security, and by capabilities of the instruments of that time, such measurements gradually faded away. They were revitalized as the topic reappeared in literature (Nakamura, 1989) and with first acquisitions of digital instruments in the 1990's. However, only after truly portable dedicated instruments were obtained for the measurements of ambient vibrations within the NATO SfP 980857 project (2005–2008) (see `www.nato.int/science/studies_and_projects/nato_funded/pdf/980857.pdf`), the quality and quantity of measurements increased to the level required by today's standards.

2. Free-Field Measurements

Although a large number of measurements were done in the last years all over Croatia, only the cases of Zagreb and Ston will be presented here. Zagreb is the capital of Croatia, a city with population approaching one million, and a history of large earthquakes. The most important one that occurred in 1880 (epicentral intensity VIII–IX° MCS) beneath the NE flanks of Mt. Medvednica damaged all houses in the city, and practically defines seismic hazard in Zagreb. In spite of rich seismic history, Zagreb still does not have an official seismic microzonation, which prompted the city officials to launch a large, long-term project of comprehensive geotechnical, geological and seismological investigations. One of the goals of the project is to produce a map of seismic microzonation, and microtremor measurements are expected to provide important information.

The city of Ston is located about 50 km to the NW of Dubrovnik, in southern Dalmatia. It is a small town with rich history, known for the ancient salterns and the third longest fortification walls in the world. As a part of the Dubrovnik Republic, Ston was also one of the first townships in this part of the world that developed according to the strict urban code enforced from the 14th century onward. Earthquakes are frequent there. The most recent devastating one occurred in 1996 (M = 6.0, intensity VIII° MCS in Ston), and Ston needed almost a decade to recover. Detailed damage reports exist in the archives, so we hoped that they could be compared to the noise measurements.

2.1. MEASUREMENTS IN ZAGREB

The Zagreb metropolitan area encompasses over 640 km^2. It consists of thick (100 m or more) alluvial sediments (clays, sands, gravel) in the Sava river valley, that gradually get thinner towards the Medvednica mountain to the north, which mainly consist of green slates, shales, and limestones. In order to asses the usefulness of the horizontal-to-vertical spectral ratio (HVSR) technique, we have made over 150 measurements, most of them in the proluvial

Podsljeme area where the Sava valley meets the mountain flanks (see Fig. 1). HVSR is defined as

$$\mathrm{HVSR} = (H_1 H_2)^{1/2}/V,$$

where H_1 and H_2 are amplitude spectra of the two orthogonal horizontal components, and V is spectrum of the vertical component of ambient noise.

The instrument used was the portable Tromino (produced by Micromed, Italy, `www.tromino.it`), a small all-in-one package with 3-component geophones, digitizer, GPS-timing, batteries and 512 Mb flash memory for storage. Measurements were done in the first half of 2007. They lasted for 20 min, and were processed uniformly: each trace was divided into non-overlapping 30-s long segments, spectra for all three components were computed for each of them. The three spectra were then smoothed with a 5% triangular smoothing function, and HVSR was computed as the ratio of the geometrical mean of spectra of the two horizontal components and the spectrum of the vertical one.

Figure 1 presents typical HVSR spectra obtained in an approximately 3×3 km area in the Podsljeme zone. In the south spectra are characterized by relatively broad low-frequency peaks (ranging from 0.85 to 2 Hz), indicating thick alluvial deposits. More to the north, the HVSR peaks shift towards the higher frequencies (about 3–6 Hz), as the sedimentary cover gets thinner. Reaching the foothills of Mt. Medvednica, the bedrock gets very close to the surface, as indicated by HVSR peaks at frequencies above 10–20 Hz.

The HVSR profile (as shown in the map in Fig. 1) is 2,500 m long. This is a spatial spectrogram constructed of all HVSR spectra measured within 500 m from the profile trace. It clearly shows systematic increase of the fundamental frequency as one moves from the left (south) to the right (north). Assuming an average S-wave velocity of 300 m/s in the whole sedimentary layer above bedrock, the observed variation of fundamental frequency maps into thickness variation from over 100 m in the south to only a few meters in the north, which is in agreement with (very few) available geotechnical data.

Podsljeme is today one the most prestigious residential areas in Zagreb, where houses are typically 2–4 storeys high, with expected fundamental frequencies of about 5 Hz or higher. According to our measurements so far, it is also the only place in Zagreb where such buildings are in danger due to soil-structure resonance during earthquake shaking.

2.2. MEASUREMENTS IN STON

In Ston ambient noise measurements were done in 2005 and 2006. A total of 70 free-field points were measured, as shown in Fig. 2. They were processed as described in the previous section. The town itself is situated between the

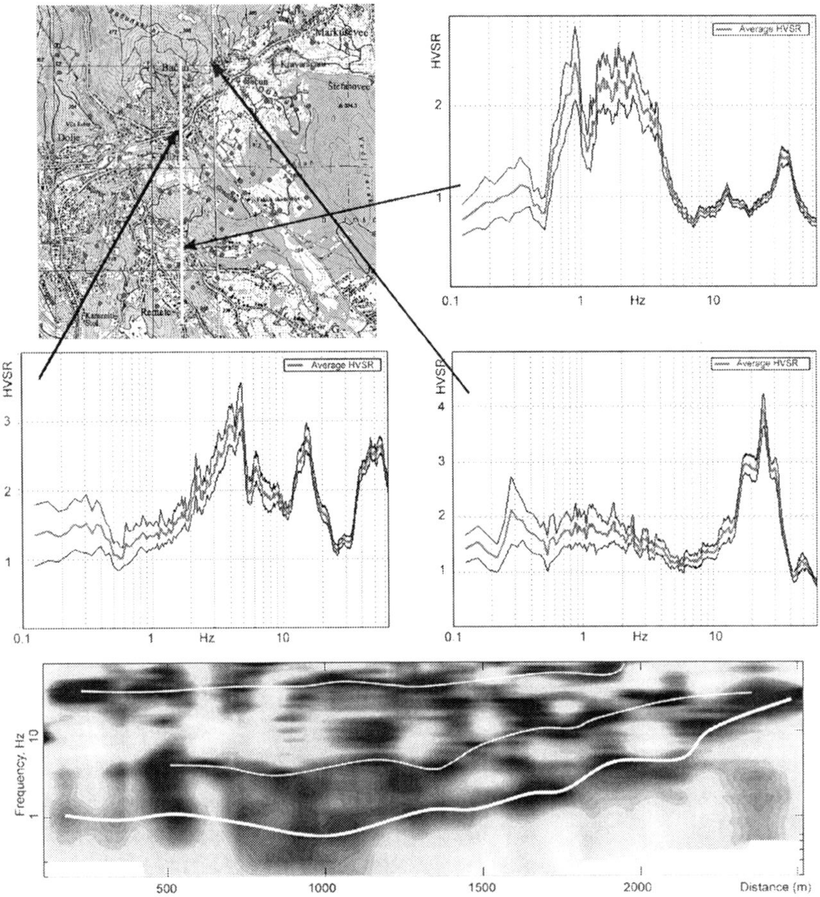

Figure 1. *Top left*: Map view of the Podsljeme area in Zagreb which was chosen as a test neighbourhood for the HVSR measurements. Black dots show the locations of measurement points. The foothills of the Medvednica Mt. are in the northern and north-western part. The white AB-line shows the location of the profile in the bottom. *Top right and middle*: Examples of the measured HVSR spectra (mean ±1 standard deviation), showing how the dominant frequency shifts towards higher values as we move along the profile from A to B. *Bottom*: HVSR profile AB (see the map on the top). Only measurements within 500 m from the profile line are considered. Dark shades correspond to high HVSR values. The lines are drawn to emphasize features, and have no direct geological interpretation, although clear systematic increase of the fundamental frequency with the thinning of the sedimentary cover close to B is evident.

Stoviš hill, and the shallow Ston channel. Limestones prevail. According to a few boreholes, there is about 15–30 m of weathered weak material (mostly sands) above the bedrock beneath the town. Our measurements confirm this as the fundamental frequency all across the plane beneath Stoviš vary

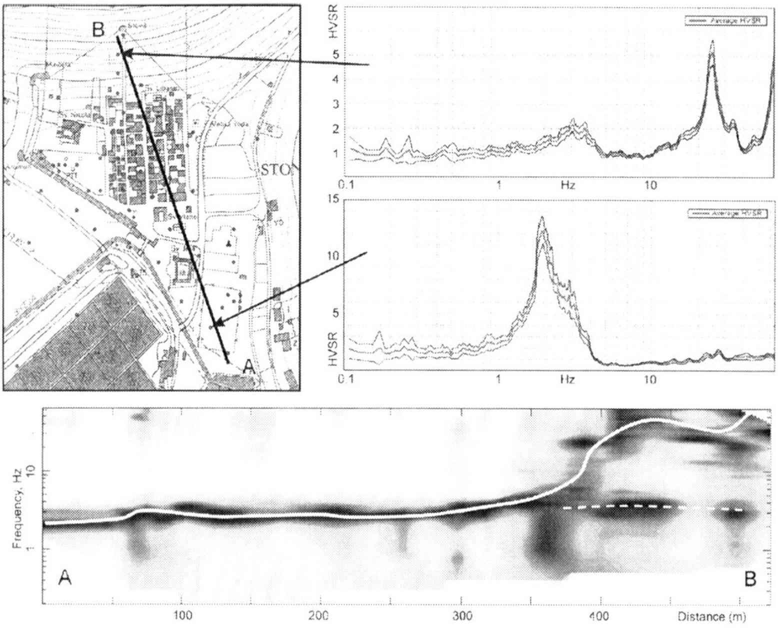

Figure 2. *Top left*: Map view of the town of Ston (southern Dalmatia). Black dots show the locations of measurement points. The Stoviš hill is in the north. The AB-line shows the location of the profile in the bottom. *Right*: Examples of the measured HVSR spectra (mean ±1 standard deviation), showing typical results obtained at the hill and in the plane beneath. HVSR spectra in the southern part are all characterized by a single pronounced peak in the range 2–4 Hz. The locations up the hill, situated practically on the bedrock are characterized by much higher dominant frequencies, well above 10 Hz. *Bottom*: HVSR profile AB. Only measurements within 80 m from the profile line are considered. Dark shades correspond to high HVSR values. The lines are drawn to emphasize features, and have no direct geological interpretation.

between 2 and 4 Hz. As we start climbing up the hill, the layers' thickness rapidly decrease and the dominant frequency increases to over 20 Hz.

Fundamental frequencies of stone houses in Ston vary between about 3 and 6 Hz, depending on their height, shape and position. This frequency interval coincides well with the dominant soil frequencies beneath the town centre and especially at the Stoviš foothills, which, together with high amplification, may explain severity of the damage (VIII° EMS) caused by the 1996 earthquake whose epicentre was 20 km away.

Results obtained in Zagreb and Ston confirmed applicability of the HVSR technique in cases of both thick and thin sedimentary covers, especially in determining the soil fundamental frequency. Based on preliminary measurements described above, ambient noise measurements are officially adopted to be used in the course of microzonation of Zagreb.

3. Measurements in Buildings

The bulk of measurements, about 75% of total, were done in Zagreb. Those results will be given later in this section. At least one measurement (16–20 min long) was done in each building as close as possible to the top floor. In some cases, additional observations were performed at intermediate levels too. Each measurement inside was accompanied by a free-field one, taken immediately before or after the one in the building, using the same instrument. Whenever possible the free-field measurements were performed at least one building height away from the building itself.

As resonance effects will depend also on the damping of the structure, it was of interest to compile a program that will simultaneously estimate periods of free vibrations and their respective damping from the records of ambient vibrations in buildings and in the free field. These efforts resulted in a collection of `Matlab` (`www.mathworks.com`) routines, assembled together in a graphical user interface (GUI) FREDA (FREquency-Damping Analyses). The GUI is shown in Fig. 3. Main features of FREDA include:

- Plain ASCII files of ambient vibration time-histories as input
- Instrument corrections (for displacement and velocity)
- Correction for the reference spectrum (excitation signal)
- Five modes of analyses:
    1. Time domain:
        - Slightly modified nonparametric analyses (NonPaDAn, Mucciarelli and Gallipoli, 2007), based on log-decrement approach
        - NonPaDAn on a suite of bandpass-filtered signals
    2. Frequency domain:
        - Spectral single-peak transfer function analyses
        - Spectral sweep transfer function analyses
        - HVSR
- Each mode may also be used with the random decrement (Cole, 1971) signature of the signal as input
- Uses real or synthetic signals
- Output graphics (`eps`, `jpg`)

The first two of the spectral methods are based on matching the theoretical amplitude response of a SDOF oscillator to the observed building response.

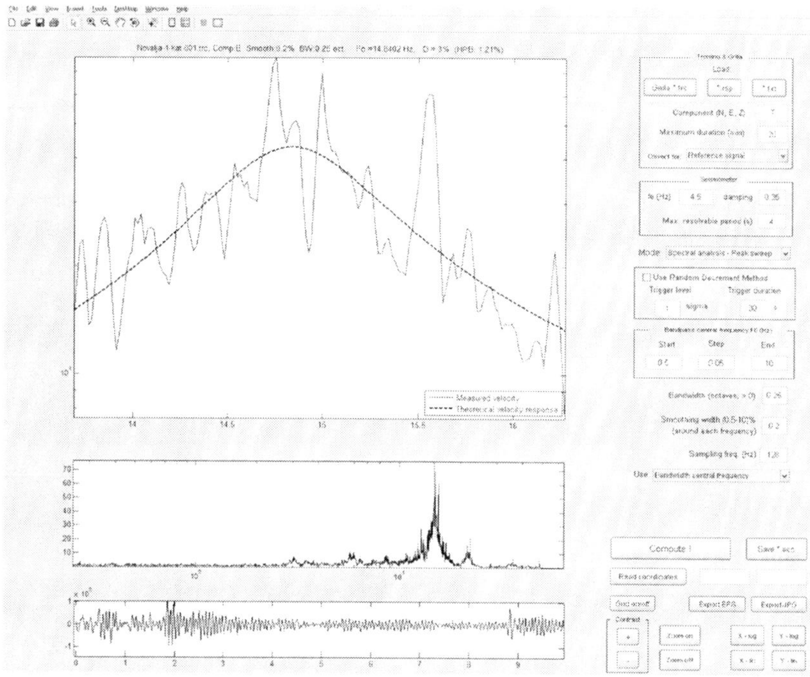

Figure 3. FREDA graphical user interface (GUI), showing an example of the spectral single-peak transfer function analyses. The bottom subplot shows the first 10 s of the 20 min long measured noise time series of the transversal horizontal component of the building vibrations induced by ambient noise. Above it is its Fourier spectrum divided by the spectrum of the free-field noise. The main window shows the blow-up of the selected peak and the best fitting SDOF theoretical response. All controls for choosing the mode of analyses and various parameters are in the right part of the GUI.

The program has been tested extensively using synthetic signals as well as measurements on many building types. The comparison of results reveals that estimates of frequencies and damping obtained by spectral methods are in general more robust and less dependent on parameters of the respective algorithm, than the results based on time-domain analyses. Spectral algorithms are also much better in resolving higher modes. The random decrement method is in most cases found to be inferior to spectral or band-pass procedures using the original signal.

In particular, the use of HVSR is not recommended (see an example of the bell-tower in Zadar in Fig. 4), although it may yield reasonable frequency estimates in some instances. However, there is no theoretical basis for its application as we can not safely assume that horizontal and vertical spectra do not differ at the ground level. This is especially dangerous if soil amplification is significant (with prominent HVSR peaks), in which case the free-field

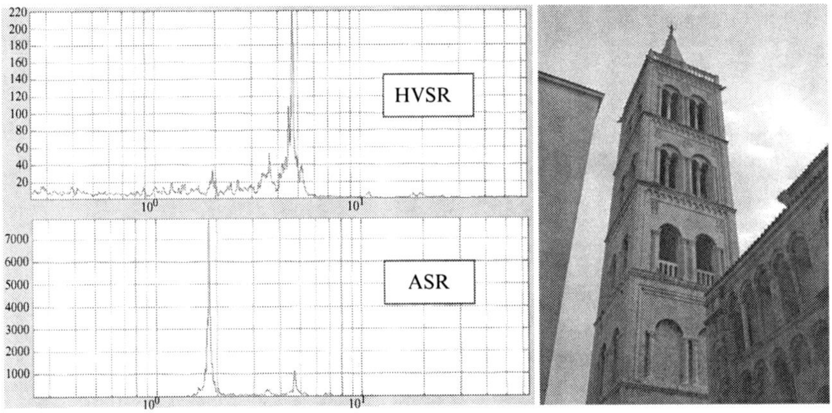

Figure 4. Analyses of response of the 60 m high, stone bell-tower of the St. Anastasia church in Zadar (*right*). Ambient vibrations were recorded at the top terrace at the height of about 50 m. Left, top: HVSR. Left, bottom: Amplitude Spectral Ratio (ASR) between horizontal spectrum recorded on the tower and in the free-field. Note the large relative difference between the amplitudes of modes of oscillation. Based on HVSR, one might be mislead and declare the 4.8 Hz the fundamental mode. The abscissa is frequency in Hz. Analyses of spectral ratio in the bottom yield (for the fundamental mode): frequency $f_1 = 1.81$ Hz, damping $D_1 = 0.5\%$ of critical.

Figure 5. *Left*: The Palagruža lighthouse; *Right*: Spectral ratios of the lighthouse transversal (top) and longitudinal (bottom) response to microtremor excitation measured at five levels relative to the free-field spectrum.

HVSR may contaminate building response, leading to false identification of possible resonance. All subsequent analyses were done using amplitude spectral ratios (ASR), i.e., spectra of vibrations measured in the building, divided by the corresponding spectrum of the excitation signal (microtremors recorded in the vicinity of the building).

Figure 5 shows spectra of microtremor-induced vibrations of the stone masonry Palagruža lighthouse building located on a small island in the centre of the Adriatic Sea, measured at five levels within the building. The spectra clearly exhibit a strong peak at about 5.6 Hz, the amplitude of which increases

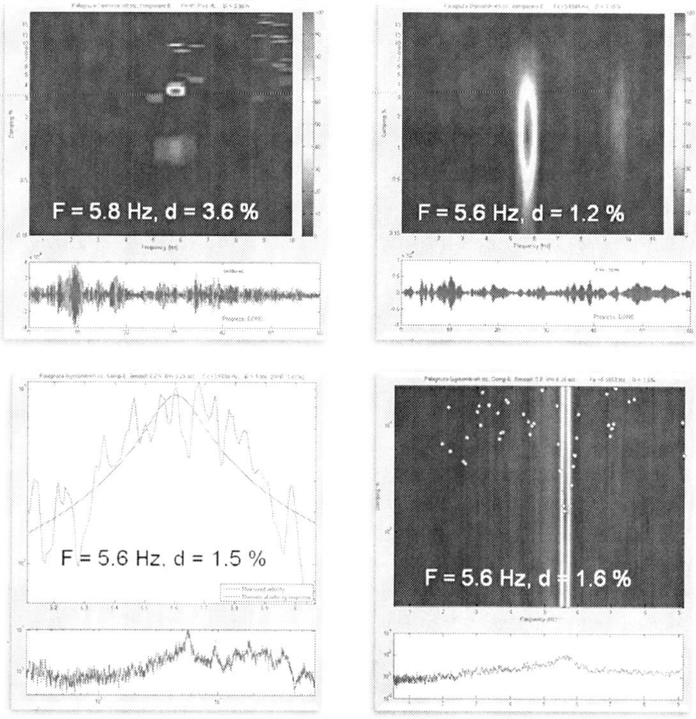

Figure 6. Results of FREDA analyses of the Palagruža lighthouse top floor recording using 4 modes of analyses: NonPaDAn (*top left*), bandpass NonPaDAn (*top right*), spectral single–peak transfer function analyses (*bottom left*), and spectral sweep transfer function analyses (*bottom right*). F is frequency (Hz), d is damping (% of critical).

with the height of the floor. The recording at the top level (on the tower) has been analysed in FREDA, and results obtained using four modes of analyses are presented in Fig. 6. Clearly, in this case, all methods yielded nearly the same frequency of the building's fundamental mode, whereas damping estimates vary between 1.2 % and 3.6% of critical.

Further example of measurements in stone masonry buildings is the one of the City Hall in Ston (greater Dubrovnik region), which was heavily damaged in the $M = 6.0$ earthquake in 1996 (epicentral distance about 16 km). As shown in Fig. 7 the building response to ambient vibrations is characterized by two distinct peaks at about 3 and 4 Hz, the first of which closely corresponds to the soil fundamental frequency as revealed by HVSR of microtremors recorded in front of the building. We therefore believe that pronounced soil-structure resonance must have occurred, causing damage in Ston which was significantly higher than in the nearby village of Slano, located much closer to the earthquake epicentre.

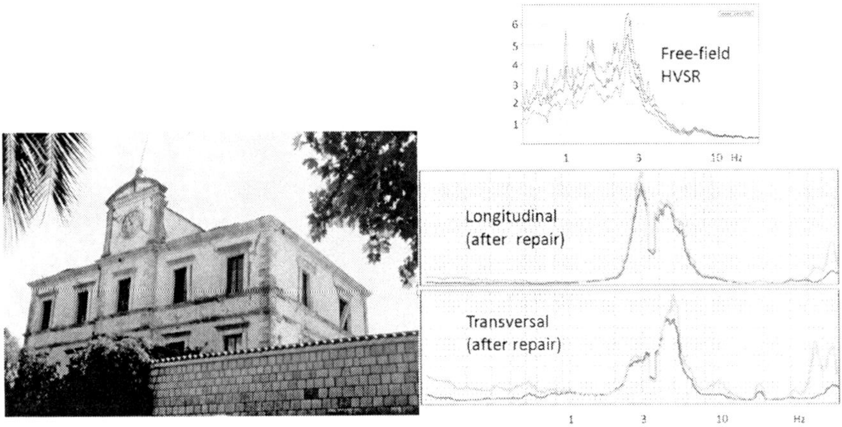

Figure 7. *Left*: The City Hall of Ston, damaged in the Ston-Slano earthquake of 1996. *Right, top*: HVSR of microtremors measured in the park in front of the building; *Right, centre and bottom*: Spectra of the building response to microtremor excitation (after building repair, corrected for the input spectrum). Note that frequencies of the first mode match the fundamental soil frequency, indicating soil-structure resonance may have been responsible for heavy damage the building suffered.

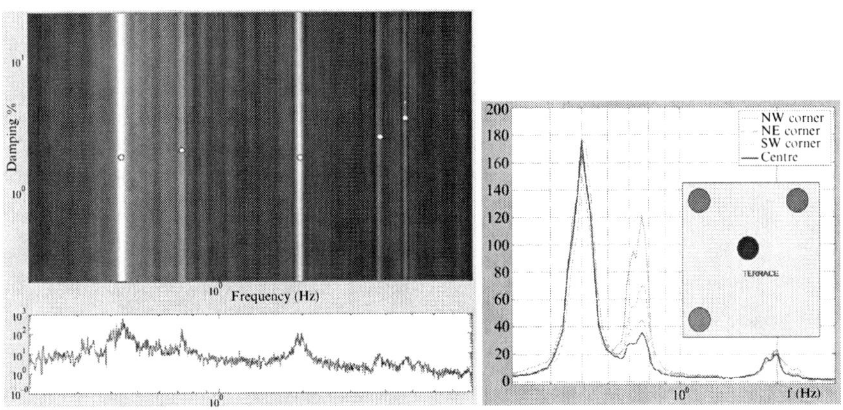

Figure 8. *Left*: Spectral sweep analyses of vibrations measured on top of the 26-storey skyscraper in Zagreb. The bottom plot is amplitude spectrum of horizontal component of recorded velocity after deconvolving the excitation signal as measured in the free-field. The top plot presents the results, marking at least five vibrational modes. The local maxima of the frequency-damping surface are accentuated by white circles. *Right*: Spectra of horizontal vibrations, measured simultaneously at three corners and in the middle of the terrace. Notice how the peak at 0.73 Hz is the only one whose amplitude varies with the location (and is the smallest in the centre), indicating a torsional mode.

Application to a RC building is illustrated by an example of one of the highest skyscrapers in Zagreb (26 floors, 96 m). Figure 8 (left) shows FREDA analyses of recordings of the ambient-noise induced vibrations in the centre

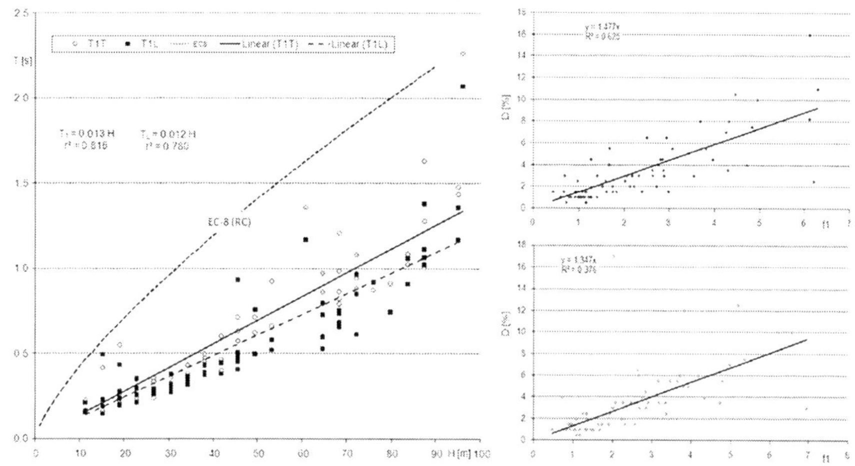

Figure 9. *Left*: Period-height relationship for RC-buildings in Zagreb. Circles—transversal direction (perpendicular to the building long-axis); Squares—longitudinal direction. EC-8 curve is shown by dotted line. *Right*: Damping-frequency relationship for the same data set (top: transversal, bottom: longitudinal direction). Dampings in % of critical.

of the terrace at the top of the building. At least five modes are discernible, with the following frequencies (f) and dampings (D, % of critical): f_1 = 0.44 Hz, D_1 = 1.0%; f_2 = 0.73 Hz, D_2 = 1.4%; f_3 = 1.95 Hz, D_3 = 1.6%; f_4 = 3.83 Hz, D_4 = 2.0%; f_5 = 4.72 Hz, D_5 = 3.0%. Comparing the spectra of vibrations simultaneously recorded at the corners and in the centre (Fig. 8, right), suggests that frequency of 0.73 Hz corresponds to a predominantly torsional mode, as its amplitudes at corners are notably larger than in the centre.

Data collected in Zagreb indicate that period-height relationship for RC buildings (Fig. 9) differs substantially from the one provided by Eurocode-8 which yields periods that are, on the average, about twice too large compared to the observed ones. Dampings increase with fundamental frequency in a way very similar to observations from Japan.

We have checked the stability of measured spectra on various buildings by repeated measurements during different times in a day, seasons, and weather conditions. Typical variation of estimated frequencies was found to be within a few percent. For damping, the values varied no more than ±1–2% of critical damping.

4. Conclusions

Recent measurements of free-field microtremors in Croatia proved to be valuable in providing additional insight into the geotechnical properties of the soil, especially in estimating the fundamental frequency of sedimentary deposits. Measurements of ambient vibrations in buildings were shown to be efficient and quick, yielding reliable, accurate and temporally stable estimates of frequencies and damping of the building vibrational modes. Combining the free-field measurements with those within houses and other structures can point to constructions likely to exhibit soil-structure resonance. The measurements done so far form an initial nucleus of the building inventory, a collection of fundamental periods, damping, spectral shapes, and other data which can prove important in documenting building structural integrity and assessing the degree of possible damage in future earthquakes.

References

Cole, H. A. (1971) Method and apparatus for measuring the damping characteristic of a structure, United State Patent No. 3,620,069.

Kanai, K. (1957a) Semiempirical formula for the seismic characteristics of the ground, *B. Earthq. Res. Inst.* **35**, 309–325.

Kanai, K. (1957b) The requisite conditions for predominant vibration of ground, *B. Earthq. Res. Inst.* **35**, 457–471.

Mucciarelli, M. and Gallipoli, R. M. (2007) Non-parametric analysis of a single seismometric recording to obtain building dynamic parameters, *Ann. Geophys.* **50**, 259–266.

Nakamura, Y. (1989) A method for dynamic characteristics estimation of subsurface using microtremor on the ground surface, *Quart. Rep. Railway Tech. Res. Inst.* **30**, 25–33.

ANALYSIS OF DYNAMIC IMPACT ON A GROUND SLOPE DURING DESTRUCTION OF AN EMERGENCY HOUSE

Ivan O. Sadovenko, Dmytro V. Rudakov (dmi3rud@mail.ru)*,
and Vasyl' I. Timoschuk
Dept. of Hydrogeology and Eng. Geology, National Mining University, 19, K. Marx Av., 49005 Dnipropetrovsk, Ukraine

Abstract. A procedure to determine the static load equivalent to the mechanical impulse after hypothetical destruction of a many-storied emergency house is developed and applied to estimate plastic deformations of the hosting slope. The correctness of the analytical estimations and the finite-element modeling was confirmed by comparison with the available monitoring data.

Keywords: dynamic impact, compressible soil, shear deformations, static load, landslide, finite elements, numerical modeling

1. Introduction

Unfavourable engineering and geological conditions in urban areas may often lead to destruction of many-storied houses after long-term exploitation. Uncontrollable falling down of emergency house constructions poses a risk to neighboring buildings, may initiate landslides and cause micro-seismic effects. Thus, the reliable estimation of dynamic impact on soil during building demolition and related processes is of increasing importance for built-up urban areas.

The typical case of such situation is the emergency many-storied house on the slope of the clough "Rybal'ska" in Dnipropetrovsk (Ukraine). The geotechnical monitoring has revealed threatening accumulation of irreversible soil deformations during the last years. The total settlements from 1974 till 1988 reached 312 mm, the maximal inclination made 780 mm, and fracture widths on upper stories became 80 mm. Attempts to diminish the dangerous inclination have resulted in increase of deformation rates due to cyclic loading caused by heavy equipment and pressure redistribution below the basement. Uncontrollable stripping down of the house leads to local

*Corresponding author.

T. Schanz and R. Iankov (eds.), *Coupled Site and Soil-Structure Interaction Effects with Application to Seismic Risk Mitigation,* NATO Science for Peace and Security Series C: Environmental Security,

destruction threatening of its stability in general. The probable collapse of the house is expected to result in moving the landslide slope, with damaging near-by constructions.

Soil reaction on dynamic impact consists in forming the compression-extension and shearing zones due to seismic waves followed by the displacements of building parts. Such processes were studied numerically (Wolf, 1988; Wegner et al., 2005) with applications to soil-structure interaction below tall buildings due to seismic excitation. The attention was paid mostly on factors determining wave propagation including frequency content, wave pattern, etc.

Meanwhile, there are a number of situations in civil engineering with short soil-structure interaction followed by essential ground changes and local seismic effects. After such kind of influence the process intensity usually declines and a soil reverts to the almost previous state under equilibrium of energy and work spent for deformations. Alternatively, dynamic impact may cause irreversible deformations threatening the stability of landslide slopes and embedded constructions.

The aim of this paper is to develop an approach to estimate post-effects of local and short-term soil-structure interaction after dynamic influence on compressible ground.

2. Analysis of the Dynamic Impact on the Soil

Analyzing the state of emergency-house constructions and the dynamics of the possible destruction allows drawing the conclusion that the most possible scenario could be unpredictable in time collapse of the overhead covers accompanied by the consecutive falling down walls and plates. To increase forecast reliability a simultaneous destruction of all house sections is assumed. In view of physical reasons this process can be simulated as transmitting an impulse just after a direct strike on the ground considered as a construction underlying the house. This corresponds well with the known approach in seismology to estimate earthquake force by the energy transmitted from waves to rocks and buildings. Usually acceleration in the soil below the building is evaluated using the registered amplitude and period of oscillations; which allows scaling the damage of affected houses.

The strength and deformation behavior after dynamic and static loading of rocks and soils were generalized by Gzovsky (1975). The so-called conditionally instantaneous deformations make prevailing contribution to the total soil compression in relatively short periods. The distinctions between dynamic and static loading and their impact on the compressibility of soil considered as a three-component system were discussed by Liakhov and Poliakova (1967).

The aforementioned studies make it possible to predict ground slope state applying an equivalent static load on the emergency house basement and the near-by buildings by modifying the static behavior parameters. The total specific work spent for soil deformations on the unit area under the building basement can be written as

$$A_s = p_s \int_{V_1}^{V_2} dV_s, \tag{1}$$

where p_s is static pressure, V_1 and V_2 are the limits of soil volume change. Decreasing the volume due to deformations leads to additional settlements. On the other hand,

$$A_s = E_s K_a, \tag{2}$$

where E_s is the specific kinetic energy of impact on the ground distributed over a unit area accounting for the bearing surface, K_a is the attenuation factor of a compression-extension wave between the emergency house and the surrounding buildings during the destruction.

The parameter E_s is calculated according to the data about the building mass, the interaction area and the falling velocity. The estimated value of specific energy is 626 kN/m. The generally accepted value of 0.05 for K_a is used (Dynamic Calculation of Buildings and Constructions, 1984), whereas for buildings located on the other side of the clough K_a decreases to 0.02.

The possible value of the static pressure influencing the closest buildings can be estimated using Eq. (2). In accordance with the building standards, a difference of 10 cm between settlements on the opposite side of a construction is considered as critical, which corresponds to the value $V = 0.1\ \text{m}^3$ in Eq. (1) so that $p_s = E_s K_a / V = 313\,\text{kN/m}^2$.

The estimated static load equivalent while destructing the emergency house enables numerical modeling of the stress–strain state of the slope taking into account kinetic energy attenuation. The forecast technique consists in re-iterated solving of the boundary value problem using the finite element method (Fadeev, 1987; Zienkiewicz, 1967). It implies step-by-step loading the ground slope with the static equivalent load corresponding to the dynamic influence of the collapsing house.

3. Modeling of the Soil Deformation

Stresses and strains are related by the Hooke's law written for the 2D case (Fadeev, 1987) if the critical stress limit is not exceeded. Deformations are restricted by the strength $T < 0$ in the tension zone; and failure in the compression zone is governed by Mohr-Coulomb's criterion. The rock behaviour in the elastic-plastic state is described by the equations

$$\sigma_1' = S + \text{Ctg}\,\varphi\,\sigma_3', \tag{3}$$

$$\sigma_3' = [E_n(\varepsilon_1 + \varepsilon_3) + S(\nu_n - 1)]/(1 - \nu_n \,\mathrm{Ctg}\beta + \mathrm{Ctg}\beta - \nu_n), \qquad (4)$$

where σ_1' and σ_3' are the principal stresses; the parameters $E_n = E/(1 - \nu^2)$ and $\nu_n = \nu(1 - \nu)$ are related to the Young's modulus E and Poisson's ratio ν; $S = 2C\,\mathrm{Ctg}(45° - \varphi/2)$, $\mathrm{Ctg}(\varphi) = (1 + \sin\varphi)/(1 - \sin\varphi)$, C and φ are cohesion and friction angle respectively. The angle β in equation (4) determines the plasticity flow mode that is considered associated at $\beta = \varphi$ for this model, which allows accounting for dilatancy of rocks under the post-failure deformations. The principal stresses σ_1' and σ_3' can be treated as theoretical limits corresponding to actual deformations such as the numerical iterative solution tends to that values.

The loss of rock strength for the post-failure deformation mode is described by the stability criterion written in principal stresses as:

$$\sigma_c = \sigma_1 - (2\lambda + 1)\sigma_3, \qquad (5)$$

where $\lambda = \sin\varphi/(1 - \sin\varphi)$, σ_c is the strength limit in compression.

The soil state after elastic and post-failure deformations can be determined by analysing the equations proposed in Pustovoitenko and Timoschuk (2001) (Table I). Here ε_1 is the elastic deformation, $\varepsilon_{1,e}$ is the elastic deformation limit, E_d is the decline modulus.

The loss of strength beyond post-failure deformations is simulated by replacing Eq. (3) with relationships 2 and 3 in Table I. Besides, the used model and program (Fadeev, 1987) include the option to study the fault and rupture zones by means of an analysis of equation (4).

The enlargement of plastic deformation zones is controlled numerically during calculation of the stress–strain state. As a criterion of transition to the plasticity state the exceeding of the critical value of shear deformation within the bounds of the landslide slope is accepted. The soil state is estimated along the critical directions depending on the intensity and the value of the shear deformation. They characterise the relative displacements of points on the slope that depend on convergence of the iterative algorithm with respect to the theoretically determined pressure.

TABLE I. Equations for analysing soil deformations

N	Mode	Governing equation
1	Elastic deformations $\varepsilon_1 \leq \varepsilon_{1,e}$	$\sigma_1' = (2\lambda + 1)\sigma_3 + \sigma_c$
2	Exceeding the strength limit under condition $\varepsilon_{1,e} \leq \varepsilon_1 \leq \varepsilon_{1,e} + (\sigma_c - \sigma_r)/E_d$	$\sigma_1' = (2\lambda + 1)\sigma_3 + \sigma_c - E_d(\varepsilon_1 - \varepsilon_{1,e})$
3	Decreasing strength to the residual σ_r value $\varepsilon_1 \geq \varepsilon_{1,e} + (\sigma_c - \sigma_r)/E_d$	$\sigma_1' = (2\lambda + 1)\sigma_3 + \sigma_r$

4. Numerical Study and Results

The key points in the numerical modeling were exact reproducing the site geometry and the dynamic impact on the soil. The section corresponding to the critical direction regarding possible landslide development was studied in details.

Soil deformations were simulated within the area of about 500 m length and of maximal height of 60 m. The considerable size of the domain and the imposed zero deformations on the lower and lateral side boundaries minimised the influence of external effects on the processes inside the slope. Refining the mesh to 5 m of the element length was proved sufficient for the basement having several tens of meters size. The main parameters accepted with the 95% level of confidence for artificial, and loess-loamy soils and quarterly sediments are the following; $E = 2.5–24$ MPa, specific weight $\gamma = 16.78–20.04\,\mathrm{kN/m^3}$, $C = 7–191$ kPa, $\varphi = 8–20°$; the values for C and φ are estimated, besides, by laboratory testing. Soil changes were quantified by the intensity of deformations and pressure, and the convergence of the iterative process to the theoretical values as well. The results are demonstrated in Fig. 1.

The numerical analysis has shown that the ground slope within the studied area is now still before the limit equilibrium state, which is confirmed by the absence of large plastic deformation zones below the basement and a good convergence of the algorithm. The maximum values of shear deformations are concentrated in the bottom part of the slope above the depth of 25–30 m. Consecutive increase of the loading on the edges of the initial cracks in the slope intensifies irreversible deformations and enlarges plastic flow areas both along the clough main direction and above the depth of 25–30 m.

The difference between settlements of some blocks under the nearest building reaches the critical value of 10 cm when the equivalent load on

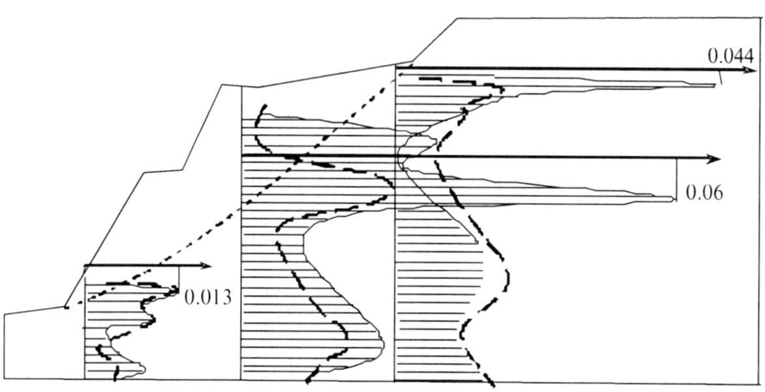

Figure 1. Distribution of shear deformation along the vertical profiles: the unloaded slope (dashed curves), the slope loaded by $p_s = 0.5$ MPa (solid curves).

the crack border equals 0.3 MPa. The load p_s corresponding to the maximal settlement difference for this building was estimated within the range of 0.1–0.6 MPa. If $p_s > 0.5$ MPa the plastic deformations grow unlimitedly, which points to landslide development accompanied by irreversible deformations above the potential sliding surface.

Regarding the hypothetical character of the predicted soil-structure interaction the numerical results can not be examined directly *in-situ*. However, it was possible to match measured and calculated displacements on the same slope for the section located on the distance of about 100 m from the emergency house where geometry, static load, and rock properties are quite similar (Fig. 2). The numerical model identified on the example of the analogous

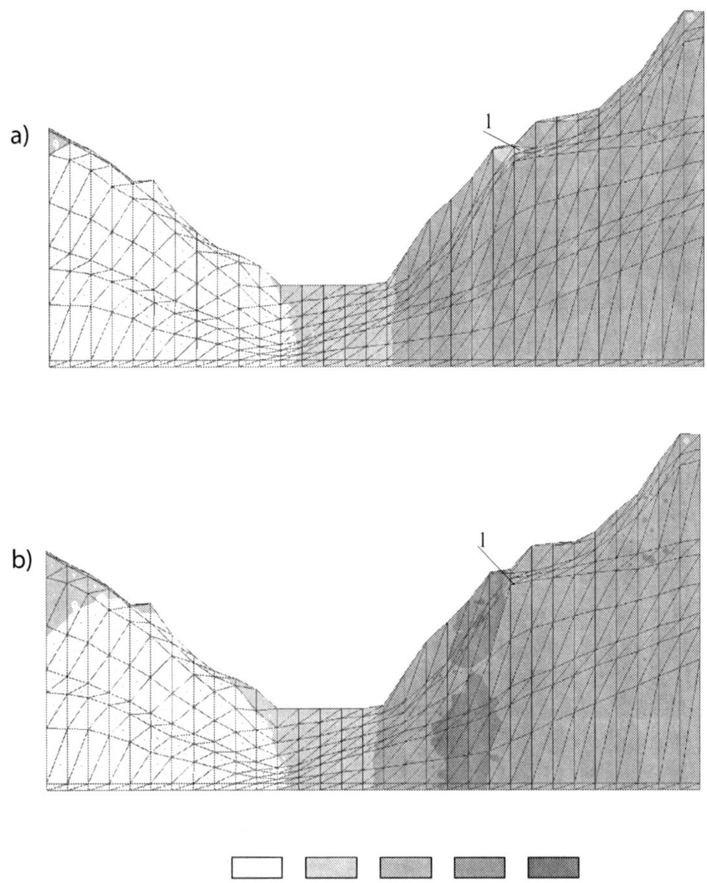

Figure 2. Relative shear deformation in the ground slope on the landslide site: 1—location of the house analogous to the emergency one; (a) undisturbed state, (b) load $p_s = 0.5$ MPa.

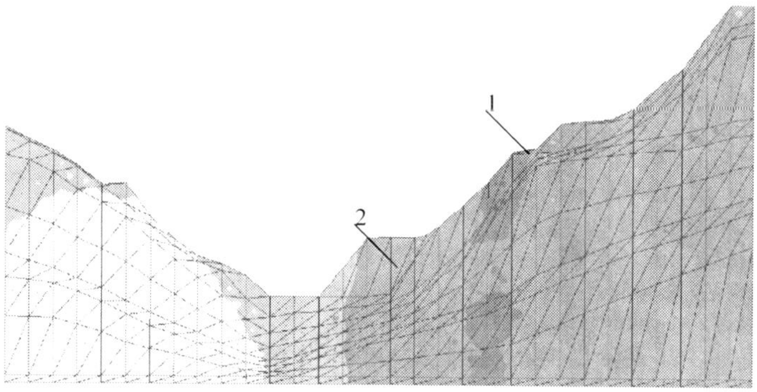

Figure 3. Shear deformation in the slope at $p_s = 0.5$ MPa: 1—location of the house analogous to the emergency one, 2—the counter-banquet. For colors see Fig. 1.

section has confirmed the limit equilibrium state of the slope and the arising fault zones found in the upper layer of sediments having 5–7 m thickness.

The range of calculated settlements around the contour of the nearest building basement of 0.393–0.430 m corresponds well to the measurements interval in the same points 0.332–0.450 m. The satisfactory convergence of the numerical algorithm up to the monitoring date shows the reliability of the input parameters, robustness of the modeling results and conclusions concerning the emergency building.

According to the numerical modeling results the slope stability can be increased by installation of additional load in the form of a counter-banquet placed in the talweg direction (Fig. 3). The estimated efficiency of such preventing measure becomes the highest when the load on the slope is about 0.5 MPa and the loading zone width equals 50 m. Additional weight of counter-banquet reduces the plastic deformation area by 15–20%.

5. Conclusion

It was proposed a procedure to determine the static load equivalent of the dynamic impact during short-term soil-structure interaction. This procedure was applied to model plastic deformations of the landslide slope and to calculate the settlements of the buildings on this slope while possible destruction of the many-storied house takes place. It was found out that the additional uneven settlements of the near-by building will exceed the critical values whereas plastic deformations will lead to reduction of the slope stability and, eventually, to landslide triggering.

References

Dynamic Calculation of Buildings and Constructions. Designer's Handbook (1984) Moscow, Stroyizdat.

Fadeev, A. V. (1987) *The Method of Finite Elements in Geomechanics*, Moscow, Nedra.

Gzovsky, M. V. (1975) *Fundamentals of Tectonophysics*, Moscow, Nauka.

Liakhov, G. M. and Poliakova, N. I. (1967) *Waves in Solid Media and Loads on Constructions*, Moscow, Nedra.

Pustovoitenko, V. P. and Timoschuk, V. I. (2001) *Geotechnogenic Processes in Unstable Soils*, Kyiv, Naukova dumka.

Wegner, J. L., Yao, M. M., and Zhang, X. (2005) Dynamic wave–soil-structure interaction analysis in the time domain, *Computers and Structures* **83**, 2206–2214.

Wolf, J. P. (1988) *Soil Structure in the Time Domain*, Englewood Cliffs, NJ, Prentice-Hall.

Zienkiewicz, O. C. (1967) *The Finite Element Method in Structural and Continuum Mechanics*, London, McGraw Hill.

AUTHOR INDEX

Albers, Bettina, 65
Apostolska, Roberta, 239
Athanatopoulou, Asimina M., 209

Bala, Andrei, 101
Blagov, Dilyan, 239
Bonev, Zdravko, 239, 253

di Prisco, Claudio, 139
Dikmen, S. Umit, 271
Dineva, Petia, 43, 53

Florin Balan, S., 101
Folić, Boris, 179
Folić, Radomir, 179

Galli, Andrea, 139
Gatmiri, Behrooz, 89
Georgiev, Valeri, 253
Gicev, Vlado, 151

Hannich, Dieter, 101
Herak, Davorka, 291
Herak, Marijan, 291

Iankov, Roumen, 253

Kappos, Andreas J., 195
Kiymaz, Guven, 271
Kouteva-Guentcheva, Mihaela, 33

Manolis, George D., 209

Necevska-Cvetanovska, Golubka, 239

Panza, Giuliano F., 33
Paskaleva, Ivanka, 33
Psarropoulos, Prodromos N., 77, 127

Rangelov, Tsviatko, 43
Ritter, Joachim, 101
Rudakov, Dmytro V., 303

Sadovenko, Ivan O., 303
Schanz, Tom, 53, 253
Sextos, Anastasios G., 195, 225
Snieder, Roel, 281

Taskari, Olympia, 225
Timoschuk, Vasyl' I., 303
Todorovska, Maria, 169
Trifunac, Mihailo D., 3
Tsompanakis, Yiannis, 115, 127
Turk, A. Murat, 271

Vaseva, Elena, 239
Vecchiotti, Mauro, 139

Wuttke, Frank, 53, 253

Zania, Varvara, 127

Breinigsville, PA USA
25 August 2009
222927BV00006B/96/P